THROUGH THE LENS
OF ANTHROPOLOGY

THROUGH THE LENS OF ANTHROPOLOGY

An Introduction to Human Evolution and Culture

Robert J. Muckle, Laura Tubelle de González, and Stacey L. Camp

THIRD EDITION

UNIVERSITY OF TORONTO PRESS

Toronto Buffalo London

© University of Toronto Press 2022
Toronto Buffalo London
utorontopress.com

ISBN 978-1-4875-4014-2 (cloth)   ISBN 978-1-4875-4017-3 (EPUB)
ISBN 978-1-4875-4015-9 (paper)   ISBN 978-1-4875-4016-6 (PDF)

Library and Archives Canada Cataloguing in Publication

Title: Through the lens of anthropology : an introduction to human evolution and culture / Robert J. Muckle, Laura Tubelle de González, and Stacey L. Camp.
Names: Muckle, Robert James, author. | González, Laura Tubelle de, author. | Camp, Stacey Lynn, author.
Description: Third edition. | Includes bibliographical references and index.
Identifiers: Canadiana (print) 20220189234 | Canadiana (ebook) 20220189250 | ISBN 9781487540142 (cloth) | ISBN 9781487540159 (paper) | ISBN 9781487540173 (EPUB) | ISBN 9781487540166 (PDF)
Subjects: LCSH: Anthropology – Textbooks. | LCGFT: Textbooks.
Classification: LCC GN25 .M83 2022 | DDC 301 – dc23

We welcome comments and suggestions regarding any aspect of our publications – please feel free to contact us at news@utorontopress.com or visit us at utorontopress.com.

Cover design: Michel Vrana
Cover image: Charlotte Corden © 2021

We wish to acknowledge the land on which the University of Toronto Press operates. This land is the traditional territory of the Wendat, the Anishnaabeg, the Haudenosaunee, the Métis, and the Mississaugas of the Credit First Nation.

University of Toronto Press acknowledges the financial support of the Government of Canada and the Ontario Arts Council, an agency of the Government of Ontario, for its publishing activities.

ONTARIO ARTS COUNCIL
CONSEIL DES ARTS DE L'ONTARIO
an Ontario government agency
un organisme du gouvernement de l'Ontario

Funded by the   Financé par le
Government   gouvernement
of Canada   du Canada

Canadä

Our book is dedicated to the domestication of coffee, barley, and hops. The writing of this book was fueled by coffee and will be celebrated with beer.

LOCATIONS MENTIONED IN THE BOOK

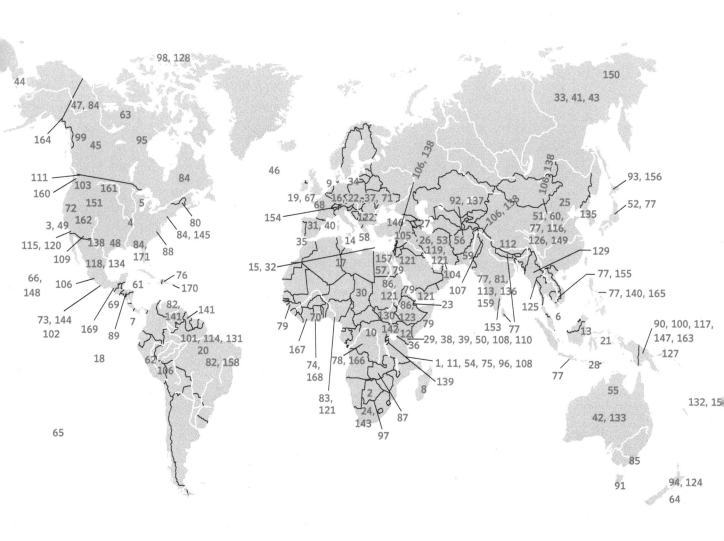

BRIEF CONTENTS

CONTENTS

Chapter 5

CULTURAL DIVERSITY FROM THREE MILLION TO 20,000 YEARS AGO 107

Chapter 6

CULTURAL DIVERSITY FROM 20,000 TO 5,000 YEARS AGO 133

Chapter 7

ARCHAEOLOGY OF THE LAST 5,000 YEARS

Chapter 8

STUDYING CULTURE

Chapter 9

LANGUAGE AND CULTURE

Chapter 10

Chapter 11

Chapter 14

ANTHROPOLOGY AND SUSTAINABILITY

ILLUSTRATIONS

FIGURES

TABLES

MAPS

BOXES

ACKNOWLEDGMENTS

The authors acknowledge the contributions of many. We appreciate our families for their patience, and for not complaining too much while we were writing. Laura would like to express her love and gratitude especially for the support of her husband, Luis; her daughters, Maya and Lirén; and her three dogs, Mochi, Ginger, and Jake, who kept her company in the office while writing. Stacey is grateful for the unflappable support of her husband, Ben; her children, Lana and Tyson; her loyal terrier, Mimsy; and the squirrels and birds in her backyard who served as a constant source of amusement while writing. Bob expresses appreciation for the support of his wife, Victoria, and his yours-mine-and-ours children: Miriam, Esther, Jonathan, Cody, Tomas, and Anna. He does not feel he received any support from his cat, Whisky; dog, Rosie; or any of the finches, all named Darwin.

We are very appreciative of all those friends, colleagues, and former students who have contributed photos at no cost in order to keep the cost of production low and ultimately make the book affordable to students. This includes Dr. Gillian Crowther, Emma Kimm-Jones, Nadine Ryan, Sashur Henninger, Luis A. González, Dr. Julie Lesnick, Andres Guerra, Daniel Chit, Dr. Tad McIlwraith, and Robert Gumpert. We are also indebted to our good friend and colleague Barry Kass, owner of imagesofanthropology.com, who has provided many photos for each edition.

We are grateful for the original artwork in this edition. Our cover image was drawn by Charlotte Corden, an illustrator and fine artist with an MA in anthropology, who designed a wonderful custom cover to illustrate some of the people, primates, and artifacts talked about in the book. There are new images within the pages created by Dr. Katherine Cook, who in addition to being a professor at the University of Montreal is also an artist. We recognize Katherine's talent for conceptualizing and creating illustrations to match our thinking and writing.

This third edition is built upon the suggestions of many reviewers of the proposals and previous editions. Most of these reviewers have been anonymous, but of those who have identified themselves, we are especially grateful for the many positive suggestions from Dr. Barbara J. King, professor emerita at the College of William and Mary; Dr. Erin McGuire, a professor at the University of Victoria; and Dr. Jason Antrosio, a professor at Hartwick College.

Of course, we are also indebted to the good people at the University of Toronto Press. From the initial proposal through the publication of the second edition, we were guided by the wonderful Anne Brackenbury. For this third edition, we have been guided by editor Carli Hansen. Both Anne and Carli have been invaluable assets to the quality of our books. We also thank our copyeditors for making us look like better writers than we really are, including Samantha Rohrig of The Editing Company for her work on the third edition.

ABOUT THE AUTHORS

Each author had primary responsibility for multiple chapters, although there was considerable overlap. Bob Muckle had primary responsibility for the introductory chapter and the chapters on primates, evolutionary theory, and human biological evolution (Chapters 1–4). Bob Muckle and Stacey Camp shared responsibility for the archaeology chapters (Chapters 5–7), and Laura González had primary responsibility for the chapters on cultural anthropology and sustainability, integrating linguistic anthropology throughout (Chapters 8–14).

Robert (Bob) Muckle is a professor in the anthropology department at Capilano University in North Vancouver, Canada. He has been practicing, teaching, and writing about anthropology since the 1980s. His primary teaching and research interests are in archaeology, biological anthropology, and the Indigenous peoples of North America. He also is interested in applying archaeology to help solve issues related to sustainability. He has worked on dozens of field projects in North America and Africa, including working collaboratively with Indigenous peoples of North America. Publications include *Introducing Archaeology*, third edition (co-authored with Stacey Camp, 2021), *Indigenous Peoples of North America: A Concise Anthropological Overview* (2012), and *The First Nations of British Columbia: An Anthropological Overview*, third edition (2014). He also edited a volume of readings for students titled *Reading Archaeology* (2008). If he could live anywhere at any time in the past, he would choose North America about 10,000 years ago. This is because he prefers low population densities, enjoys using an atlatl, and would like to taste fresh mammoth. Alternatively, he would like to be hanging out around Stonehenge about 4,000 years ago, drinking a drink we now call beer. As long as he remembers, he's had a thing for Stonehenge.

Laura Tubelle de González is a professor of anthropology at San Diego Miramar College in Southern California. She has taught cultural and biological anthropology courses for 20 years and is a campus advocate for social justice, equity, and inclusion. She specializes in cultural anthropology, having done fieldwork in Mexico and India. Laura is the author of *Through the Lens of Cultural Anthropology* (2019), another book in the "Through the Lens" series from University of Toronto Press. She is a past president of the Society for Anthropology in Community Colleges (SACC) and the recipient of the American Anthropological Association/ Oxford University Press Award for Excellence in Undergraduate Teaching of Anthropology in 2018. If Laura had limitless funds, she would travel the world with the express purpose of tasting every local cuisine, preferably carrying with her a book, a camera, and some dancing shoes. Alternatively, she would like to pilot a dirigible and be addressed as "Captain."

Stacey L. Camp is an associate professor in the anthropology department and director of the Campus Archaeology Program at Michigan State University in East Lansing, Michigan. She has taught cultural anthropology, history, and archaeology courses for the past 15 years. She has worked on archaeology projects in Ireland, China, and the Western United States. Her primary research interest concerns the archaeology of migrants and diasporic communities living in the Western United States in the nineteenth and twentieth centuries. She has also published on gender equity and parenting in archaeology; the politics of heritage and memory; the archaeology of contemporary waste and sustainability efforts; and managing and digitizing archaeological collections. She currently serves as an associate editor of the journal *Historical Archaeology* and as the archaeology representative on the American Anthropological Association's Committee on Ethics. She is the author of *The Archaeology of Citizenship* (2013) and the second author of the third edition of *Introducing Archaeology* (2020) with Bob Muckle.

PREFACE

The ultimate objective of this book is to provide students with an appreciation of what we call the lens of anthropology, or in other words, the way in which anthropology frames and views the world. *Through the Lens of Anthropology* introduces the perspectives, methods, and ideas of anthropology, as well as some of its theories. The book also contains highlights of some anthropological research in order to provide students with concrete examples. We hope that this book will contribute in a positive way to students learning about anthropology in its many forms and applications.

We realize that for many students, a single introductory anthropology course may be their only formal education in anthropology, and this book may be the only one they ever read that is explicitly devoted to the subject. While keeping the text readable and user-friendly, we strive to provide enough information so that students can appreciate the broad and detailed fields of the discipline. It's not important to us that 20 years from now students still recall specific details, such as the number of australopithecine species. Rather, we hope that students will come to understand and appreciate the anthropological lens and how it is applied.

Several key themes set this book apart from others. In particular, we have written the book with an emphasis on food and sustainability, topics in which all authors have an interest. Where Box Features relate to food issues, we have identified them with the heading "Food Matters." As well, most chapters have additional examples related to aspects of food and sustainability, and we have also devoted an entire chapter to anthropology and sustainability (Chapter 14), to emphasize the connections between them.

Secondary themes of the book include the Indigenous peoples of North America and how anthropology is embedded in popular culture. Finally, although Chapter 9 focuses especially on linguistic anthropology, more ways to talk about language are embedded in all of the chapters in the second half of the book, which emphasizes cultural anthropology. "Talking About" boxes focus on one aspect of language as it relates to the themes in each of these chapters.

New for the Third Edition

Bob Muckle and Laura González are very pleased to welcome Dr. Stacey Camp as a co-author. Her experience in archaeology and teaching and her written contributions are great assets to this book.

For this third edition, we have made a conscious effort for the text to remain highly readable, and for it to engage students in thinking about current issues through the lens of anthropology. We have updated research relating to each of the four fields of anthropology and to major social movements of the past several years.

Significant changes include discussing the work of and adding more citations for anthropologists identifying as female, Indigenous, Black, or people of color. We increased the number of examples of the work of anthropologists in the context of contemporary events and issues such as the Black Lives Matter movement, Indigenous issues, the #MeToo movement, gender issues, and the COVID-19 pandemic. We discuss the anthropology of pandemics, past and present, with a focus throughout the book on COVID-19 and its repercussions.

We have upgraded the box features, replacing or making significant revisions to 11 of them. Entirely new boxes include The Archaeology of Pandemics; The Archaeology of Food Insecurity; Outer Space: "The Final Frontier"; Ancient Aztec Foodways; Gender in Politics; How COVID-19 Reveals Class Divisions; and Islamophobia and the "Chinese Virus."

We have made changes to our use of pronouns, expressing gender neutrality throughout, and we have expanded subjects like marriage and family to better represent a wider spectrum of possibilities. In order to support and validate those authors to whom we refer in the text with doctorate degrees (PhDs), we have made sure to use the title "Dr." with their names. With these changes, we hope to further normalize these practices.

Finally, we have upgraded the art in the book, replacing 22 figures and adding original art by Dr. Katherine Cook.

In consequence of these updates and additions, several dozen sources, many written between 2018 and 2021, have been added to the References section.

NOTE TO INSTRUCTORS

We have designed this book so that topical coverage is not restricted to the sequence of the chapters as laid out. Following the introductory chapter, the sequence leads to the primate background (Chapter 2), followed by evolutionary thought (Chapter 3), human biological evolution (Chapter 4), archaeology (Chapters 5–7), and then cultural anthropology and linguistic anthropology (Chapters 8–14). We know that some instructors prefer to begin with cultural anthropology. Similarly, some may choose to cover evolutionary thought before primates. One of the ways we have accommodated alternative coverage sequences is to bold important glossary terms multiple times in the book – for instance, the first time a glossary term appears in the archaeology section and again the first time it appears in the cultural anthropology section. In this way, the book doesn't assume the student already knows the term when it appears in the latter half.

Although different parts of the book have different foci (i.e., biological anthropology, archaeology, and cultural anthropology with linguistics), we strive to integrate the four-field approach where appropriate. Our primary themes of food and sustainability, for example, are one way of showing this integration, and the incorporation of linguistics into various contexts is another. In addition, when a topic is addressed multiple times in the book, we cross-reference the chapters.

Each chapter includes a list of learning objectives, a chapter summary, and questions to guide students' reading and provide a framework for thinking about the issues covered. In addition, we and the University of Toronto Press provide a full set of ancillaries for instructors adopting the text. These ancillaries include an instructor's manual with chapter outlines and key points, lecture suggestions, activities and assignments, answers to the review questions found in the book, lists of key terms with page references, suggested readings, Weblinks, PowerPoint slides, and a test bank. For more information about the instructor's manual, PowerPoint slides, and test bank, and to download the images, maps, figures, and tables from the book, instructors should visit www.lensofanthropology.com.

We wanted to keep this book in a concise format for two main reasons. First, the concise format brings the cost of the book down for students; the high cost of textbooks is a concern for many instructors, including us. Second, we know that you, the instructor, have your own knowledge, experience, and goals for your course. We have provided the basic structure upon which we hope you will build, utilizing examples of your own that will make your course come to life.

NOTE TO STUDENTS

More than 300 words in the text appear in bold, indicating that they can be found in the glossary. You are encouraged to check the glossary to see how these words are defined. Many of them have a particular meaning in anthropology that may be different from the way they are used in common conversation (words such as *theory* or *gender*, for example). Many words are boldfaced two or three times in the text. This accommodates courses that follow a different sequence of chapters than that laid out in this book. We also encourage you to use the learning objectives to guide your reading, as well as the review and discussion questions to reflect and apply what you have read to your own experiences.

The term *North America* is used frequently throughout this book. We recognize that there are multiple ways of defining North America, but when we use North America we are primarily referring to the United States and Canada.

There are free online learning resources that can greatly enhance your engagement with the book's content. Visit www.lensofanthropology.com for self-study questions, chapter outlines, Weblinks, further reading, and to download the images, maps, figures, and tables contained in the book.

1

INTRODUCTION: VIEWING THE WORLD THROUGH THE LENS OF ANTHROPOLOGY

LEARNING OBJECTIVES

In this chapter, students will learn:

- *about the nature and scope of anthropology, including its use as a lens or framework, its major branches and subfields, and the anthropological perspective.*

- *about the two key themes in this book: (1) anthropology and food, and (2) anthropology and sustainability.*

- *what it means to be human, and about the nature of culture, with particular attention given to its definition, components, interconnectedness, and ever-changing dynamic.*

- *why anthropology has broad appeal.*

- *about the history of anthropology in North America.*

- *about the contemporary challenges of anthropology.*

- *how anthropology is situated within the contexts of academia, the business world, and popular culture.*
- *how anthropology is important in an increasingly connected world.*

Anthropologists are interested in all aspects of humanity – both past and present, biological and cultural. Anthropology is a lens through which we can understand human diversity and address contemporary challenges.
#AnthropologyIsImportant

INTRODUCTION

How, why, and when did **humans** come to be? Why do various groups of people around the world have different physical characteristics, and why do so many people speak, think, and do things differently than those you are most familiar with?

The short answer is that **science** in general and **anthropology** in particular tell us that humans have been around a very long time. Habitually walking on two legs was the first important thing to occur in the human lineage, and the growth of intelligence the second. Differences in languages, thoughts, and customs arise from many things but are usually the result of humans' successful adaptation to the environment and other people around them. A longer, more thorough answer requires some knowledge of anthropology. This text is designed to give students some of that fundamental knowledge and the anthropological thinking skills to address these questions and others like them.

To make sense of the world around us and our place in it, it's useful to have a framework within which to work. Frameworks help organize thoughts and guide our understanding of both the natural and cultural worlds. Many frameworks exist. For example, traditional mainstream religions offer one major kind of framework, and Indigenous ideologies another. Science and academic inquiry are frameworks as well. Just as there are many different religious and Indigenous frameworks, there are also different kinds of scientific and academic frameworks.

It is important to have a framework for understanding phenomena. Frameworks help people work toward explanations and narrow down the kinds of information to consider. What constitutes a framework is a set of principles, methods, and theories, along with the knowledge to investigate, understand, and explain phenomena.

Anthropology is one such framework – a way of understanding how humans came, and continue, to be. It isn't the only way, but it is one way. In practical terms, this means that anthropology has a particular set of characteristics that distinguish it from other ways of knowing. These characteristics are clarified later in this and subsequent chapters.

In this book we refer to the anthropological framework as the **lens of anthropology**. As with any framework, it provides a basic structure to help organize our thoughts. It provides focus and clarity. The lens is a particular set of ideas, methods, theories, ethics, views, and research results.

This book introduces the discipline of anthropology, mostly in the context of the academic world but also from the perspective of practical applications. Mostly, it informs readers about the human world as seen through the lens of anthropology. This opening chapter outlines the nature of anthropology, the **anthropological perspective**, the history of anthropology, how anthropology is situated within the world today, and the importance of anthropology in an increasingly connected world. Subsequent chapters turn the lens toward our place in the world of **primates**, human biological evolution, cultural evolution, and the wide range of cultural diversity among the world's populations today. Where appropriate, the chapters include references to underlying anthropological methods and theories as well as research findings and insights.

Besides offering an overview of the basic framework of anthropology, this book highlights two themes of contemporary interest in anthropology: food and sustainability.

Food is an area of considerable interest in anthropology, including such topics as the emergence of meat eating, the origins of cooking, processes involved in food production, cooking and gender, patterns of eating across cultures, food security, and food taboos. Explicit references to food are made in most of the chapters, in both the main text and the Box Features.

Sustainability is another area of considerable interest in anthropology, including such topics as nonhuman primates' contributions to sustainable environments, the identification of sustainability (or lack thereof) in the human past, and maintaining environmental sustainability in contemporary times. This book understands sustainability as having three main aspects: environmental, social, and economic. As with food, explicit references are made to sustainability in most chapters in both the main text and some of the Box Features. Sustainability is also the entire focus of Chapter 14.

BOX 1.1   **Food Matters: Anthropology and Food**

It is hard to think of a more essential thing in human existence than food. People need nourishment to survive. Therefore, much of the social and cultural life that embeds humans in their daily activities results from finding, distributing, preparing, consuming, and disposing of food. For this reason, anthropologists across the four fields have addressed the questions surrounding the human relationship to food since the beginning of the discipline.

There are many ways in which food issues are similar across cultures. All humans may consume a wide variety of foods to support health. As omnivores, people choose from foods available in the environment (which may vary greatly from one ecosystem to another) and receive the nutrients the human body requires. In addition, humans prepare food by cooking it. This is grounded in an ancient legacy in which hominins' bodies and communal life were greatly affected by the cooking process. Finally, the activities around food procurement are deeply embedded in a complex system of social, economic, political, and religious norms and expectations. People's daily lives are limited, supported, and enriched by eating as a cultural and symbolic act.

Anthropologists interested in food issues may identify themselves simply as biological, cultural, archaeological, or linguistic anthropologists with an emphasis on foodways. They may also choose a more specialized subfield, such as nutritional anthropology (which takes a biocultural approach), ethnoecology (which examines traditional foodways), gastronomy (which combines cooking, food science, and cultural meanings, especially of fine foods), or food studies (which tends to focus on issues of culture, history, and identity).

Studying peoples' foodways has become more important than ever in light of the impacts of climate change and globalization. Due to unpredictable weather patterns, farmers may lose entire harvests during a severe storm or heat wave. Inuit ice fishermen can no longer reliably read the sky to know if they should undertake a fishing expedition. Because of a globalized economy that opens up access to land and water, companies and nations claim ownership of resources that are outside their boundaries. Water, privatized by a bottling company, vanishes in underground aquifers, leaving less for farmers seeking to irrigate crops. The World Bank funds massive food aid programs, while African farmers' harvests rot in granaries, owing to lack of demand.

Food issues are sustainability issues and also anthropological issues. What is more central to people's lives than food?

THE APPEAL OF ANTHROPOLOGY

The appeal of anthropology is broad. For some, it is the subject matter; for others, the sense of discovery, becoming familiar with the strange, the methods of inquiry, the explanatory power, or the lifestyle.

BOX 1.2 **Anthropology and Sustainability**

It has been clear to scientists for some time that life on our planet is becoming unsustainable. Louder and more urgent calls to action emerge every few years. In 1992, the scientific community published the "World Scientists' Warning to Humanity." In 2000, the United Nations Millennium goals were published; they included specific targets to reach by 2015 in social, environmental, and economic realms to mitigate some of the most detrimental effects. Seeing that we hadn't yet reached these targets, in 2014, the Intergovernmental Panel on Climate Change (IPCC) published their synthesis report, which stated bluntly that if we didn't act immediately to cut carbon pollution, there would be severely damaging and "irreversible impacts" that would hamper our ability to survive as a species.

Clearly, the sustainability of people and life on earth is an issue that we cannot ignore. Why include anthropology in a discussion about sustainability? Anthropology, across its fields, is uniquely positioned as a field of study to provide the kinds of broad and deep understandings about people in their environments – understandings that can lead to solutions. Since the beginning of the discipline, anthropologists have sought to learn about the long-term interactions of people in their environments. From the time of hominin evolution, through more recent prehistory, to the rise and fall of civilizations, anthropologists have sought to understand the reasons for successes and failures in all of the ecosystems on earth.

Today, anthropologists study some of the most marginalized people living on the planet, including Indigenous people living in small-scale, traditional societies. As the world becomes more connected through globalization and industry, these groups are often the most oppressed. At the same time, these groups hold vast amounts of traditional knowledge about the ecosystems in which they live. This knowledge appears to be more important than ever to save the biodiversity of the planet. As these small-scale cultures disappear, so do their knowledge, languages, understanding and use of flora and fauna, and preindustrialized ways of making a living.

It is here among the people who are on the edges of the modern Western world that anthropologists discover the kinds of connections with nature that modernization has largely discarded in the quest for status and power that characterizes the contemporary world. Therefore, anthropology can not only provide holistic and long-term views of why cultures succeed or fail in their ecosystems but also illuminate the kinds of human connections to one another and to the natural world that characterize our species.

The appeal of anthropology to the general public is perhaps best indicated by the widespread use of anthropology in **popular culture** and news media. Depictions of anthropology and anthropologists are manifest in films of all sorts, ranging from serious documentaries about peoples past and present to ficitionalized accounts of anthropologists at work. Programs on television, streaming services, and various

Figure 1.1

MAASAI CEREMONIAL WALK. Images like this play into the popular or public appeal of anthropology, which often focuses on people and places that are remote in time and/or place, and are highly visual. Pictured here is a Maasai ceremony taking place in Tanzania in 2019.

Credit: Kairi Aun/Alamy

internet-sharing platforms similarly embed anthropology in both the serious scholarship of humans as well as in fictionalized accounts.

News media tend to support the broad appeal of anthropology, often reporting on important discoveries, especially if there are strong accompanying visuals such as mummies, skeletons, artifacts perceived to be treasure, ruins of an ancient civilization, or anything that is likely to be considered odd or unusual behavior to their regular audience. For example, an image of a group of Maasai participating in a ceremony in Tanzania in 2019 (see Figure 1.1) meets many of the criteria for generating and maintaining the appeal of anthropology, insofar as it focuses on a people who, from a North American perspective, look different, act different, dress different, and perform what, to an outsiders' view, appears to be odd and unusual ceremonies. With its bright pops of color and angled perspective, the image itself is also visually appealing; it catches the eye and draws the audience in. As you will learn in this book, anthropology is much more than this, but the appeal of images like that of the Maasai remains.

The appeal of anthropology is further supported by the burgeoning growth in anthropology-related tourist adventures, such as those to well-known archaeological sites or to witness and participate in the daily lives and ceremonies of people in distant and remote places.

For many who have made a career in anthropology, it was the subject matter (i.e., humans) that initiated their interest, and the methods that have convinced them that their choice was the right one. For others, it is the lifestyle that is the draw – whether that be the life of a professor and the world of academia, that of a researcher out in the field, or perhaps that of someone immersed in the world of business, where anthropologists can take what they know and apply it in the interests of people, organizations, and governments everywhere.

It would be remiss to not mention that for some, the appeal of anthropology is not even the study of humans; rather, it is the prospect of working with nonhuman primates. As is covered in more detail in Chapter 2, the focus of many anthropologists is on the study of monkeys and apes.

Of course, the appeal of anthropology is not universal. In many ways, anthropology is a product of colonialism, which continues to be perpetuated by many people and institutions today to the detriment of marginalized groups around the world. Many anthropologists, however, are actively working toward decolonization and are active in using anthropology to address issues related to social justice, political upheaval, civil unrest, Indigenous empowerment, racism, food security, and sustainability. That is appealing as well.

DEFINING ANTHROPOLOGY, DEFINING HUMAN, AND DEFINING CULTURE

Three of the most important words a student of anthropology should become familiar with, at least in regard to how they are used in anthropology, are *anthropology*, *human*, and *culture*. There is a good chance that most students already have some idea of the meaning of these words, but they may not be aware of their specific meanings or how these meanings may vary depending on context. There is little consensus, in fact, even among anthropologists, about what these words mean, but this is not necessarily a bad thing. There are some general understandings of what each word means in anthropology, but it isn't necessary that everyone use the same definition in every context.

There are many definitions of anthropology. The one constant in all valid definitions is that anthropology involves the study of humans. Beyond this, however, how one defines the discipline depends on context. Outside North America, anthropology is often considered to focus on peoples and cultures of contemporary times (or the very recent past). In North America, however, anthropology is usually considered to include studies of human cultures and human biology, past and present.

In many instances, such as when general distinctions between various fields of study are being made, a simple definition such as "anthropology is the study of humans" may be sufficient. There are many disciplines that focus on humans, however, so it is often desirable to distinguish between the particular kinds of things about humans that anthropologists are interested in, such as human culture and human biology, or methods and perspectives. The inclusion of the "evolutionary, comparative, and holistic perspectives," for example, distinguishes anthropology from other fields of human study. Other uniquely anthropological perspectives are included later in this chapter.

Human has a distinct meaning in anthropology and may be used in ways that are unfamiliar. For many people, human equates with **Homo sapiens**, the genus and species to which we all belong. For anthropologists interested in the human past, however, *Homo sapiens* is insufficient. As will be clarified and expanded upon in subsequent chapters, many anthropologists equate human with a certain kind of primate whose normal means of moving around is walking on two legs (**bipedalism**) and who emerged several million years ago. Those who accept that bipedalism is the distinguishing characteristic of humans therefore equate human with the biological family **Homininae**, which includes the genus *Homo* as well as other genera (plural of genus) existing between seven million and one million years ago. Some anthropologists, especially those focusing on past cultures, equate human with the genus *Homo*, for the simple reason that it is with the emergence of the genus *Homo* more than two million years ago that we first have undeniable physical evidence of human culture. In sum, some anthropologists equate human with *Homo sapiens*, some equate human with the genus *Homo* (which includes *sapiens* as well as other now extinct species), and some equate human with the biological family Homininae.

Culture, too, has a distinct meaning in anthropology, which may not correspond to its usage in other contexts. Culture is a core concept in anthropology and is covered more fully in Chapter 8. As with the definitions of anthropology and human, there are many different ways to define culture, even within the discipline of anthropology. Some definitions focus on the mental templates that govern peoples' behavior; others focus on customs. We consider culture to include aspects of ideology and behavior, as well as the products of those thoughts and behaviors (i.e., material culture). Thus, we define culture as the learned and shared things that people think, do, and have as members of a society. "Things that people think" refers to ideology, which includes belief systems and values. "Things that people do" includes behaviors that are commonly referred to as customs. "Things that people have" is commonly referred to as material culture.

Defining Anthropology

Anthropology is ...

- the study of humans, in all places and at all times.

- the study of human culture.

- the study of human biology.

- the study of humans, focusing on the description and explanation of human cultures and human biology, and including the scholarly collection, analysis, and interpretation of data related to humans, past and present.

- the scholarly study of humans through evolutionary, comparative, and holistic perspectives.

There are many components of culture, including those relating to **subsistence**, diet, technology, communication, settlement patterns, economic systems, social and political systems, belief systems and ideology, art, and health. All cultures have these components, and each is discussed more fully in later chapters. When anthropologists speak of subsistence, food procurement, or foodways, they are usually referring to the methods by which people get their food, such as **foraging** (also known as hunting and gathering), **pastoralism**, **horticulture**, **agriculture**, or **industrialism**. Diet refers to the specific kinds of food eaten. **Technology** refers to the way people have made or used things, including such things as making and using tools, cooking, harvesting, and building. Communication refers to all ways people have of communicating with each other, including speech, sounds, gestures, art, and writing. Settlement patterns refer to the movements of people within their territories, and the ways in which they create their living spaces. Economic systems focus on the way people obtain and distribute resources. Social systems include the methods by which order is maintained within a community or group, and political systems involve the processes by which order is maintained with other groups. **Ideology** includes shared beliefs and values, art includes both visual and performing arts, and health and healing refer to physical and mental health, illnesses, and methods of treatment.

It is important to understand that culture is dynamic, fluid, and ever-changing. All components of a culture do not change at the same rate, and the components do not change in the same order across cultures, but they all do eventually change.

It is possible to be part of multiple cultures and subcultures at the same time. On the basis of language alone, there are several thousand distinct cultures in the world today. This is based on the notion that where languages are distinct, other aspects of ideology, customs, and material culture are typically distinct as well. Most elements of any one culture are shared with other cultures as well. It is the suite of characteristics that distinguishes distinct cultures and subcultures. Consider, for example, that readers of this text may identify with North American culture, which has several distinct values, behaviors, and other qualities unique to North Americans but not shared widely with people outside North America. Beyond that, people may also identify with other cultures associated with their heritage, country, or geography. For example, although they share many aspects of North American culture, there are distinct differences between American and Canadian cultures, east coast and west coast subcultures, southern and northern subcultures, Indigenous and non-Indigenous cultures, and corporate and non-corporate subcultures.

Components of Culture
• Subsistence/food-getting/ food procurement
• Diet
• Technology
• Communication
• Settlement patterns
• Economic systems
• Social systems
• Political systems
• Belief systems and ideology
• Art
• Health and healing

Many people are often able to operate within multiple cultures. It is common, for example, for some Indigenous peoples in North America to move between their traditional Indigenous culture, especially as it exists on their reserves or reservation, and the typically more dominant non-Indigenous cultures in urban areas.

THE FOUR FIELDS AND APPLIED ANTHROPOLOGY

Anthropology in North America is usually considered to have four academic fields, sometimes referred to as branches or subfields: **cultural anthropology, archaeology, biological anthropology**, and **linguistic anthropology**. As illustrated in Figure 1.2, anthropologists often also apply their skills outside academia in a fifth branch called **applied anthropology**, which utilizes skills and methods of each of the other four branches.

Cultural anthropology, also known as social anthropology or socio-cultural anthropology, focuses on cultures of the present and recent past. Training in this branch often involves immersing oneself within a culture for several months or more and then producing an **ethnography**, which is a written description of that culture. Immersion in a culture is often called **ethnographic research**, and the method itself, whereby one both observes and participates in a culture, is known as **participant observation**. Thousands of ethnographies have been written by cultural anthropologists over the past 150 years, and they provide much of the raw data of anthropology. Beyond simply describing peoples' lifeways, cultural anthropologists also seek to interpret and explain larger patterns of culture. Cultural anthropologists often work in the academic world; however, there are also many who find themselves working in the world of business, for governments, and in the not-for-profit sector. Insight into the world of contemporary cultures through cultural anthropology is the focus of Chapters 8 through 14.

Archaeology (also spelled archeology) may be defined as the study of humans through their material remains, which essentially means the physical evidence of their activities. Most archaeology is focused on **prehistory** and the historic period, but some archaeologists focus on the contemporary world. The primary raw data of archaeology includes **archaeological sites** and **artifacts**, which are usually found during fieldwork. The primary objectives of archaeologists are to describe and explain the human past and to document the rapidly disappearing physical record of the human past. About 90 percent of archaeologists working in North America are involved in a kind of archaeology known as

Figure 1.2
THE BRANCHES OF ANTHROPOLOGY.
This illustration represents the four main branches of anthropology. Each branch also includes an "applied" component.

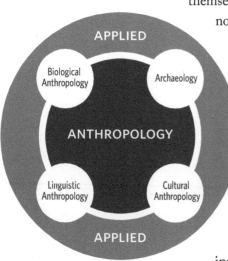

cultural resource management (CRM), or **commercial archaeology**, which essentially involves looking for and recording archaeological sites in advance of development projects. The results of archaeological research, especially in regard to outlining the development of human culture over the past two million years, are the focus of Chapters 5, 6, and 7.

Biological anthropology, also known as **physical anthropology**, focuses on human biology, past and present. This includes the study of human biological evolution as well as the study of contemporary biological variability. Biological anthropologists study skeletal material (e.g., bones and teeth) as well as DNA and other molecular substances.

Primatology, which is the study of nonhuman primates within a framework of anthropology, is usually considered a subfield of biological anthropology. Biological anthropologists are interested in the study of other primates in their own right to help us better understand our place in the animal world from a biological perspective, to provide us with models of how early humans may have behaved, and to help conserve endangered populations. Primate taxonomy, evolution, and behavior are covered in Chapter 2.

Paleoanthropology, which involves the study of early human biology and culture, is often considered to be part of biological anthropology, although it tends to involve the recovery, analysis, and interpretation of both biological and cultural evidence of early humans. The study of human biological evolution, mostly based on studies by paleoanthropologists, is the focus of Chapter 4. The study of early human cultural evolution is the focus of Chapter 5.

Most biological anthropologists work in the academic world, but some are employed elsewhere in fields such as **ergonomics** or forensics. Forensic anthropologists use their expertise in biological anthropology mostly to identify victims of crimes or disasters; they typically ascertain an individual's sex, age at death, and cause of death, but their duties may also include such things as identifying the ancestral population of the individual and their medical and nutritional history.

Linguistic anthropology is the study of human languages within the framework of anthropology. This includes classifying languages (such as putting them into taxonomic categories); examining languages to help determine past human migrations and interactions; and studying language change, the influence of language on other elements of culture and cultural influences on it, and language usage. Most

Figure 1.3
ANTHROPOLOGISTS AT WORK.
This image depicts a sampling of the kinds of things anthropologists do. Beginning at the bottom left and moving counterclockwise, the image depicts an anthropologist interviewing someone; a laptop displaying a visual language program using symbols of the Innu language; a primatologist; an archaeologist; a tablet with an open access symbol; anthropologists contributing to public activism; a biomedical anthropologist, represented by DNA, the health symbol, and scientific analysis; and finally, a camera to represent visual anthropology.
Credit: Katherine Cook

Figure 1.4
INDIGENOUS PEOPLE IN THE KALAHARI.
Cultural anthropologists study peoples and cultures all over the globe. Pictured here are Indigenous people in the Kalahari Desert region of Botswana, Africa. Anthropological studies of Indigenous peoples of this area, sometimes known as the San, typify anthropological interests in what is foreign and unfamiliar to North Americans. It would be incorrect to think, however, that it is only the foreign and unfamiliar that anthropologists study.
Credit: Hyserb/Shutterstock

linguistic anthropologists work in academic institutions, but there are also career opportunities elsewhere, such as with **Indigenous** groups and not-for-profits seeking to document the rapidly disappearing languages of the world.

Most anthropologists in North America have some training in at least three, and often all four, of the major branches of anthropology. It is common, for example, for one who primarily identifies as an archaeologist to also have training in and a good understanding of cultural anthropology and biological anthropology. Likewise, a cultural anthropologist usually has some training in and understanding of linguistic anthropology and archaeology.

While each branch has an applied anthropology component, it is clearly in the field of archaeology that the largest percentage of anthropologists practice the discipline outside the purely academic arena. That being said, many cultural anthropologists are employed by Indigenous groups to help them document their cultures, especially in the areas of **Traditional Use Studies (TUS)** and **Traditional Ecological Knowledge (TEK)**. Corporations also employ cultural anthropologists to help them with their own internal organizations, as well as to learn how to better deal with consumers or people in other countries. Some cultural anthropologists have even found employment working for the US military in conflict zones, on the assumption that the expertise a cultural anthropologist has is likely to lead to better-informed decisions by military personnel.

THE ANTHROPOLOGICAL PERSPECTIVE

It is primarily the anthropological perspective that constitutes the lens of anthropology. This perspective has several elements, including being holistic, evolutionary, comparative, qualitative, focused on linkages, focused on change, and based on fieldwork. Taking a **holistic** perspective means that anthropologists view all aspects of human biology and culture as being interrelated. That is, for a thorough understanding of any one component of human biology or culture, anthropologists recognize that a full understanding involves studying the links between all components. In the study of early human evolution, for example, anthropologists understand that intelligence is correlated with brain size, making tools is linked with dexterity, meat eating is correlated with digestive enzymes, and so on. They also recognize that all components of culture are intricately interrelated, so that an anthropologist primarily interested in art recognizes that art may influence or be influenced by politics, social systems, ideology, technology, and more. Anthropologists also understand that a change in one component of a culture invariably causes changes in other components.

Taking an evolutionary perspective means that a good understanding of biological and cultural traits is best considered in regard to long-term evolutionary changes. Anthropologists use a database that extends millions of years into the past. They understand that changes rarely occur in a vacuum, and while some changes can occur quickly across time and space, it is at least worth considering the record of the past.

Taking a comparative perspective means that anthropologists often compare things in their research. For biological anthropologists, this may mean that when they find an ancient bone that looks as if it may be human, they compare it to known human bone matter to determine the species to which it may best be classified. In the same way, when archaeologists find an ancient tool of unknown function, they may compare it to similar-looking things in use today to make inferences about its purpose. Linguistic anthropologists often compare the vocabularies of languages to draw inferences about interactions in the past, and cultural anthropologists compare various components of multiple cultures to reveal how people may adapt to similar circumstances.

Taking the **qualitative** perspective means that anthropologists tend to focus on descriptive research rather than **quantitative** data. Anthropologists do use statistical analysis, but it is rarely the primary method of research. Rather than have people complete surveys and then quantify the results, for example, anthropologists tend to seek deeper meaning and insight by focusing on fewer individuals for longer periods.

Focusing on linkages relates to the holistic perspective. Much more so than other disciplines, anthropology tends to focus on linkages – the linkages between human biology and culture, the linkages between various parts of the human body, and the linkages between the various components of a culture. Biological anthropologists understand, for example, that walking upright efficiently is linked to changes in the skull, back, pelvis, legs, and feet. Cultural anthropologists may not know as much about human political systems as political scientists, nor as much about

Key Elements of the Anthropological Perspective

- Holistic
- Evolutionary
- Comparative
- Qualitative
- Focused on linkages
- Focused on change
- Based on fieldwork

settlement patterns as human geographers, but they are likely best situated to understand the connection between politics, settlement, economics, religions, and other factors.

Anthropologists often focus on change. They understand that both human biology and human cultures are undergoing constant change, and this ties in with holistic and evolutionary perspectives and with focusing on linkages. Anthropologists are interested in how and why change occurs, both within groups and over time.

Another key element of anthropology is its focus on fieldwork. Anthropologists tend to collect their own data. Biological anthropologists want to find bones of early humans or personally extract the DNA from bones already in collections. Archaeologists want to find and excavate archaeological sites themselves. Linguistic anthropologists usually prefer to work directly with native speakers. Cultural anthropologists often immerse themselves in another culture to make their own observations and collect their own data. This sets anthropology apart from the many other disciplines that use data collected by governments, agencies, or other groups.

Cultural relativism is an important concept in anthropology. As described by Matthew Engelke (2017), it underpins all of anthropology: "Put simply, cultural relativism is critical self-awareness that your own terms of analysis, understanding and judgment are not universal and cannot be taken for granted.... It is an approach, a styling. It is what helps anthropologists guard against the dangers of assuming that their common sense or even informed understanding – about justice or affluence or fatherhood or the elementary forms of religious life – is self-evident or universally applicable" (pp. 16–18).

Other characteristics of anthropology that may be considered to fall within the anthropological perspective include the following:

- Anthropology tends to be more interested in populations than individuals.
- Anthropology is interested in big-picture questions. What makes us human? Why and how did we evolve the way we did? How are some traits adaptive?
- Anthropology is also interested in small things, such as how people greet each other.
- Anthropology recognizes that most (but not all) traits, both biological and cultural, are adaptive in some way.
- Anthropology recognizes that biological and cultural characteristics are not perfect.
- Anthropology recognizes that there are multiple ways of adapting, and that one way is not necessarily better than another.
- Anthropology recognizes that similar problems can be solved in different ways.

HISTORY OF ANTHROPOLOGY, MOSTLY IN NORTH AMERICA

Anthropology emerged globally as a widely recognized academic discipline in the 1800s, primarily in Europe. It developed out of an interest in the observation of cultural diversity around the world, as well as a largely European fascination for ancient archaeological sites and artifacts.

It wasn't until the late 1800s that anthropology took hold as a scholarly discipline in universities, but there was certainly considerable activity that could broadly be classified as anthropology before this time. In the 1700s, for example, Thomas Jefferson (who would later go on to become the third president of the United States) excavated one of the thousands of large earthen mounds that dotted the landscape of the eastern and central parts of the United States. Jefferson's objective was to draw some conclusions about who had created the mounds. The hypothesis that the mounds were created by the ancestors of Indigenous peoples still living in the area was confirmed.

Interest in the mounds continued throughout most of the 1800s, particularly in regard to studying them before they would be destroyed by colonial settlement, ranching, farming, and other activities. Almost all the mounds have since been destroyed. A notable exception is **Cahokia**, near St. Louis, Missouri, which is now a **World Heritage Site** (see Chapter 7). Much of the work on the mounds was funded by the American Ethnological Society and the Smithsonian Institution.

Figure 1.6

INDIGENOUS WOMEN DOING ARCHAEOLOGY.

The profession of anthropology includes an increasing number of Indigenous peoples. Pictured here are two First Nations women from Canada working on an archaeological project.

Credit: Nadine Ryan

The beginning of pure scholarly or theoretical work in anthropology in North America is often associated with Lewis Henry Morgan (1818–81). Morgan made significant contributions to both ethnography and theory, but is best known for developing the **unilinear theory** of cultural evolution, outlined in *Ancient Society, or Researches in the Lines of Progress from Savagery through Barbarism to Civilization* (1877). In this book, Morgan proposed that every society in the world started out in a state of savagery. Some progressed to barbarism, and others then progressed to civilization. Savagery and barbarism each had three stages, making for seven stages in total. Classification into any one stage was based primarily on subsistence strategy and technology. If a people did not have pottery, for example, they were savages. According to Morgan, cultural diversity around the world could be explained by some societies failing to progress as quickly as others. Although some anthropologists liked this theory, many did not support it, and it was largely discredited by anthropologists within a few decades. Many anthropologists knew then, and all know now, that there are usually multiple ways of adapting to environments and other groups, and no one way is necessarily better than another.

There were many other interesting developments for anthropology in North America beginning in the late 1800s. One such development included the establishment of the Bureau of American Ethnology, which was created to collect information on the Indigenous peoples of the continent and supported both archaeological and ethnographic research projects. The rapidly declining populations of Indigenous peoples (due largely to disease and conflict with those of European descent) and rapidly changing cultures (as traditional lifeways were changing in response to **colonialism**) gave rise to a sense of urgency to document these cultures. This created a kind of anthropology known as **salvage ethnography**, which became the most common kind of anthropology in the late 1800s and early 1900s. Another development with implications for North American anthropology was the golden age of museum collecting in North America, beginning in the late 1800s and continuing into the early 1900s. Anthropologists and others collected millions of objects from Indigenous peoples for museums in North America and Europe. Some were negotiated and paid for, others were not. Hundreds of thousands of human skeletons were included in the collections.

The most dominant figure in the history of North American anthropology is Franz Boas (1858–1942), who moved to the United States from Germany in the late 1800s. Boas's own fieldwork focused on the Indigenous peoples of the continent, especially in the Pacific Northwest region. However, Boas made many other significant contributions, including being an outspoken critic of Lewis Henry Morgan's unilinear model of cultural evolution and developing the notions of cultural relativism and historical particularism, which became foundations of the discipline in North America. Boas is also widely credited with many important developments in the four-field approach in anthropology as it is practiced in North America, and he trained many of the most prominent North American anthropologists of the early 1900s (including Dr. Alfred Kroeber, Dr. Margaret Mead, and Dr. Edward Sapir), encouraged women to become anthropologists, and formally trained and collaborated with Indigenous peoples (including Ella Deloria and George Hunt).

The history of anthropology in North America has been intricately intertwined with the Indigenous peoples of the continent. Since the late 1800s, some anthropologists have had good relations with Indigenous peoples, but for many, the relationship can be characterized as having been exploitative on the part of anthropologists. Serious and widely published criticisms by Indigenous peoples of anthropology in North America began to become well known in the 1960s, and since that time, relations can generally be characterized as better. Most anthropological work involving Indigenous peoples, for example, is now done only with the consent of Indigenous peoples, and with the anthropologist providing something of value, including knowledge, to Indigenous peoples.

Over the past few decades, anthropologists trained and working in North America have disentangled the relationship between the discipline and the Indigenous peoples of the continent. Indigenous peoples are not as central to North American anthropology as they once were. Anthropologists still work with Indigenous peoples in the traditional areas of research, such as ethnography,

BOX 1.3   **The Indigenous Peoples of North America and Anthropology**

Anthropology in North America has a long history of entanglement with Indigenous peoples of the continent. From the late 1800s to the late 1900s, the overwhelming focus of anthropological study was on Indigenous groups in the territories now known as Canada and the United States. Anthropologists saw the rapid rate at which Indigenous populations were declining, traditional lifeways were changing, languages were disappearing, and archaeological sites were being destroyed. This led to many anthropologists undertaking what is known as salvage ethnography, recording as best they could what life was like before the influence of Europeans. There was some specialization, but many anthropologists were practicing four-field anthropology, meaning fieldwork for them usually included studying the Indigenous peoples in their own territories, undertaking ethnography (cultural anthropology), learning and recording languages (linguistic anthropology), measuring the physical attributes of the people (biological anthropology), and excavating archaeological sites.

Although there were certainly some good relationships between anthropologists and Indigenous peoples, it is justifiable to state that until the later part of the twentieth century, the relationship was largely exploitative. Anthropologists would often take much from the Indigenous peoples in regard to their cultural knowledge and beliefs, as well as hundreds of thousands of human skeletons and millions of artifacts, while providing nothing or very little in exchange. Anthropologists were advancing their own careers, filling museums, and making contributions to the discipline of anthropology at the expense of Indigenous peoples. Anthropologists began to be called out by some Indigenous peoples in the 1960s. One of the most prominent voices, Vine Deloria Jr. (Dakota Sioux) published a scathing criticism in his book *Custer*

archaeology, and linguistics, but their interests in Indigenous peoples also include many other areas, including Indigenous identity and **cultural appropriation**.

There are many other threads of interest in contemporary anthropology in early twenty-first-century North America including, but certainly not limited to, corporate culture, youth culture, popular culture, militarization and warfare, terrorism, food, sustainability, disease, education, queer culture, and gender. Many anthropologists now work among the voiceless and disenfranchised in North America, such as the homeless in urban areas and undocumented migrants, often challenging widely held misconceptions about their lives. Many anthropologists also address the concept of **race**, covered more fully in Chapter 8.

Over the past few decades there has also been a change in the makeup of those in the profession of anthropology. As in most academic disciplines, there has been a long history of White male dominance in North American anthropology. Male

Died for Your Sins: An Indian Manifesto (1988), which includes the following excerpt:

> INTO EACH LIFE, it is said, some rain must fall. Some people have bad horoscopes; others take tips on the stock market ... but Indians have been cursed above all other people in history. Indians have anthropologists.... Over the years anthropologists have succeeded in burying the Indian communities so completely beneath the mass of irrelevant information that the total impact of the scholarly community on Indian people has become one of simple authority.... The implications of the anthropologist ... should be clear for the Indian. Compilation of useless knowledge "for knowledge's sake" should be utterly rejected by the Indian people.... In the meantime it would be wise for anthropologists to get down from their thrones of authority and PURE research and begin helping Indian tribes instead of preying on them. (78-100)

The relationship between Indigenous peoples and anthropologists has significantly improved in recent decades. Many Indigenous people have entered the profession, and anthropologists who continue to work with Indigenous peoples in North America do so largely with their permission and on their behalf. Linguistic anthropologists often work with Indigenous groups in efforts to record and revitalize languages; archaeologists often work in support of claims of Indigenous rights and territories; and cultural anthropologists are often involved in assisting with Traditional Use Studies (TUS) and documenting Traditional Ecological Knowledge (TEK). Anthropologists are also often involved in supporting Indigenous peoples in addressing stereotypes, misconceptions, cultural appropriation, and commodification of their heritage. In many ways, the relationship that anthropologists now have with Indigenous peoples can be characterized as supportive rather than exploitative.

dominance has decreased in recent decades, and ethnic diversity is increasing – but still has a long way to go. Numbers of anthropologists may be roughly equitable in regard to gender, but in regard to ethnicity, people of European descent and light skin color remain a significant majority.

One of the biggest challenges for anthropology today is confronting its legacy of racism. While some have recognized anthropology's imperial origins, many others have been hesitant to acknowledge how racism continues to impact the discipline and create barriers for Black, Indigenous, and people of color (BIPOC) scholars. Among those who do recognize anthropology's racist legacy are a number of BIPOC archaeologists who have been voicing these sentiments for quite some time (e.g., Agbe-Davies, 2002; Battle-Baptiste, 2011; Franklin, 1997a, 1997b). The recent **COVID-19** pandemic, which has disproportionately affected Black and Latinx communities, and the **Black Lives Matter** movement have created a fresh resolve among some anthropologists to face the reality that racism is embedded in the fabric of contemporary society, including within the subfields of anthropology (e.g., Franklin et al., 2020).

Other ongoing challenges for anthropology include sexism, gender-based discrimination, and sexual harassment and violence. As with racism, anthropologists have long recognized the reality of these problems in society but have been slow to address these issues within their own field. Starting in the 1980s, archaeologists began speaking up about gender-based discrimination in the discipline, as well as in depictions of women in the past (e.g., Conkey & Gero, 1991; Gero, 1985). In more recent years, anthropologists have similarly challenged heterosexuality as the norm in the past and have observed that many sexual identities exist and have long existed (e.g., Schmidt & Voss, 2000). Since 2017, the rapid growth of the **#MeToo** movement has also inspired many people to share their stories of sexual harassment, violence, exploitation, and abuse (e.g., Hodgetts et al., 2020), and in light of recent surveys documenting cases of sexism and gender-based discrimination among anthropologists (e.g., Clancy et al., 2014), there has been a renewed vigor in recognizing and addressing these issues within the field of anthropology.

Anthropologists have also recently begun to seriously consider how people with disabilities can be involved in anthropology, including field projects and lab work. Despite the existence of anti-discrimination laws that prevent people with disabilities from being excluded from workplaces and higher education, anthropologists historically have not been very accommodating (Enabled Archaeology Foundation, n.d.; Fraser, 2007; Phillips & Gilchrist, 2012).

Of course, in these early decades of the twenty-first century, anthropologists face many practical challenges as well, including how to work during times of natural disasters, climate change, political upheaval, civil unrest, and pandemics, all of which affect both the people and places that anthropologists study, and the anthropologists themselves. In some cases, planned research is put on hold or canceled, or alternate methods and ways of communication are used. For example, while many anthropological studies have been put on hold during the COVID-19 pandemic, many archaeologists engaged in CRM projects have been deemed **"essential workers,"** meaning that their jobs are considered critical to the maintenance of infrastructure needed for society to function. In such cases, archaeologists have had to adapt their field methods and establish new protocols to keep everyone on site safe (see Figure 1.7).

The early decades of the twenty-first century have also seen anthropology increase its public voice though podcasts and writing for mainstream media. The open-access anthropology magazine *SAPIENS* (sapiens.org), for example, regularly publishes articles and essays by anthropologists, providing insight into humans for a wide, public audience.

ANTHROPOLOGY, COLONIALISM, AND DECOLONIZATION

Anthropologists have been described as agents of colonialism, utilizing anthropological studies of Indigenous peoples in North America to support the systemic structures oppressing Indigenous peoples and disassociating them from lands, resources, and cultures. This includes, for example, the work of cultural anthropologists treating Indigenous peoples and cultures as objects and giving little in return, and archaeologists using government and legal systems to excavate archaeological sites and collect skeletons and artifacts.

A recent movement in North America called decolonization has the goal of dismantling the historic and ongoing systemic oppression of Indigenous peoples in the United States and Canada. This is most apparent in Canada. The movement is led by Indigenous peoples, but many anthropologists are now supporting the movement and acting as agents of decolonization. Besides speaking out against the historical and ongoing impacts of colonialism, anthropologists are also publicly valuing Indigenous ways of knowing as being at least as important as scientific or academic ways of knowing. Archaeologist Dr. George Nicholas (2018) writes, "As ways of knowing, Western and Indigenous Knowledge share several important and fundamental attributes. Both are constantly verified through repetition and verification, inference and prediction, empirical observations and recognition of pattern events." As noted by Nicholas, Western and Indigenous ways of knowing are not necessarily antithetical, and using a combination of both systems can provide mutual support and lead to unanticipated insights.

Some anthropologists are also working toward dismantling colonialism by prioritizing the wishes of Indigenous peoples over the goals of academic studies or government policies, and by speaking out against the systemic oppression of

Figure 1.7
ARCHAEOLOGISTS WORKING IN CHALLENGING TIMES.
Anthropologists often find it necessary to adapt to events occurring around them, both natural and cultural. The COVID-19 pandemic resulted in some archaeologists being considered "essential workers." Pictured here are Dr. Duane Quates, Jack Biggs, and Jeff Burnett doing archaeology on the campus of Michigan State University during the pandemic, taking necessary precautions such as wearing masks and practicing physical distancing.
Credit: Stacey Camp

Indigenous peoples long supported by governments and large segments of White settler society. This also includes stepping away from the centering of Europeans in the narratives of North America's past, such as the use of "prehistory" to refer to the time before Europeans arrived and "New World" to the refer to the Americas, terms that neglect some 15,000 years or more of Indigenous history and occupation. As part of the decolonization process, some anthropologists are also consciously changing their vocabulary, using the preferred Indigenous name for a given community, nation, or tribe rather than the anglicized name often imposed upon them and using terms preferred by Indigenous peoples such as "ancestors" rather than "skeletal remains."

SITUATING ANTHROPOLOGY

Anthropology can be found in multiple contexts, including the academic world, the business world, and popular culture. Perhaps the most common perception of anthropology is that of an academic discipline, operating primarily out of colleges, universities, and museums. Indeed, most colleges and universities have anthropology departments with professors who teach, research, and write. It is usually the results of anthropological work done in the context of the academic world that make their way into mainstream media.

Anthropology has also found a home in many kinds of businesses, however. For example, some companies seek advice from anthropologists in order to better understand the dynamics of their own businesses. Other companies seek anthropological knowledge so as to better understand their clients or business partners and learn how not to offend those they do business with in other parts of the world. Others seek anthropological research and data to better market to target groups.

Some examples of applied anthropology in business are described in Chapter 8, including those of anthropologists working for General Motors. Many other well-known companies, including Google, Microsoft, and Intel, either have their own anthropologists on staff or hire anthropologists as consultants. Another example of applied anthropology is the work of Dr. Robin Nagle, a professor at New York University, who is also the anthropologist-in-residence at the New York City Department of Sanitation. Nagle, whose research focuses on the labor and infrastructure necessary to deal with garbage, wrote an ethnography of the Department of Sanitation called *Picking Up* (2013).

Anthropology can also be considered in the context of popular culture. Anthropologists study popular culture, and both anthropologists and the discipline of anthropology are firmly embedded in popular culture. Of all the fields, archaeology seems to get the most attention in popular culture, with archaeologists commonly portrayed as adventurers and stories revolving around the past. Reports of discoveries of human fossils make their way into mainstream media quickly, and in recent years several successful television programs have been based on the work of forensic anthropologists. Box 1.4 considers both the study of popular culture and the portrayal of anthropology in it.

THE IMPORTANCE OF ANTHROPOLOGY IN AN INCREASINGLY CONNECTED WORLD

As the world becomes increasingly connected, the importance of anthropology also increases. For example, in recent times, anthropologists have been able to make important contributions by helping people suffering from epidemics, natural disasters, and conflict. They do this in multiple ways, including by using their cultural knowledge to help those suffering as well as by educating those seeking to provide aid. This is especially important, for example, in areas where Indigenous peoples may mistrust or not understand modern medicines and health facilities, and where people may have a general mistrust of governments or foreigners. Anthropologists can work in educating or serving as mediators between those providing and those receiving aid. Anthropologists can mitigate potential misunderstandings, and they also recognize, through the holistic perspective, that even emergency aid can have profound effects on other aspects of a culture.

Anthropologists have much to offer in discussions and planning for a sustainable future for people on the planet. They can use research on primate ecology, for example, to help sustain forest environments and support the people who live there. They can use examples from archaeology to demonstrate what has and has not worked for past societies in regard to the long-term sustainability. Importantly, by working with contemporary populations, cultural anthropologists develop both a local and a global view of deficiencies and successes in terms of sustainability.

Other areas where anthropologists make useful contributions are climate change and **food security**. Archaeologists, for example, can cite multiple cases of how people have adapted to changing environments in the past, for instance by building smaller houses in colder times. The field of biological anthropology points to biological markers of stress or malnutrition in the diet resulting from dietary changes or food insecurity. Cultural anthropologists can cite examples of how various communities from around the world are able to maintain food security, or what factors impede it.

A relatively new and important anthropological focus is on the anthropology of outer space. Areas of interest include documenting orbital debris (also known

BOX 1.4 **Anthropology, Popular Culture, and News Media**

Anthropology has an interesting relationship with popular culture and news media. Anthropology and anthropologists are firmly embedded in popular culture, and popular culture is a topic of interest that anthropologists study. When it comes to anthropology in the news, however, there are some problems.

Real anthropological work, featuring the work of real archaeologists, is often featured in semi-scholarly publications such as *National Geographic*, *SAPIENS*, *Archaeology*, and *Smithsonian* magazine.

Anthropology has become firmly embedded in movies, television, and video games. Popular examples include the series of *Indiana Jones* films and the *Lara Croft: Tomb Raider* video game and movie franchise.

Anthropologists are occasionally involved in the creation of movies as well. Primatologist Dr. Michael Reid, for example, served as a consultant on ape behavior for the Hollywood production of *Rise of the Planet of the Apes* (2011), and linguistic anthropologist Dr. Christine Schreyer created the Kryptonian language for the Superman movie *Man of Steel* (2013). Keeping with the theme of artificially created languages, Schreyer also studies the community of contemporary speakers who have learned the Na'vi language created for the movie *Avatar* (2009).

Sometimes anthropology is associated with popular culture through its link with celebrities and politics. For example, Dr. Ann Dunham and her work became popularized after the election of Barack Obama, the 44th president of the United

Figure 1.8 INDIANA JONES.
Fictional anthropologists are embedded in popular culture including movies, television, novels, comic books, and video games. One of the best-known fictional anthropologists is Indiana Jones.
Credit: Courtesy of the Everett Collection

States. Obama is the son of Dunham, who was an anthropologist.

Many anthropologists focus on popular culture as a scholarly area of interest. Anthropologist Shirley Fedorak, for example, has authored a book called *Pop Culture: The Culture of Everyday Life* (2009), which explores such topics as television, music, the internet, folk and body art, sports, food, and wedding rituals through the lens of anthropology.

Anthropological research and perspectives occasionally reach mainstream news media, but they are often filtered through a journalistic lens and reinforce historical stereotypes of the discipline. To provide some insight on how anthropology is portrayed in mainstream news media, anthropologist Dr. Hugh Gusterson (2013) researched mentions of anthropology in the *New York Times*. Gusterson discovered that rather than writing the stories themselves, anthropologists are most often mentioned in articles written by others. In other words, according to Gusterson, "They are ventriloquized to the public by reporters who turn them into stories or use their quotes to enliven other stories" (p. 12). Gusterson also observed that the *New York Times* overrepresented archaeology and biological anthropology and that, in regard to cultural anthropology, "[j]ournalists help perpetuate a notion of anthropologists as guardians of the savage-slot" (p. 12). Those who decide which anthropology-related pieces get published are evidently biased by their own perceptions and assumption of the discipline. Gusterson writes, "readers of the *New York Times* are left with a picture of anthropology as an enterprise in the salvage and preservation of marginal peoples or commentary on the fluffy bits of human behavior that don't interest economists and political scientists," and "anthropologists are constructed in the public sphere as having little to say about some of the most urgent and pressing political and economic controversies of the day" (p. 13). It is suggested that while economists and political scientists have been able to extend their expertise beyond the confines for which they were trained, and in some cases become general or public intellectuals, this hasn't been the case for anthropologists, who have been relegated to the marginal areas of interest. This is a shame, most anthropologists would probably agree, since anthropologists have much to offer in global debates about many significant issues, including conflict, food security, gender, migration, refugees, and sustainability. The holistic, comparative, and evolutionary approaches of anthropology and the significant amount of anthropological research undertaken on these and other topics are valuable, but for many are underappreciated. Rather than relying on mainstream media to publicize their research and perspectives, however, some anthropologists are increasingly turning to social media, blogging, or writing columns for popular outlets such as *SAPIENS* (sapiens.org).

Figure 1.9
ASTRONAUT ON THE MOON.

Outer space is an emergent area of interest in anthropology, with anthropologists of each of the four fields having much to contribute in regard to understanding the impact of being in space on the human body and the importance of biological and cultural diversity for future colonies in space.

Credit: Courtesy of NASA

as space junk) and remains of human activity on Earth's moon and on other planets. Biological anthropologists have roles to play in studying the impact of outer space on human bodies and recognizing the importance of biological diversity on future colonies in space. Cultural anthropologists, too, have important roles to play in future colonies in space as well as in studying the culture of those involved in space exploration. Archaeologists Dr. Justin Walsh and Dr. Alice Gorman (also known as Dr. Space Junk) are involved in studying the culture of the International Space Station, primarily through examining the close to one million photographs that have already been taken aboard the space station and doing ethnographic observations of space station materials as they return to Earth (issarchaeology.org). NASA has recently published an open-access book featuring some of the contributions of anthropologists to space exploration (Vakoch, 2014).

Anthropologists, perhaps more than most people, recognize the value of diversity, both biological and cultural. It is important that as the world becomes increasingly connected, both biological and cultural diversity be appreciated. Ultimately, it may ensure the survival of our species.

SUMMARY

This chapter has provided an overview of anthropology. One objective has been to clarify the nature and scope of the discipline, including important terminology, concepts, and perspectives. Another objective has been to prepare students for what lies ahead in subsequent chapters. Mirroring the Learning Objectives stated in the chapter opening, the key points can be summed up as follows:

- Anthropology is the scholarly study of humans. This includes human biology and human cultures, past and present. The perspectives, methods, theories, and research results of anthropology provide a good framework (or lens) through which to view and understand humans. The four major branches of anthropology are cultural anthropology, archaeology, biological anthropology, and linguistic anthropology. Each branch includes an applied component.
- The anthropological perspective includes holistic, evolutionary, comparative, and qualitative approaches. This perspective also recognizes the importance of examining links between various components of human biology and culture, collecting one's own data, and focusing on understanding how and why things change.
- Culture can be defined as the learned and shared things that a group of people have, think, and do. The principal components of culture include subsistence strategies, diet, social and political systems, communication, technology, art, and ideology. Components are interrelated and influence each other. Cultures are constantly changing.
- The history of anthropology in North America has largely been focused on the Indigenous peoples of the continent, although anthropological interests have broadened significantly in recent decades. Franz Boas is widely recognized as a very influential figure in anthropology in North America for both his own research contributions and his training of future anthropologists.

- Contemporary challenges for anthropologists include overcoming the discipline's legacy of racism, sexism, and other forms of discrimination, as well as the challenges of addressing decolonization and working in an age of global pandemics, climate change, and civil and political upheaval.
- Anthropology may be considered in the contexts of academia, business, and popular culture. Most pure research is undertaken by those in academia, although there are applied components (mostly in business applications in each of the subfields). Many corporations hire anthropologists to research and provide insight into their own employees as well as their partners and clients. Anthropologists study popular culture and are embedded in it.
- Anthropology has an important role to play in an increasingly connected world. Anthropologists can make important contributions to helping those in need in times of disaster, and they can offer examples and suggestions of how to cope with the problems of living in the twenty-first century, including issues related to food security, sustainability, and climate change.

Review Questions

1. How do anthropologists define *anthropology*, *human*, and *culture*?

2. What is the appeal of anthropology?

3. What are the main components of culture?

4. What are the key elements of the anthropological perspective?

5. What are the main branches of anthropology?

6. What is the history of anthropology in North America?

7. What are the contemporary challenges facing anthropologists?

8. How is anthropology situated in the contexts of academia, the business world, and popular culture?

9. Why is anthropology important in an increasingly connected world?

Discussion Questions

1. What might be some of the advantages of using an anthropological perspective to view and understand the world?

2. What might be some of the disadvantages of using an anthropological perspective to view and understand the world?

3. Should anthropologists be considered "essential workers"?

4. How has anthropology been complicit in perpetuating colonialism and its enduring legacy of oppression?

Visit **www.lensofanthropology.com** for the following additional resources:

SELF-STUDY QUESTIONS **WEBLINKS** **FURTHER READING**

2

WE ARE PRIMATES: THE PRIMATE BACKGROUND

LEARNING OBJECTIVES

In this chapter, students will learn:

- *why it is important to understand humans as part of the primate world.*
- *the basics of primate taxonomy.*
- *about key events in primate evolution.*
- *the methods anthropologists use to study primates in the wild.*
- *the principal research interests and findings of anthropologists studying nonhuman primate behavior.*

The first thing to understand is that humans are primates. The second is that studying the evolution and behavior of other primates helps us understand ourselves.
#WeArePrimates

- Studying primates provides an understanding of our place in the world. Recognizing the similarities in biology and behavior is important to understanding that humans are not quite as unique as many think they are.

- Understanding humans as primates allows anthropologists to make inferences about the conditions under which evolutionary changes occur.

- Understanding our common evolutionary history helps to explain the various ways of adapting to environments biologically. This includes understanding that there is not necessarily a single best way for adaptation to occur. In terms of biology, similar problems can be solved in different ways.

- It is important to see how various primates adapt to similar circumstances in different ways, behaviorally. This is important to understanding how humans came to be and the behavioral strategies they adopted.

- Studying contemporary primates is useful for providing models of how early humans may have lived. This includes such things as group size, subsistence and settlement strategies, diet, and social and political systems.

- Studying contemporary primates can provide models for understanding how human culture, including language and tool use, may have evolved.

- Studying primates is useful to other areas of inquiry, including evolutionary biology.

- Studying primates today is important for sustainability. Primates are integral to many natural ecosystems. With the knowledge obtained from studying primates, anthropologists may educate others and help maintain the sustainability of the natural environments where primates live.

- Studying primates is important for understanding issues related to diet. Anthropological knowledge of primate diets can be used to provide education about the relationships between diet and biology, and the correlation of biological changes with dietary changes.

- Studying primates is important for aiding the rehabilitation of primates that have been removed from their natural environments.

- Studying primates is important for aiding their protection and conservation. Anthropologists know that diversity, both biological and cultural, is fundamentally important. Preserving primate diversity and their habitats is important for their survival.

- Studying primates is important for being able to critically evaluate popular, pseudoscientific, and anti-scientific ideas about humans, other primates, and the past.

INTRODUCTION

To fully understand human biology and human culture, it is important to understand humans as primates. Anthropologists know that humans are not quite as unique as many believe us to be. An anthropologist's comprehensive understanding of humans involves not only people from around the world today, but also those from the past. It also includes fundamental knowledge of how we are related to the other animals most like us, typically in the taxonomic order **Primates**. Anthropologists know that a comprehensive understanding of both human biology and human culture includes knowledge of the evolution of primates and the behavior of nonhuman primates.

Anthropologists recognize that there are problems with using primates as models for ancestral humans. Anthropologists are usually careful to explicitly recognize the limitations of their research in formulating models of how early humans may have been, but the research remains useful for suggesting possibilities. Basically, if behaviors are observed in contemporary populations or among groups of people from the recent past, and similar behaviors are observed in our closest relatives, then there is a good chance that our human ancestors behaved in the same way as well.

PRIMATE TAXONOMY

Primates are a **taxonomic order** belonging to the class Mammalia, commonly known as mammals. Biologists currently recognize about 5,500 different species of mammals and divide them into more than a dozen orders – an order being a major subdivision of the class. Placement in the order Primates is dependent on having most or all of a specific set of characteristics that distinguish the order from all other orders.

Anthropologists generally accept that there are about 500 species of primates. Discrepancies occur as new species are identified, and distinctions between species and subspecies are often unclear. More than 100 new species and subspecies have been described since 1990.

The principal characteristics that distinguish the order Primates are outlined below. The characteristics are not necessarily unique to primates, and not all primates necessarily have all the characteristics listed. If an animal has most of the characteristics listed, however, it is almost certainly a primate. The characteristics listed are those that are most frequently used, but as with many things in the sciences, definitions are often fluid and subject to debate. Some scientists identify more than two dozen distinguishing criteria of primates (see, for example, Tuttle, 2014).

With respect to primates, prehensibility (from **prehensile**) refers to the ability to grasp things with the digits of the hand and/or feet and, in some cases, the tail. Grasping is enhanced by

Distinguishing Characteristics of Primates

- Prehensile hands and feet
- Nails instead of claws
- Forward-facing eyes/ vstereoscopic vision
- Large brains (both in actual size and relative to the rest of the body)
- Single offspring
- Long period of infant dependency
- Diurnal
- Arboreal
- Movement in a variety of ways (quadrupedal, knuckle-walking, climbing, clinging/jumping, brachiation, bipedal)
- Social
- Nonspecialized diets

flexibility in fingers and toes, allowing the digits to separate and bend, and by the opposability of thumbs and, in many primates, toes.

Just as we have fingernails and toenails, so do the other approximately 500 species of primates. Primates have forward-facing eyes, which allow an overlapping field of vision from both eyes, which in turn provides for excellent depth perception. Primates tend to have large brains, especially in relation to the rest of their body. Although there are exceptions, most primate species have one offspring at a time, and there is a lengthy period of dependency on the mother. With rare exceptions (e.g., orangutans), primates are very social and live in groups, often with established hierarchies, important family relationships, and friendships. The overwhelming majority of primates are diurnal (active during the day), although some species are most active during the night (nocturnal). Similarly, most primates are arboreal (spending most of their time in trees), although there are several exceptions including, of course, humans.

Primate diets are diverse (see Box 2.1). As an order, primates may be considered to be omnivorous, but in practice particular **taxa** tend to specialize in fruit, leaves, or insects. Meat eating is rare but does occur among some ape and monkey populations.

Primates are often characterized as having a generalized body plan, which means they can do a lot of different things that many other animals cannot. No other mammal, for example, has the flexibility of limbs seen in primates. With this flexibility, primates can run, jump, move sideways, move on two legs for at least short distances, climb, leap, swing, and brachiate (or swing from arm to arm).

Primate dentition is also considered generalized. Incisors, canines, premolars, and molars enable primates to eat a wide variety of foods, unlike other mammals that typically have teeth specialized for either hunting and meat eating (e.g., carnivores) or grazing (e.g., herbivores).

There are many categories within the order Primates, including suborders, infraorders, superfamilies, families, genera, and species. These are illustrated in Figure 2.1. What follows is a brief overview of the various categories, focusing on those that include humans as a member.

There are two suborders of primates: **Strepsirhini** and **Haplorhini**. The distinguishing characteristics of each are mostly relative, as shown in Table 2.1. Most of the species that people would easily recognize as primates are members of the suborder Haplorhini, including all monkeys, apes, and humans. Strepsirhini include many different species, but the most common are known as lemurs.

Strepsirhini include those primates that are less obviously "primates" to nonspecialists, meaning they may lack the full set of distinguishing characteristics, or the characteristics are in less typical form. Compared to most other primates, Strepsirhini tend to have a greater reliance on **olfaction** (sense of smell). Associated with this, they tend to have a larger snout than other primates, and a **rhinarium**.

BOX 2.1   **Food Matters: Primate Diets**

Primates, as a taxonomic order, may be considered **omnivorous**. In practical terms, this means that primates eat a wide range of foods including plants, insects, and, in some cases, small mammals. It would be wrong to think, however, that all primate species or populations are omnivorous. Some species focus on fruits, some on rougher foliage, and others on insects.

The three principal dietary strategies of primates are **frugivory**, **folivory**, and **insectivory**. Frugivory (having a diet focused on fruits) is the most common, although it is not unusual for frugivorous primates to supplement their diets with leaves and insects. Folivory (having a diet comprised mostly of leaves and other rough foliage) and insectivory (having a diet comprised of insects) tend to be linked to body size. Smaller primates, including many of the strepsirhines and some of the smaller haplorhines, may be considered to be primarily insectivores, although they also eat plants. The larger primates, such as gorillas and orangutans, tend to be, at least in some contexts, folivores. Orangutans, for example, tend to prefer fruit but also eat leaves, bark, and insects. Likewise, gorillas favor fruit but will often eat foliage. Chimpanzees also tend to prefer fruit but also eat foliage, insects, and, in some cases, small mammals they hunt.

Based on numerous primate studies on diet, Chapman and Chapman (1990) wrote an article called "Dietary Variability in Primate Populations," published in the journal *Primates*. They were able to demonstrate that primate populations often switch between frugivory, folivory, and insectivory. They note, for example, that orangutans in their study had been observed to be primarily frugivorous one month (with 90 percent of their feeding time spent on fruit and the other 10 percent split evenly between leaves and insects), and primarily folivorous another month (with 75 percent of their feeding time spent on leaves, 15 percent on bark, and 10 percent on fruit). In another example, they showed that a population of spider monkeys in Costa Rica switched strategies as well, one month eating only fruit, one month eating mostly leaves, and spending significant time in another month eating insects.

Primate diet is linked with biology. The teeth of primates that are primarily insectivores, for example, have molars with pointed cusps that allow the efficient piercing of insect exoskeletons. Teeth of folivorous primates, on the other hand, are effective for slicing through leaves and other rough foliage, while the teeth of primates that are mostly frugivorous tend to have rounded cusps, enabling effective crushing of fruit. Depending on their diet, primates will have differing kinds of microorganisms to aid in digestion, and the features of their digestive tracts will also differ. Chimpanzees, for example, can digest foods that humans cannot due to different kinds of microbiota and the length of their intestines.

Other kinds of adaptations to accommodate diet include stomach size and the presence of cheek pouches. Some primates, such as baboons and macaques, have large cheek pouches, which allow them to store food temporarily. Chimpanzees, on the other hand, have relatively large stomachs.

In some areas, it isn't unusual for multiple species of primates to occupy the same environment. One of the reasons for this is that since they have different diets, there is little or no competition for food between the different species.

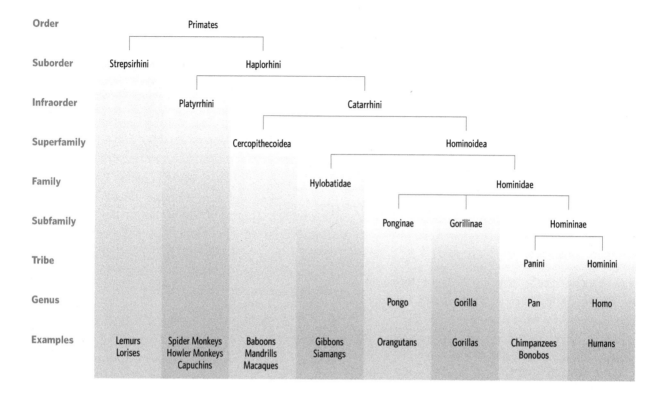

Order — Primates
Suborder — Strepsirhini, Haplorhini
Infraorder — Platyrrhini, Catarrhini
Superfamily — Cercopithecoidea, Hominoidea
Family — Hylobatidae, Hominidae
Subfamily — Ponginae, Gorillinae, Homininae
Tribe — Panini, Hominini
Genus — Pongo, Gorilla, Pan, Homo
Examples — Lemurs, Lorises / Spider Monkeys, Howler Monkeys, Capuchins / Baboons, Mandrills, Macaques / Gibbons, Siamangs / Orangutans / Gorillas / Chimpanzees, Bonobos / Humans

Figure 2.1
PRIMATE TAXONOMY.
This figure presents one of several ways of classifying primates, including humans. The diagram is simplified to focus on major categories of apes and humans.

Relatively few species of primates are nocturnal, but most of those that are belong to this suborder. Those primate species that retain a claw in addition to fingernails and toenails are strepsirhine. In general, compared to other primates, Strepsirhini tend to have more limited prehensibility, and some lack color vision. Although some species of Strepsirhini inhabit regions of tropical Asia and mainland Africa, most species live on the island of Madagascar off the southeast coast of Africa. One of the most well-known kinds of Strepsirhini is the ring-tailed lemur.

The Haplorhini, compared to Strepsirhini, have no rhinarium, worse olfaction, and better vision. Nocturnalism and multiple births occur among some species of Strepsirhini, but both characteristics are rare among Haplorhini.

There are about 40 species of Strepsirhini and several recognizable taxa, such as families, genera, and species. Some anthropologists focus their research among Strepsirhini, especially in Madagascar, but because humans belong to the suborder Haplorhini, there is more anthropological interest in Haplorhini than Strepsirhini.

There are two infraorders of Haplorhini that anthropologists study, the **Platyrrhini** and the **Catarrhini**. Major differences between the two are outlined in Table 2.2. Platyrrhini is roughly synonymous with "New World monkeys," which in practical terms means the monkeys of Central and South America. Catarrhini includes "Old World monkeys," meaning monkeys of Africa, Asia, and Europe, as well as apes and humans.

There are several distinguishing characteristics of Platyrrhini. The term *Platyrrhini* itself refers to the characteristics of the nose – basically flat, with nostrils flaring outward.

TABLE 2.1

Differences between Strepsirhini and Haplorhini

	Strepsirhini	Haplorhini
Sense of Smell	Better	Worse
Prognathism	More	Less
Rhinarium	Present	Absent
Sense of Vision	Worse	Better
Nocturnal	Some species	Rare
Multiple Births	Some species (twins)	Rare
Brain Size	Smaller	Larger

TABLE 2.2

Differences between Platyrrhini and Catarrhini

	Platyrrhini	Catarrhini
Location	Central and South America	Africa, Asia, and Europe
Body Size	Smaller on average	Larger on average
Arboreal	All species	Most species
Prehensile Tail	Some species	No species
Dental Formula	2-1-3-3	2-1-2-3
Nostrils	Widely spaced, flaring outward	Closely spaced, facing down
Sexual Dimorphism	Relatively little	Often pronounced

Their natural habitat includes the tropical and subtropical forested regions of Central and South America. All monkeys have tails, but it is only among some species of Platyrrhini that the tail is prehensile. If one observes a monkey hanging by its tail, it is a Platyrrhini. All Platyrrhini are primarily arboreal, tend to be smaller than the monkeys of Africa, Asia, and Europe, and exhibit relatively little sexual dimorphism. Widely known Platyrrhini include spider monkeys, squirrel monkeys, howler monkeys, and capuchins.

Catarrhini are distinguished from Platyrrhini in several ways. Catarrhini include the Old World monkeys (**Cercopithecoidea**), as well as all the apes and humans.

Figure 2.2
LEMUR.
Neither monkey nor ape, lemurs belong to the primate suborder Strepsirrhini.

Credit: Kevin Schafer/Alamy

Most Catarrhini are arboreal, but some, like baboons and gorillas, spend considerable time on the ground. Forests are their most common habitat, but some Catarrhini (e.g., baboons) occupy savannah-grassland environments as well. The term *Catarrhini* refers to the narrow nose, which has closely spaced, downward-facing nostrils.

Dental formula refers to the kind and number of teeth in the mouth. It is usually expressed as a series of numbers for each quarter of the mouth, going from front to back. All Catarrhini, including humans, have the same dental formula, commonly expressed as 2–1–2–3, where the first "2" indicates incisors, the "1" represents canines, the second "2" represents premolars, and the "3" represents molars. Among Catarrhini, this dental formula is the same for each quarter of the mouth – upper right, upper left, lower right, lower left. In total, adult Catarrhini have 32 teeth. Many find it interesting that humans have the same dental formula as chimpanzee, gorillas, baboons, and more than 100 other species of monkeys and apes from Africa, Asia, and Europe. The third molar is the tooth in humans commonly known in North America as the wisdom tooth. Many people now have the wisdom tooth removed before it erupts; sometimes it simply never grows. The lack of the third molar also occurs on occasion among other primates.

Catarrhini comprise two **superfamilies**: Cercopithecoidea and **Hominoidea**. Cercopithecoidea are the Old World monkeys. Hominoidea include apes and humans. Principal differences are listed in Table 2.3. Cercopithecoidea, or Old World monkeys, have tails, but these are not prehensile. Old World monkeys also tend to be larger than their New World counterparts, and they have more sexual

Figure 2.3
MACAQUE.
Macaques are monkeys, belonging to the primate infraorder Catarrhini.
Credit: Nadine Ryan

Figure 2.4
CAPUCHIN.
Capuchins are monkeys belonging to the primate infraorder Platyrrhini.
Credit: Steve Taylor ARPS/Alamy

TABLE 2.3

Differences between Cercopithecoidea and Hominoidea

	Cercopithecoidea	Hominoidea
Common Names	Old World monkeys	Apes and humans
Tail	Present	Absent
Brain Size	Smaller (less developed)	Larger (more developed)
Body Size	Smaller	Larger
Ontogeny	Shorter	Longer
Shoulders	Less developed	More developed

dimorphism. Hominoidea are tailless, tend to be larger than Cercopithecoidea, and have extended **ontogeny** (i.e., increased length of dependency), larger and more developed brains, and more developed shoulders, enabling **brachiation**.

There is no consensus on the subdivision of the Hominoidea. One popular classification system (and the one used in this text) recognizes three families of Hominoidea: (1) **Hylobatidae**, which includes gibbons and siamangs of Southeast Asia, sometimes known as the "lesser apes"; (2) **Pongidae**, which includes three genera – **Pongo** (the genus to which orangutans belong), Gorilla, and **Pan** (which includes two species: *Pan troglodytes* [chimpanzees] and *Pan paniscus* [bonobos]); and (3) Homininae, the family to which humans belong, as do all ancestors since the split from the common ancestor of chimpanzees about seven million years ago. In this book, we consider Homininae – and its informal name hominin – to equate with human and to include the **genus** *Homo* and all other genera of bipedal primates descended from the common ancestor of humans, chimpanzees, and bonobos. This means that there is only one genus of hominin today (i.e., *Homo*), but hominins also include genera of the past, such as *Australopithecus*, *Ardipithecus*, *Kenyanthropus*, *Sahelanthropus*, *Orrorin*, and *Paranthropus*.

It is important to recognize that other classification systems do exist. When some refer to hominin, they may be referring to a form of classification that also includes chimpanzees and bonobos. Most biological classification systems simply use the categories of species, genus (a group of similar species), and family (a group of similar genera), but when describing humans and apes, some use additional categories such as subfamily (a subcategory of family), tribe (a subcategory of subfamily), and subtribe (subcategory of tribe).

Table 2.4 summarizes the place of humans in the primate world.

TABLE 2.4

Human Taxonomy

Taxonomic Group	Includes	Examples
Class Mammalia	29 orders, about 5,500 species	Cats, dogs, bats, rats
Order Primates	About 500 species	Monkeys, apes
Suborder Haplorhini	About 300 species	Old and New World monkeys, apes
Infraorder Catarrhini	About 200 species	Old World monkeys, apes
Superfamily Hominoidea	All apes and humans	Chimpanzees, bonobos, gorillas
Family Homininae	Only humans	*Homo sapiens* and ancestral humans

PRIMATE EVOLUTION

Primate evolution has occurred primarily, and perhaps entirely, in the Cenozoic era, which began 65 million years ago and continues to the present. The Cenozoic is one of four geological eras, the others being the Mesozoic, from roughly 250 to 65 million years ago, in the age of the dinosaurs; the Paleozoic, from about 540 to 250 million years ago; and the Precambrian, from the origins of the earth about 4.5 billion years ago to 540 million years ago.

In the latter stage of the Mesozoic, mammals were in existence but exhibiting nowhere near the diversity and abundance they would come to express in the Cenozoic. Conventional thinking is that when the dinosaurs became extinct about 65 million years ago, likely due to environmental change, the new environments and lack of dinosaurs opened up new ecological niches, which mammals quickly filled. This diversification is a good example of **adaptive radiation**. It is likely that many new kinds of mammals, including primates, evolved to fill these niches.

Those who identify early remains as belonging to the order Primates and the further subdivisions all the way down to **species** necessarily make choices about which remains belong to what species. When assigning various remains to different species, we are making assumptions about the ability of different animals to mate and prodvuce fertile offspring based on bones that are sometimes tens of millions

Figure 2.5
ORANGUTAN.
Orangutans are critically endangered apes whose natural habitat is in Indonesia and Malaysia.
Credit: Kabir Bakie/CC BY-SA 2.5

of years old. We really don't know the variability within various species, genera, and other categories. A relatively small sample size also affects our ability to classify.

Understanding and describing primate evolution is simplified through the use of time periods called epochs, created by geologists. Epochs are based in part on observable changes in the geological record, including the fossil record of plants and animals. There has been discussion about naming a new epoch – the **Anthropocene** – based on significant changes to the earth caused by human activities, but there is as of yet no agreement on a start date for the Anthropocene, with some suggesting about 12,000 years ago and others suggesting a start date in the 1800s. The principal developments in primate evolution, by epochs of the Cenozoic era, are outlined in Table 2.5.

The fossil record of primates in the Paleocene is sketchy. There is some evidence of mammals that appear well suited for life in trees (e.g., features indicate climbing capabilities) and that have primate-like teeth, but whether this is enough to classify them as primates is debatable. One potential primate from this period is *Purgatorius*, discovered in Montana. Some suggest it may be ancestral to all later primates, but it should be remembered that its classification as a primate is itself dubious.

Primates were certainly well established by the Eocene. Dozens of primate species, at the least, lived during this time period; some suggest there may have

BOX 2.2   Rafting Monkeys

One of the most intriguing areas of interest in the study of primate evolution has to do with the origin of the Platyrrhini, commonly known as the New World monkeys. They begin to appear in the fossil record of Central and South America about 30 million years ago, but their antecedents are unknown.

One common hypothesis suggests there was ongoing evolution from some of the earliest forms of primates existing in North America during the Eocene. A problem is the lack of supporting evidence in the fossil record. While there were animals best described as primates in the region during the Oligocene, there is no evidence of primates at all in the fossil record for at least 10 million years before the Platyrrihini, and those early forms did not resemble monkeys.

Another hypothesis is that the earliest Platyrrhini rafted over from Africa. The rafts were presumably floating islands of natural debris. To some it seems like a far-fetched idea, but for many it seems the most probable scenario. This hypothesis is covered by Dr. Alan de Queiroz (2014) in his book *The Monkey's Voyage: How Improbable Journeys Shaped the History of Life*. Several kinds of evidence are used in support of this hypothesis, including the knowledge that primates made it hundreds of miles from Africa to the island of Madagascar tens of millions of years ago (presumably by rafting on floating vegetation) and that, in more recent times, early populations of humans apparently traveled to some of the islands of Southeast Asia. Other support includes the observation of very large natural rafts, including one described to be as large as Belgium and another with living trees growing to a height of 30 feet. Such natural rafts have been observed to occur at the mouths of rivers, where fallen trees and other kinds of vegetation accumulate and on occasion float into the ocean.

One of the biggest problems to deal with in accepting the rafting hypothesis is the sheer distance of the journey from Africa across the ocean to Central or South America. But as pointed out by de Queiroz, while the distance today is about 1,800 miles, it was probably half that about 40 million years ago. It has also been suggested that there were likely multiple islands in the ocean during the voyage.

Models factoring distance, currents, and winds suggest that the voyage may best be measured in days rather than weeks. Further support is provided by the recognition of molecular studies suggesting that other animals also made the journey from Africa to the Americas via natural rafts, although these other animals were mostly lizards and snakes.

The fossil evidence may also be interpreted as supporting the rafting hypothesis. There are some primate fossils from the Oligocene in Africa that appear to be good ancestral candidates for both the Platyrrhini and Catarrhini, with evidence including, for example, the dental formula most commonly associated with Platyrrhini (2-1-3-3).

TABLE 2.5

Overview of Primate Evolution 65 Million Years Ago (MYA) to Present

Geological Epoch (Time Period)	Key Events
Paleocene (65–55 MYA)	Probable emergence of primates
Eocene (55–34 MYA)	Proliferation of primates
Oligocene (34–23 MYA)	Establishment of Catarrhini and Platyrrhini
Miocene (23–5.3 MYA)	Probable emergence of Hominoidea and Homininae
Pliocene (5.3–2.6 MYA)	Multiple genera of Homininae, emergence of genus *Homo*
Pleistocene (2.6 MYA–11,700 years ago)	Many species of humans
Holocene (11,700 years ago to present)	*Homo sapiens* dominates

been as many as 200. Evident primate characteristics include those related to an emphasis on vision and prehensibility. The diversity of remains suggests that there was already an evolutionary split between Strepsirhini and Haplorhini, meaning the last common ancestor that monkeys, apes, and humans had with lemurs lived more than 34 million years ago.

The Oligocene is the time period when Catarrhini and Platyrrhini become evident. Many believe that around 30 million years ago a population of a monkey or monkey-like primate floated on a natural raft of vegetation from Africa to Central or South America and became the founding population of all subsequent primates there.

The Miocene is associated with the emergence and proliferation of Hominoidea. The last common ancestor of orangutans, gorillas, chimpanzees, and bonobos probably lived about 20 million years ago. The last common ancestor of gorillas, chimpanzees, bonobos, and humans probably lived about 10 million years ago. The last common ancestor of chimpanzees, bonobos, and humans probably lived about seven million years ago.

It is important to understand that all contemporary primates have evolved and continue to do so. This includes humans. Humans did not evolve from chimpanzees, but they have a common ancestor with them. Chimpanzees (*Pan troglodytes*) and bonobos (*Pan paniscus*) have continued to evolve as well. The last common ancestor of chimpanzees and bonobos lived about two million years ago. It is likely that

Figure 2.6
GORILLA.
Gorillas are the largest living primates, with adults sometimes exceeding 400 pounds (180 kg). They are a member of the superfamily Hominoidea.
Credit: David Cantrille/Alamy

groups of the common ancestor to both became separated by the Congo River in Africa at about that time.

PRIMATE BEHAVIOR

As outlined in the opening section of this chapter, there are many reasons for anthropologists to study nonhuman primate behavior. These include providing models of how early humans may have lived, an understanding of the advantages of certain kinds of behaviors exhibited by all primates, including humans, and investigating the various ways our closest relatives have of solving problems.

Methods

Like studying people, studying primates in the wild requires a long-term commitment on the part of anthropologists. The primates being studied need to be able to trust the human observers, and the observers need the primates to act as normally as possible – as if the observers were not there.

Most primatologists with a background in anthropology prefer to study primates in the wild rather than in zoos or research facilities. This is mostly because anthropologists are interested primarily in how primates behave in the wild, without human interference. Also, primatologists trained in anthropology tend to be more interested in what primates actually do in the wild, rather than what they may be capable of, which is often the focus of zoo or lab-based studies. Anthropologists know, for example, that some nonhuman primates can be taught to use lighters to start fires and to drive golf carts, but this has relatively little value to anthropology.

The primary method of primatology is direct observation of primate groups in the wild. Researchers use scientific methods, including hypothesis testing, and collect data using specific techniques such as focal-animal sampling, in which one individual is followed for a specific period. Evidence concerning that individual's behaviors is recorded, and then the next "subject" on a prepared list is observed. This method, plus the use of filming, reduces unintentional bias that may otherwise occur when the researcher's attention is drawn by the biggest, flashiest animals in a group. Collection of dung for fecal analysis may supplement observations, allowing for analysis of diet, stress, or reproductive hormones.

Three of the most well-known primatologists are Dr. Jane Goodall, Dr. Dian Fossey, and Dr. Biruté Galdikas. Box 2.3 includes samples of their reflections and observations.

Principal Research Interests

Communication

Research on primate communication includes studying vocalizations, gestures, expressions, and language. Research reveals several reasons for vocalizations, or calls. Vocalizations may be used to identify information about the sender, such as individual identity, as well as the location of individuals, food sources, and potential threats (e.g., predators). Among a number of monkey species, for example, there are specific alarm calls for specific predators. Gestures include such activities as arm waving, hugging, and, among gorillas, chest beating. Primate communication is also covered in Chapter 9.

Social Structure

Many studies focus on the social structure of primate groups. This includes the ways in which groups form – whether, for example, it is the males or females, or both, that leave their home community to find or form a new group. Among groups of Central American monkeys, for example, capuchin males leave their home group upon maturity; among spider monkeys, it is the females that leave the home group; and among howler monkeys, both males and females leave their home group. Biologically, finding or forming

Principal Research Interests in Primatology

- Communication (vocalizations, gestures, displays, expressions)
- Social structure (how groups are formed, how dominance is achieved and maintained)
- Aggressive and affiliative behaviors (conflict, grooming)
- Subsistence and diet
- Tool use

a new group keeps the gene pool diverse and prevents inbreeding among families of primates.

Primate groups are often rigidly hierarchical. Females are equal to or dominant over males in about 40 percent of primate groups. Research is often focused on how dominance is achieved and maintained, such as through strength, bluff, cleverness, and alliances. While males are dominant in most primate groups, in some, such as lemurs, females are dominant. This is usually explained as being adaptive to the environment. Female dominance ensures that mothers obtain enough food for the survival of infants in seasons in which resources are depleted.

Aggressive and Affiliative Behaviors

Affiliative and aggressive behaviors are another area of research interest. One of the most common affiliative behaviors among primates is grooming. The hygienic aspect of grooming, such as removing bugs and dead skin, is usually understood to be a by-product of the behavior, while its real importance lies in socialization. Grooming is largely viewed as a social and political activity, reaffirming alliances, relationships, and group cohesion. Other affiliative behaviors seen among primates include hugging, patting, and kissing.

Figure 2.7
CHIMPANZEE.
Pan troglodytes.
Credit: Kjersti Joergensen/ Shutterstock

Among bonobos, sex can be viewed as an affiliative behavior. Most research indicates that bonobos are quite unlike chimpanzees and other primates. Where most primate groups have dominance structures, bonobos tend to be roughly **egalitarian**. Alpha females are often the highest ranked members of the group, leading to a much less violent society than chimp society, which is run by alpha males. Where other primates may fight or make threats to resolve conflict, bonobos have sexual relations. Sex among bonobos is as common as a handshake or hug among humans. They are one of the few primates to have face-to-face intercourse occasionally, and they commonly have male–female sex, female–female sex, and male–male sex.

BOX 2.3 · **In the Words of Three Prominent Primatologists**

Three of the most well-known primatologists are Drs. Jane Goodall, Dian Fossey, and Biruté Galdikas. Each was encouraged and assisted by Dr. Louis Leakey, one of the world's best-known anthropologists of the mid-twentieth century, who saw the need for a better understanding of apes in order to understand humans. At times, these three primatologists have been called "The Trimates" and "Leakey's Angels."

Dr. Jane Goodall began studying chimpanzees at Gombe in Tanzania in 1960. Some 30 years later she reflected on her ground-breaking observations that chimpanzees ate meat and made tools:

> From the Peak I had seen, for the first time, a chimpanzee eating meat: David Greybeard. I had watched him leap up into a tree clutching the carcass of an infant bushpig, which he shared with a female while the adult pigs charged below. And only about a hundred yards from the Peak, on a never-to-be forgotten day in October, 1960, I had watched David Greybeard, along with his close friend Goliath, fishing for termites with stems of grass. Thinking back to that far-off time I relived the thrill I had felt when I saw David reach out, pick a wide blade of grass and trim it carefully so that it could more easily be poked into the narrow passage of a termite mound. Not only was he using the grass as a tool – he was, by modifying it to suit a special purpose, actually showing the crude beginnings of tool-making. What excited telegrams I had sent off to Louis Leakey, that far-sighted genius who had instigated the research at Gombe. Humans were not, after all, the only tool-making animals. Nor were the chimpanzees the placid vegetarians that people had supposed. (Goodall, 1990, p. 5)

Dr. Dian Fossey studied gorillas in Rwanda from 1966 to 1985, when she was tragically murdered in her camp. Here she describes her first observations of gorillas in the wild:

> There they were: the devilmen of native stories; the basis of the King Kong myth; the last of the Mountain Kings of Africa. A group of about six adult gorillas stared apprehensively back at us through the opening in the wall of vegetation. A phalanx of enormous, half-seen, looming black bodies surmounted by shiny black patent-leather faces with deep-set warm

Aggressive behaviors include physical displays, bluffs, gorilla chest beating, threats, and physical contact. Some primates do get physically aggressive, and some males have been observed to fight to the death in conflicts revolving around dominance. Lethal violence has been observed in a number of chimpanzee communities. One group of male chimpanzees has been observed to systematically hunt and kill males from another group, although this is certainly not a normal kind of behavior.

brown eyes. They were big and imposing, but not monstrous at all. Somehow they looked more like members of a picnic party surprised by interlopers. (as cited in Mowat, 1987, p. 14)

Fossey describes a few other early observations:

[W]e encountered an adult male – a black-back – approximately eight to nine years old, who sat watching us, but displaying no fear. Time was 11:10. He gave small hoots, more like burps, before his chest beats (each five to eight thumps) and ended two of the chest beats with branch grabbing. (as cited in Mowat 1987, p. 3)

I take another few steps back so as to be in the clearing when the animals see me, and I almost bump into the blackback male. I measure later: we were six feet apart. He stands up, blinks his eyes, opens his mouth, screams, and runs about fifty feet through brush behind him, screaming and tearing at the undergrowth. (as cited in Mowat 1987, p. 32)

Dr. Biruté Galdikas began studying orang-utans in Borneo in 1971. Here she describes the sometimes arduous task of fieldwork during the early years of following the orangutans:

Often in the days and months ahead I would wade up to my armpits in the acidic, tea-coloured swamp water, craning my neck to catch even a glimpse of the wild orangutans who traveled the canopy created by the massive hundred-foot trees. Although we were near the equator, I shivered from the coldness of the swamp water, my fingers and toes numb, my skin shriveled from constant immersion, my body raw from allergic reactions to the tannins and toxins of the water. (Galdikas, 1995, p. 87)

Elsewhere, she describes an encounter with an orangutan mother and newborn:

Moving my hand gently, I shifted the little infant. Her bright orange hair, newly dry from the fluids of birth, was soft and fluffy and contrasted with the deep, almost mahogany red of her mother's arms. Akmad's liquid brown eyes remained expressionless. She seemed unaware of my hand on her newborn. Her arm brushed carelessly across my leg as she reached for a pineapple, almost as if I didn't exist. (Galdikas, 1995, p. 3)

Differences in chimpanzee and bonobo behavior are of considerable interest to anthropologists. Why, many question, are chimpanzees so rigidly hierarchical, male-dominated, and aggressive to the point of killing adult members within their own as well as neighboring groups, whereas bonobos are much more egalitarian, have less conflict, and appear to use sex rather than aggression to prevent or resolve conflict? The basic question for many is: Was (sometimes lethal) aggression a trait of the common ancestor of chimpanzees, bonobos, and humans and then lost in bonobos? Or did (mostly male and sometimes lethal) aggression evolve separately in chimpanzees and humans? Of course, not all would agree that humans are innately aggressive at all (see, for example, the discussion of this issue in Chapter 12). The notion that innate aggressiveness is a driving force of human evolution is covered in Box 4.1.

Subsistence and Diet

Subsistence and diet are major areas of interest in primate studies. As described in Box 2.1, primate groups can be characterized as frugivorous, folivorous, or insectivorous, although the strategy may change from month to month. The diversity of foods in the diet of primates is usually high. Studies of gorillas, for example, indicate they have about 150 different plants in their diet. Orangutans have been recorded as eating over 400 different kinds of food, including over 200 different fruits. Many primates eat insects, but relatively few species eat mammals or other small animals.

Some monkeys have been observed hunting and eating meat, but most studies of hunting and meat eating focus on chimpanzees. Hunting and meat eating is very interesting to anthropologists because our human ancestors started focusing on hunting and meat eating more than one million years ago, and we are uncertain of how and why that occurred. Chimpanzee hunting and meat eating therefore provides a potential model of similar events in human ancestry.

There are several important things to understand about hunting and meat eating among chimpanzees. For one, it tends to be opportunistic rather than planned beforehand. It is likely that a decision to capture a monkey or other small animal occurs only minutes or less before the capture begins. Hunting is also usually a cooperative activity, which is unusual among nonhuman primates. In typical foraging, it is every individual for themselves, except for mothers looking after children. When it comes to hunting, however, chimpanzees often cooperate in capturing the prey, and when an animal is captured, the meat is often shared. Sometimes the meat is shared only among those who participated in the hunt; other times it is shared with others in the group. Sharing does not occur like this in any other subsistence activity. It is also males that do the hunting. Females have been observed to participate, but it is most often the males. Interestingly, all these characteristics of hunting demonstrated by chimpanzees are also common among groups of modern human hunter-gatherers, except that hunting by humans is often planned in advance.

Besides considering primate hunting and meat eating, anthropologists are also interested in primates as prey. Monkeys are often the prey of chimpanzees, but there are other predators to consider as well. Of course, the kind of predator depends on the region. Leopards and tigers are predators of primates in Asia, for example, and raptors and other large birds feed on primates as well. Primates are usually cautious around watering holes, since that is where many predators, including snakes and other reptiles, lie in wait. It is often the lower-ranking primates that will first come out of the trees to approach the water. Primates frequently will use alarm calls to warn others of predators, increasing the danger for the individual making the call. It should also not be forgotten that humans are predators of primates as well, hunting monkeys and apes for food, pets, and zoos.

Tool Use

Tool use among primates is another interesting area of research. Until the 1960s, it was widely thought that only humans made and used tools. Now there are several recognized instances of tool use among a variety of primates. Chimpanzees modify blades of grass and sticks to obtain termites in their mounds, some monkeys and apes use rocks to crack open nuts, and chimpanzees use sticks to poke small animals hiding in trees. Leaves are used as sponges, and gorillas have been observed using sticks to test the depth of water and to assist with walking. Before Dr. Jane Goodall first observed tool making among chimpanzees, the idea of "Man the Tool Maker" distinguished humans from other primates. When Goodall reported seeing a chimp in her observation group stripping leaves from a twig to fashion a tool, Dr. Louis Leakey famously commented, "Now we must redefine tool, redefine Man, or accept chimpanzees as humans." Instead, we have accepted a fourth option: that many other primates make and use tools as well. We also no longer use "Man" to refer to humans or people.

PRIMATES IN CRISIS: ECOLOGICAL STABILITY AND CRITICAL THINKING

About half the approximately 500 species of primates are endangered, some critically so. The reasons for this include the destruction of primate habitats, the viewpoint that primates are pests to be eradicated, the illicit trade in primates for pets and zoos, and the idea that primates can be considered human food (issues more fully discussed in Box 2.4). This is why so many primatologists work toward conserving primate habitats and rehabilitating rescued primates.

A Note on Primates and Ecological Sustainability

Nonhuman primates in the wild are integral to maintaining ecological sustainability. This is especially important in tropical and semitropical forests. Primates have an important role in seed dispersal, which promotes the ongoing growth and

BOX 2.4 **Primates in Crisis**

Many primatologists start their field studies firmly embedded in an anthropological framework, with the ideal of contributing to the methods, theories, and discoveries of the discipline. Once in the field, however, it is not unusual for researchers to shift some of their focus toward additional objectives, including rehabilitation and conservation.

The Primate Specialist Group (PSG) of the International Union for Conservation of Nature Species Survival Commission (IUCN/SSC) is one of the key organizations keeping track of newly discovered primate groups and primates in crisis, and frequently updating and publishing lists of endangered species. They report, for example, that all the great apes – orangutans, gorillas, chimpanzees, and bonobos – are endangered, some critically. They regularly identify the 25 most endangered primates. The 2019 list included a variety of lemurs from Madagascar, several species from the Americas, and more than a dozen species of monkeys and apes from Africa and Asia.

The following is an excerpt from *Stolen Apes: The Illicit Trade in Chimpanzees, Gorillas, Bonobos, and Orangutans*, a publication of the United Nations that documents the nature and severity of the primates in crisis due to trade:

> Great apes are trafficked in various ways. In many cases wild capture is opportunistic: farmers capture infant apes after having killed the mother during a crop-raid, or bushmeat hunters shoot or trap adults for food, and then collect babies to sell. However, organized illicit dealers increasingly target great apes as part of a far more sophisticated and systematic trade. They use trans-national criminal networks to supply a range of markets, including the tourist entertainment industry, disreputable zoos, and wealthy individuals who want exotic pets as status symbols. Great apes are used to attract tourists to entertainment facilities such as amusement parks and circuses. They are even used in tourist photo sessions on Mediterranean beaches and clumsy boxing matches in Asian safari parks. (Stiles et al., 2013, p. 8)

Bushmeat is a term often used in the context of primates as food. It is not unusual to see primates for sale as food in some markets in Africa. It is not the local level of bushmeat hunting that is the big problem, however; rather, it is the slaughtering of primates for the international trade in bushmeat, serving those in other regions where primate meat is considered a delicacy. Because of the close biological relationship between apes and humans, there is some risk that eating apes may result in acquiring some of the pathogens causing diseases carried by the apes. The eating of apes has consequently been blamed, often incorrectly, for epidemics including the 2014 outbreak of Ebola in Africa (bats were a far more likely host). As noted in the UN document, however, it is those who butcher the apes who are at highest risk, since they are the ones to come in contact with the blood and organs of the apes.

One of the most egregious causes of orangutans being endangered is the transformation of their natural forest habitats of Southeast Asia into plantations for the production of palm oil, which in turn ends up in thousands of products including packaged foods, detergents, and cosmetics. With the reduction of their natural forest habitats, orangutans have difficulty surviving. Many starve to death, while others are killed to prevent them from eating parts of the newly planted palm trees.

development of forests. Primates typically move from a few to several miles a day within their territory. They feed on the fruit of trees in one area and then, through their feces, deposit the seeds of those fruits in another. This is an important aspect of ecological sustainability that should not be overlooked. A reduction in numbers of primates can be devastating for ecological sustainability. This is particularly important to understand in light of the knowledge that thousands of primates are removed from their natural habitats every year.

Using Knowledge of Primates to Think Critically

Knowledge of primate biology, primate evolution, and primate behavior can be used to deconstruct popular notions of primates and primate-like beings. A good example of this is applying knowledge about primates to reports of Bigfoot, as demonstrated in Box 2.5. Based on anthropological knowledge of primates past and present, most anthropologists doubt Bigfoot is real. Those who do believe in it usually suggest that it is a remnant population of *Gigantopithecus*, the largest primate that ever lived. *Gigantopithecus* lived in Asia, but some believe it may have migrated to North America along with many other animals, including humans, during the late stages of the last ice age.

SUMMARY

This chapter has provided an overview of the primate world and the place of humans in it. This includes an understanding of what makes primates different from other orders of mammals and the distinguishing criteria of the major taxa of interest to biological anthropologists. The chapter also provides an overview of primate evolution during the Cenozoic, with a particular focus on the emergence of primate taxa to which humans belong, and an overview of the major areas of research interest in the study of nonhuman primate behavior. Mirroring the Learning Objectives stated in the chapter opening, the key points can be summed up as follows:

- There are many reasons why anthropologists study nonhuman primates. Most importantly, nonhuman primates provide models of how early humans may have adapted and evolved, both biologically and culturally.
- There are many subdivisions of primates. There is consensus that humans are members of the suborder Haplorhini, infraorder Catarrhini, and superfamily Hominoidea. There is no consensus, however, on how humans are classified within the superfamily Hominoidea. In this book, humans are considered to belong to the family Homininae, which includes all primates that are bipedal

BOX 2.5 **Assessing Bigfoot**

Many people have reported seeing a large, human-like ape in various parts of North America. Mostly it is known as Bigfoot, but there are many regional variations, including Sasquatch. There have been thousands of reported sightings over the past several decades. It is typically described as being quite large (8–10 feet tall and several hundred pounds), bipedal (walking on two legs), and hairy. It is almost always reported as being solitary, and many also report it as being omnivorous (e.g., eating stolen food from campsites and dumpsters) and nocturnal. It also typically flees when sighted. The name Bigfoot is derived from what many consider to be the most compelling physical evidence: very large, human-like footprints, usually in mud, often identified following an alleged sighting.

Most anthropologists doubt that such a creature really exists. Studies of nonhuman primate behavior simply do not support it. Of the approximately 500 species of primates, humans are the only bipeds; anthropologists have never found any kind of evidence of another primate that habitually walks upright. Reports of solitary behavior are also troubling; being social is a characteristic of almost all primate groups. Orangutans are a rare exception, with males spending a significant amount of time alone, but even then, the mother–child bond remains. In apes and humans, it is necessary for mothers to look after children for years, so one would expect that if Bigfoot did exist, it would be more likely that mothers with children, rather than solitary males, would be observed. Reports of Bigfoot being omnivorous are also puzzling; large apes tend to be primarily folivorous (i.e., eating a diet of mainly leaves and other tough plant matter). Reports of being nocturnal are similarly problematic; most primate species are diurnal, and those that are

nocturnal tend to be small and are mostly classified as Strepsirhini – far removed from apes and humans.

Moreover, there is no compelling biological evidence for the existence of Bigfoot. Claims that specimens contain Bigfoot DNA are occasionally reported, but when these are subjected to testing, they are invariably shown to be misidentified bears or other well-known animals, or hoaxes. No bones, teeth, or soft tissue have ever been discovered.

The knowledge that before the arrival of humans in North America, no other hominoid was ever here makes the notion that Bigfoot evolved here unlikely. There is virtually no fossil record of Bigfoot. Some believe that Bigfoot may be a remnant population of *Gigantopithecus*, a large ape that lived for a time in parts of Asia between about seven million and one million years ago. We have only mandibles (the lower jaw) and teeth of *Gigantopithecus*, but based on these skeletal elements, it does appear to be a match in that it is a very large ape. We have no idea if *Gigantopithecus* was bipedal, though. Analysis of particles in the teeth suggest that *Gigantopithecus* ate a diet of plants, mostly bamboo.

Some who believe in Bigfoot suggest that the lack of a fossil record may be explained by Bigfoot deliberately burying their dead. This seems unlikely, however, because burial tends to improve preservation and archaeologists can often identify places where burials occurred.

Believers should be aware that if they do sight Bigfoot, they should probably leave it alone. It probably isn't a large human-like ape. If it exists, it is probably a human who wants to be left alone. And nobody should do it harm. Recall that if a primate is bipedal, as Bigfoot is typically reported to be, it is classified as human.

since the split with the common ancestor of humans, chimpanzees, and bonobos about seven million years ago.

- Primates have been around and evolving for about 65 million years. The biological family to which humans belong emerged about seven million years ago. Monkeys of the Americas probably rafted from Africa about 30 million years ago, and the split between the group that led to contemporary monkeys of Africa, Asia, and Europe and the group that led to apes and humans probably occurred about 25 million years ago. It is important to remember that all primate species continue to evolve, although many are also critically endangered.

- Anthropological studies of nonhuman primate behavior focus on communication, social systems, aggressive and affiliative behaviors, subsistence and diet, and tool use. There are many different ways that primates organize themselves; most have rigid hierarchies, but some, like bonobos, are relatively egalitarian. Some, but not all, primate species are aggressive, and all exhibit affiliative behaviors such as grooming each other. Primates eat mostly vegetation, although some specialize in insects. Some apes and monkeys also hunt and eat meat, but hunting is never a primary subsistence strategy among nonhuman primates.

Review Questions

1. Why is it important to understand humans as primates?

2. What are the distinguishing characteristics of primates, Haplorhini, Catarrhini, and Hominoidea?

3. What were the key events in primate evolution occurring in each epoch of the Cenozoic?

4. What are the principal research interests in primatology? Give examples of discoveries for each.

Discussion Questions

1. What might be some implications for anthropology if the loss of primates continues?

2. What may be some implications of using bonobos, rather than chimpanzees, as models of how early humans may have behaved?

3. What are some of the implications of having no consensus on primate taxonomy, such as some people restricting the use of "hominin" to refer only to humans while others also classify chimpanzees and bonobos as hominins?

Visit **www.lensofanthropology.com** for the following additional resources:

SELF-STUDY QUESTIONS	WEBLINKS	FURTHER READING

3

EVOLUTIONARY THOUGHT AND THEORY

LEARNING OBJECTIVES

In this chapter, students will learn:

- *why it is important to understand evolutionary theory.*
- *about the nature of science.*
- *the history of evolutionary thought.*
- *the basics of evolutionary theory.*
- *that mutations are key to evolution in all life forms, ranging from the creation of new strains of a virus to the evolution of humans.*

To understand humans, it is important to understand how evolution works. It has much explanatory power.

#Evolution

INTRODUCTION

There are multiple reasons that an understanding of evolutionary theory is important in anthropology. Nothing really makes sense about human biology unless we understand evolutionary processes. We are far from the perfect creation, biologically speaking. Humans are more like an accumulation of quick fixes that serve a purpose. The biology that enables people to walk upright, for example, works well for a long time, but not necessarily into middle or old age – just ask an older person how their feet, knees, hips, and back are doing.

It is also important to understand that evolution is a fact. We can see evolution happening. It is often difficult to observe it happening in our own lifetimes with plants and animals we see every day, but it *is* happening. We see it in laboratories

---

BOX 3.1   **Food Matters: Evolution in Action – Lactose Tolerance**

A good example of human evolution in recent times is the ability of many people to effectively digest milk as adults, referred to as lactose tolerance. Genetic research indicates that lactose tolerance emerged independently among populations already using domestic cattle in various parts of the world (the Middle East, Europe, and Africa) over the last several thousand years. In each case, a mutation occurred (not always the same mutation) that allowed people to digest milk without problems. The mutation became a favorable variation that was subsequently selected among populations, leading to many members of the descendant populations now being lactose tolerant.

Infants and young children are able to drink milk without ill effects due to the activity of a protein called lactase, which breaks down the lactose in milk making it easy to digest. As children mature, the ability to produce lactase is switched off for about two-thirds of the world's population. The inability to digest milk after childhood is common among mammals. Lactose tolerance in adulthood may be uniquely human. Due to a mutation that was passed on to subsequent generations, lactase continues to be active in the remaining one-third of the population, allowing them to continue drinking milk into adulthood. Rather than lactose tolerance, some people prefer to use "lactase persistence" to describe the condition, since it is the persistence of lactase that allows lactose tolerance.

Adults who are not lactose tolerant suffer many different kinds of unpleasantness if they drink milk including severe cramps, diarrhea, bloating, and flatulence. Not an enjoyable experience for those trying to digest milk, or for those around them! Without lactase persistence, drinking milk as an adult is simply not a viable option.

The majority of the 35 percent of the world's population who are lactose tolerant have European ancestry, although there are small areas in Africa and

under controlled conditions, particularly with microorganisms and insects such as fruit flies. We see it in the fact that we need a new flu vaccine every year to protect us from evolving and changing viruses. We also see evolution happening as plant breeders create new varieties and animal breeders select for particular qualities or physical traits. If you have a pet, it is the result of evolution. One good example of recent evolution in humans has to do with the ability to digest milk – lactose tolerance – which is covered in Box 3.1.

It is important to recognize that one can both believe in evolution and hold onto religious beliefs. Many scientists and religious leaders attempt to convince followers that this is not possible, but for many, there is no problem. Stephen Jay Gould (1997) wrote an important essay called "Nonoverlapping Magisteria" in which he outlined how a belief in evolution and religion need not be incompatible, as long

Asia where it is common as well. For those in North America, a high proportion of people with European ancestry are lactose tolerant. A study of populations around the world indicates that in North America, more than 80 percent of those with European ancestry are lactose tolerant. Conversely, fewer than 5 percent of Native Americans and Asian Americans are lactose tolerant. The percentage of lactose tolerant African Americans and Hispanic Americans is about 30 percent and 50 percent, respectively. Most Europeans are lactose tolerant, but other than some cattle-herding groups in Africa, elsewhere it is rare.

Lactose tolerance evolved among cattle herders. It is clear that the herders were aware of the nutritional value of dairy products before becoming lactose tolerant. Considering that milk has protein and calcium, and provides a good source of vitamins, calories, carbohydrates, and milk fat, some have described it as a superfood of the times. Archaeological research indicates that herders

accessed the nutrition of dairy by breaking down the lactose through fermentation, making cheese and yogurt that were easier to digest. The mutation leading to lactose tolerance simply sped up the process and allowed immediate nutrition by drinking.

Figure 3.1 MILK.
Lactose tolerance, including being able to drink a glass of milk without suffering digestive issues, is a recent evolutionary phenomenon.
Credit: istetiana/Shutterstock

as one recognizes that **science** best covers the empirical world while religion best covers morals and values. Gould notes that even Pope Pius XII in 1950 and Pope John Paul II in 1996 accepted this notion of nonoverlapping magisteria (or separate and distinct realms of authority). More recently, in 2014, Pope Francis also confirmed the compatibility of a belief in biological evolution with a belief in a divine creator.

Just as anthropology is a framework, so too is science. It is one of many frameworks that scientists use, especially when dealing with biological anthropology and archaeology. Science is especially important for understanding the biological evolution of humans, covered in Chapter 4. This chapter covers the nature of science, provides a brief overview of evolutionary thought, and outlines basic evolutionary concepts, especially those that are important for understanding the evolution of humans.

THE NATURE OF SCIENCE

Science is a framework, consisting of principles, methods, and ways of evaluating explanations. It is often described as empirical, meaning it relies on things that can be observed, measured, and analyzed. It often requires experiments. Results from those experiments must be repeatable by others.

The principles and methods of science are listed in Table 3.1. The first basic principle is that there is a real and knowable universe. In simple terms, the universe is real, not in your or someone else's imagination. The more important part of this principle is that the universe is "knowable." In practical terms, this means that scientists accept that understanding something is not beyond their comprehension. It does not necessarily mean that it will be figured out in your lifetime. This is because most scientific advances are very small. But they accumulate. A particular scientist makes a small advance, then that advance is built upon by others, and on it goes. That is how most science works. Imagine, for example, space exploration. It wasn't so long ago that it was commonly thought that Earth was the center of the universe. We have gone from that belief 500 years ago to landing spacecraft on Mars. Much of this work can be attributed to understanding that the universe is knowable. Similar kinds of advances apply to human biological evolution. Anthropologists accept that figuring out how and why humans emerged, how they evolved, and understanding the relationships between the various taxa of hominins is knowable. We work at it. Occasionally someone makes an important discovery and we build on that. We may not have human evolution entirely figured out in our lifetimes, but we accept that it is not beyond our capabilities.

The second and third principles are that the universe operates according to understandable laws, and that these laws are unchanging. These principles are linked

TABLE 3.1
Principles and Methods of Science

Principles	Methods
There is a real and knowable universe.	Create one or more testable hypotheses to explain observations or answer questions.
The universe operates according to understandable laws.	
These laws are unchanging.	Test the hypothesis (or hypotheses).
	Accept, modify, or reject the hypothesis.
	Continually reevaluate the hypothesis as new data become available and new hypotheses are created.

to the notion that the universe is knowable. Scientists work toward understanding laws, which are as certain as we get about something in science. There are very few things that are so certain that they are called laws, but one such example is the law of gravity. Scientists also accept that these laws have stayed constant through time, meaning the laws that were active thousands, or millions, or billions of years ago are the same laws that are active today.

In science, we can start with a guess, which proceeds to a **hypothesis**. A hypothesis must be testable, meaning there must be some way to collect and analyze data to either support or reject the hypothesis.

Theory is usually as good as it gets in science. Outside science, "theory" is often used to indicate a guess. One may often hear the phrase "It is just a theory," for example. When scientists use the word *theory*, however, it has a different meaning, with a much more positive connotation. When one speaks of the theory of evolution, for example, it may mean one of two things. The term *theory* may be used to describe multiple well-supported hypotheses about evolution, including the role of natural selection, sexual selection, mutation, gene flow, genetic drift, and more. In other contexts, mostly historical, the "theory" in the "theory of evolution" refers to notions of how evolution occurs (e.g., such as by natural selection) rather than if it occurs. When one hears of the "theory of human evolution," by comparison, it is not in the context of whether human evolution occurred. Rather, it refers to the multiple hypotheses and data that support the fact of human evolution.

The scientific method includes four primary stages. The first is to develop a hypothesis (or multiple hypotheses) to explain an observation or answer a question. The next stage is to test the hypothesis, which requires the collection and analysis of data. For example, in the study of human evolution, this may include the collection and analysis of human remains. If anthropologists wanted to test the hypothesis that walking upright occurred before a significant increase in brain size, they would need to collect human skeletons of known antiquity. This would allow them to determine if walking upright preceded an increase in brain size. The third stage includes the acceptance, rejection, or modification of the hypothesis. The fourth stage involves the continual reevaluation of the hypothesis as new data become available.

In science, the best hypothesis is usually the hypothesis that fits the data today. We recognize that new research or a reevaluation of old research may bring new data, causing the rejection of a well-accepted hypothesis. In the study of human biological evolution, for example, as new data emerge from ongoing research projects, hypotheses are continually reevaluated. Some may be strengthened by the new data, and some may lose support. The discovery of a new kind of human, an unusual bone, or a different kind of strand of **DNA**, for example, may cause anthropologists to reevaluate their ideas.

Ensuring that hypotheses are continually reevaluated is a key aspect of science. There is never certainty in science. Scientists always leave room for doubt. By continually retesting and reevaluating hypotheses, science is self-correcting. Some people find it frustrating that there is no certainty in science, but this is also one of its strengths. For example, we know much more about human evolution today than we did a few years ago. This is because we never accept a hypothesis, no matter how good it may seem, if it cannot be tested and/or does not explain the data.

Science also includes methods for evaluating competing explanations. These include ensuring the hypothesis is testable, assessing its compatibility with what we already know, how many phenomena the hypothesis can explain, and how simple the hypothesis is. These are discussed further in Chapter 7.

One of the characteristics of pseudoscience is developing a series of hypotheses and then accepting one by merely eliminating the others. This is a common ploy among nonscientists. It would be inappropriate, for example, to develop three hypotheses to explain the origin of an increase in brain size, test and reject the first two, and on this basis accept the third.

HISTORY OF EVOLUTIONARY THOUGHT AND THEORY

This section has two parts: (1) the history of evolutionary thought and the theory of Darwin; and (2) the contributions of Mendel.

History of Evolutionary Thought and the Theory of Darwin

Charles Darwin is widely credited with developing the theory of evolution, upon which much of biology today is based. His book *On the Origin of Species by Means of Natural Selection, or the Preservation of Favoured Races in the Struggle for Life*, initially published in 1859 with five subsequent editions in years following, was certainly monumental. Many suggest it is one of the two or three most important books ever written.

We know a lot more about evolution now than we did in 1859, but Darwin laid the foundation in a very clear and comprehensive way. His ideas on how evolution occurred were articulated so clearly, and he used so much evidence in support of them, that much of the generally educated public was quickly convinced not only that biological evolution occurred but also that it occurred through a process called natural selection.

Darwin (1859) summarized his ideas as follows:

> As many more individuals of each species are born than can possibly survive; and as, consequently, there is a frequently recurring struggle of existence, it follows that any being, if it vary however slightly in any manner profitable to itself, under the complex and sometimes varying conditions of life, will have a better chance of surviving, and thus be naturally selected. From the strong principle of inheritance, any selected variety will tend to propagate its new and modified form.

Darwin scholar Joseph Carroll (2003) describes the importance of *On the Origin of Species*:

> *On the Origin of Species* has special claims on our attention. It is one of the two or three most significant scientific works of all time – one of those works that fundamentally and permanently alters our vision of the world. At the same time, it is one of the few great scientific works that is also a great literary classic. It is written for the generally educated reader and requires no specialized scientific training. It is argued with a singularly rigorous consistency, but it is also eloquent, imaginatively evocative, and rhetorically compelling. (p. 9)

There are several key elements or components of Darwin's theory of how biological evolution occurs. Darwin observed that all species either do or have the potential

Figure 3.2
CHARLES DARWIN.
This portrait of Charles Darwin was done about 1840, during the time he was developing his theory of evolution by natural selection.

BOX 3.2 **Darwin – In His Own Words**

On the Origin of Species (retitled *The Origin of Species* in its sixth edition) has 14 chapters. The last is titled "Recapitulation and Conclusion," which includes the following excerpt:

> Therefore I cannot doubt that the theory of descent with modification embraces all members of the same class. I believe that animals have descended from at most only four or five progenitors, and plants from an equal or lesser number.... I should infer ... that probably all the organic beings which have ever lived on this earth have descended from one primordial form, into which life was first breathed.

Darwin makes few references to humans in *On the Origin of Species*, but he recognizes the implications of his theory of evolution for the study of humankind. Near the end of the book, Darwin writes, "In the distant future I see open fields for far more important researches.... Light will be thrown on the origin of man and his history."

The last paragraph of *On the Origin of Species* reads as follows:

> It is interesting to contemplate an entangled bank, clothed with many plants of many kinds, with birds singing on the bushes, with various insects flitting about, and with worms crawling about damp earth, and to reflect that these elaborately constructed forms, so different from each other, and dependent on each other in so complex a manner, have all been produced by laws acting around us.... Thus, from the war of nature, from famine and death, the most exalted object to which we are capable of conceiving, namely, the production of the higher animals, directly follows. There is grandeur in this view of life, with its several powers, having been originally breathed into a few forms or into one; and that, whilst this planet has gone cycling on according to the fixed law of gravity, from so simple a beginning endless forms most beautiful and most wonderful have been, are being, evolved.

In his autobiography, Darwin writes of his interest in human evolution: "As soon as I had become, in the year 1837 or 1838, convinced that species were mutable productions, I could not avoid the belief that man must come under the same law" (as cited in Carroll, 2003, p. 441).

In *The Descent of Man* (1871), Darwin writes:

> The main conclusion ... is that man is descended from some less highly organised form. The grounds upon which this rests will never be shaken, for the close similarity between man and the lower animals ... are facts which cannot be disputed.... The great principle of evolution stands up clear and firm, when these groups or facts are considered in connection with others.

to produce far more offspring than can survive, and that there is variation within all species. From this, Darwin deduced that there must be some kind of competition for resources, that some of the variability within species must be advantageous, that those with the favorable variations would produce more offspring, and that the advantageous variations could be inherited. Over time, Darwin reasoned, the accumulation of advantageous variations could lead to the evolution of a new species. The phrase "theory of evolution by natural selection" is based on the idea that it is nature that determines which variations are advantageous.

The theory is quite simple, but nobody had connected the pieces quite like Darwin did. There are multiple reasons Darwin was able to develop the theory, including his university education, his five-year voyage around the world making observations on the natural world, a passion for both describing and explaining the natural world, and his interest in reading widely. It was, for example, only after reading an essay written by an economist (Thomas Malthus) that Darwin realized that, as with humans, there must be a struggle or a competition for existence among members of other species as well.

It is important to understand that Darwin did not develop the theory of evolution in a vacuum. The theory was dependent both on the development of science in general and on other scientists' thoughts and research on the natural world, in particular those having to do with geology and biology. The development of the framework of science, for example, out of which grew the fields of geology and biology, began around 1500 **CE**. This is more than 300 years before Darwin developed his ideas.

There are many prominent researchers and thinkers who made important contributions to evolutionary thought, even though that was not the intention of all. In the late 1600s, for example, John Ray, an Anglican minister with an interest in the natural world, developed the concepts of **genus** and species, ideas that were further developed in the early 1700s by Swedish botanist Carolus Linnaeus, who added the taxa of order and class to the classification system we still use today. It was also Linnaeus who placed humans in the genus *Homo*, the order Primates, and class Mammalia. Neither Ray nor Linnaeus set out to have their work used in support of the theory of evolution, but indeed it has been. For example, we now assume that placement of similar species in a genus, similar genera in an order, and similar orders in a class are a reflection of evolutionary relationships and not just similar physical features.

Some French intellectuals had significant influence on evolutionary thought before Darwin. Georges-Louis Leclerc, Comte de Buffon, a very well-known scientist in the mid- to late 1700s, made some significant contributions. Buffon recognized (and was perhaps the first to do so) that change within species could be an effect of environment, publicly supported the idea of change, refuted the need to resort to religious doctrine to explain things, and speculated that the earth was tens of thousands of years old, rather than the 6,000 years many were claiming based on interpretations of the Bible.

Jean-Baptiste Lamarck is associated with an early nineteenth-century idea of how traits are passed on, known as the theory of the **inheritance of acquired characteristics**. Lamarck proposed that an individual that acquired specific traits during its lifetime could pass those traits on to its offspring. A classic example involves giraffes. The idea was that if a giraffe continually stretched its neck to reach higher leaves, its neck would become longer. Then, once the neck was longer and the giraffe subsequently mated, its offspring would have a longer neck too. We now know that this is not the way evolution works, and the theory was never widely accepted even in its own day, but it was an attempt that led to further thought. It was one of those smaller steps that led to bigger discoveries.

A few people we would call geologists today were also influential in the development of evolutionary thought, primarily in establishing the long time frame for life on earth that was necessary for evolution to occur. In 1650 CE, Archbishop James Ussher calculated the date of the earth's origin as 4004 **BCE** using a literal interpretation of the Book of Genesis in the Bible. (Some years later, another theologian declared more specifically that creation occurred on 23 October 4004 BCE at 9:00 a.m.) In the late 1700s, James Hutton developed the principle of **uniformitarianism**, which essentially says that the same geological processes active today were also the processes active in the past. The implication here is that since many of the processes were slow, the earth must be far greater than 6,000 years old. In the early 1830s this idea was developed further and popularized by Charles Lyell, who suggested that the earth must be at least 100,000 years old. Importantly, Darwin's early readings included the writings of Lyell on this. We know now that the earth is 4.6 billion years old.

Following graduation from university, Darwin obtained a position as a naturalist on the HMS *Beagle*, a British ship on a five-year voyage of mapping and exploration (1831–6). It was on this voyage that Darwin was introduced to extreme diversity between and within species and started formulating his ideas. One of the most significant stops of the HMS *Beagle* was at the Galápagos Islands, off the west coast of South America. Here, Darwin observed that the tortoises and finches on the islands were significantly different than those on the mainland, and he further observed a significant diversity of finches between the islands. Each variety of finch had evidently evolved to suit its environment, making a connection in Darwin's mind between diversity and environment. Those living on islands with rocky beaches, for example, had short strong beaks suitable for overturning rocks. Those living in or among trees had long thin beaks suitable for poking through bark.

Once back from the voyage, Darwin set to family life (marrying and having children) and academic work from his estate south of London. Darwin was from a wealthy family, so he was basically able to devote himself to whatever pursuits most interested him. Fortunately for the rest of the world, he wanted to describe and explain diversity. He started sketching out his ideas about evolution but told few about them, fearing a backlash from the scientific community. The final piece of the puzzle for Darwin occurred in 1838 upon reading the essay by economist Thomas Malthus.

For 20 years Darwin occasionally worked on the manuscript that would eventually become *On the Origin of Species*. Then one day in 1858 Darwin received a letter from Alfred Wallace, another English naturalist who had independently developed very similar ideas. It created a quandary. Darwin had developed his ideas over 20 years but had told few. Wallace had developed his ideas quickly and shared them. It created a dilemma because being the first to develop and publish an idea is important in the field of science. An arrangement was made for both Wallace's and Darwin's papers to be read later that year at scientific meetings in London. Accordingly, they both are credited with coming up with the idea of natural selection.

Darwin spent the next year polishing his manuscript, and the book was finally released in 1859. Darwin preferred to avoid the use of the term "evolution." In its place he used "descent with modification" and "transmutation." The book does not deal with humans, but the implication was there. In 1871, he published his thoughts on human evolution in a book called *The Descent of Man*.

Darwin deservedly receives much credit for providing the foundation of evolutionary theory. Had he not developed it, however, we would probably still be in about the same place in terms of our scientific understandings, except the accolades would be going to Alfred Wallace rather than Charles Darwin.

Darwin and Wallace provide an interesting contrast. Darwin came from a wealthy family and did not have to work. Except for his five-year voyage at a relatively young age, Darwin mostly led a quiet life at home. Wallace, on the other hand, came from a lower-class family and had to work to get by. Nonetheless, like Darwin, Wallace had a passion for describing and explaining the biological world. Wallace, however, seems to have been much more adventurous. As a young man,

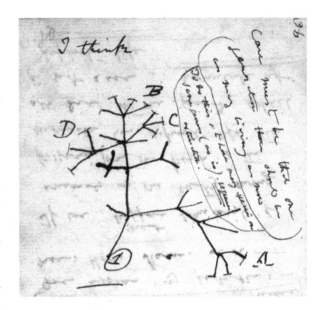

Figure 3.3
"TREE OF LIFE" DRAWING BY DARWIN.
This sketch is from one of the notebooks Darwin kept while he was developing his theory. The tips of the branches show living species, while the limbs and trunk illustrate evolutionary relationships.
Credit: Cambridge University Library

Wallace spent a few years in the Amazon River Valley in South America collecting specimens, only to lose the entire collection when the ship he was traveling home on caught fire and sank in the middle of the Atlantic Ocean. Luckily, after spending several days in a lifeboat, Wallace was rescued. Undaunted, Wallace next set his sights on Indonesia, continuing to make significant collections, and independently developed the theory of evolution by natural selection. According to some reports, Wallace came up with the theory while hallucinating during a bout of malaria.

Contributions of Mendel

Neither Darwin nor Wallace knew about the mechanics of genetic inheritance at the time they published. The foundation of **genetics** is usually attributed to Gregor Mendel, a teaching monk with a background in mathematics and botany at a monastery in what is now the Czech Republic. Mendel's contributions on understanding genetics based on experiments in the 1850s and 1860s were significant, but the work was unappreciated at the time, and Mendel died in obscurity in 1884. The importance of the work was discovered in 1900 and, combined with Darwin's theory of evolution by natural selection, provides the foundation for contemporary evolutionary theory.

Besides teaching, Mendel took it upon himself to study how inheritance works through a very careful and comprehensive program of breeding pea plants (*Pisum sativum*). He chose pea plants because they have several traits that appear in only one of two forms, as follows:

* color of flower: white or purple
* position of flower on stem: on stem (axil) or on top (terminal)
* length of stem: short or long
* shape of seed: round or wrinkled
* color of seed: yellow or green
* shape of pod: inflated or constricted
* color of pod: yellow or green

Mendel spent years studying the pea plants in the gardens of the monastery where he lived. He spent the first two years ensuring that the peas were true-breeding varieties, meaning, for example, that those that had white flowers always had white flowers through multiple generations. Mendel then spent several more years crossbreeding the plants and counting the occurrence of the various traits in each plant.

Mendel found that when plants with opposing traits were crossbred (e.g., crossbreeding those that only produced round seeds with those that only produced wrinkled seeds), only one of the traits showed up in the offspring (i.e., all the plants produced only round seeds). When then crossbreeding that generation, the trait that had disappeared always appeared again, but in a 3:1 ratio (i.e., three round seeds for every wrinkled seed).

Part of Mendel's genius was that he ensured that the plants were true-breeding varieties, and he recognized the importance of large numbers. Mendel examined approximately 30,000 pea plants over the years. Without these large numbers, it is unlikely the consistent ratio of 3:1 would have been recognized. Others had used pea plants to investigate inheritance and noticed that some traits disappear and then reappear again, but they did not recognize the ratio.

From these observations, Mendel was able to establish a very basic understanding of genetics. One of the things Mendel concluded was that traits are determined by specific units or factors in individuals, which are passed on to offspring. We now call these factors **genes**. Another thing Mendel concluded was that an individual inherits one unit from each parent. We now call these **alleles**. Another conclusion was that even though an individual has a specific trait, it may not be observable but can still be passed on. Besides understanding genes and alleles, Mendel is responsible for figuring out *dominant* and *recessive* alleles, and the differences between **genotype** (what the genes code for) and **phenotype** (the physical expression of the genes).

Relatively few human traits are passed on by simple Mendelian genetics; however, Mendel's conclusions laid the foundation for further work, which was important. Prior to the discovery of Mendel's work, there was really little understanding of how inheritance worked. Many thought that blending of traits was somehow occurring, but nobody knew how. The observations that some traits disappeared and then appeared again – rather than blending – were baffling. Most people had dismissed Lamarck's ideas, and when it came to humans, some went so far as to suggest that children looked most like the parent who dominated during sex.

Figure 3.4
STATUE OF ALFRED WALLACE.
Wallace shared credit with Darwin for developing the theory of evolution.
Credit: Robert J. Muckle

Mendel's Laws

	Flower color	Seed shape	Seed color	Pod color	Pod shape	Plant height	Flower position
DOMINANT	Purple	Round	Yellow	Green	Inflated	Tall	Axial
RECESSIVE	White	Wrinkled	Green	Yellow	Constricted	Short	Terminal

MODERN EVOLUTIONARY THEORY

Darwin's theory of evolution by natural selection and Mendel's pioneering work on genetics laid the foundation for evolutionary theory, but many advances have occurred since the late 1800s, which are outlined in this section.

Key concepts in evolutionary theory include mutation, natural selection, sexual selection, gene flow, genetic drift, and adaptive radiation. **Mutations**, which are errors in the replication of DNA, are the ultimate source of variation within populations. They can be neutral, beneficial, or harmful. We often hear about the bad mutations that are passed on via sperm or egg, since they can cause diseases, deformities, and death. DNA is complex and errors occur all the time. Your hair and fingernails grow, for example, by replication of cells. If you have a mutation in those cells, it is no concern. If, however, there is a mutation in the sperm or egg, it can be a big deal.

Mutations usually occur simply as copying errors. They can also occur, however, from exposure to radiation, chemicals, and viruses. All human variability ultimately derives from mutations. As anthropologist Dr. Eben Kirksey (2021) states in his book *The Mutant Project: Inside the Global Race to Genetically Modify Humans*, "strictly speaking, we are all mutants" (p. 7). He elaborates further:

At a molecular level, each of us is unique. Each of us starts life with forty to eighty new mutations that were not found in our parents…. During the course of a normal human life we also accumulate mutations in our bodies, even in our brains. By the time we reach age sixty a single skin cell will contain between 4,000 and 40,000 mutations … These genetic changes are the result of mistakes made each time our DNA is copied during cell division, or when cells are damaged by radiation, ultraviolet rays, or toxic chemicals. Generally, mutations aren't good or bad, just different. (p. 7)

Indeed, it is because of mutations that our ancestors had variability from which bipedalism and larger brains emerged. It may have been a mutation, for example, that allowed one person to walk upright for a few meters farther than the average person could walk. Under specific conditions, that ability may have been favorable and may have led to that individual having more offspring, who also had the favorable trait. It is likely that the mutations that were selected for since the split of our common ancestor with chimpanzees and bonobos number in the millions.

In recent years, many people have been engaged in the study of mutations among viruses. Viruses that cause the flu in humans, for instance, regularly mutate, which is why new flu vaccines are created every year. Similarly, the virus that causes **COVID-19** has undergone several mutations that have led to the development of multiple new variant strains.

Gene flow and **genetic drift** are important concepts. Gene flow is when genes move between populations that are members of the same species but that do not normally mate with one another. Because of separation, new alleles may have formed in one population that may then be passed on to another. Regarding human evolution, for example, it was almost certainly the case that people moving out of Africa in the distant past mated with pre-existing populations in the Middle East and Asia. Gene flow would have occurred. Genetic drift, on the other hand, is a random factor in evolution. It is when changes in **allele frequency** occur by chance; for instance, when a small group leaves its parent population and begins a new population elsewhere. The smaller the group, the larger the changes may be in allele frequency.

Adaptive radiation occurs when a species rapidly adapts to an ecological niche, often expanding its population quickly and diversifying into multiple species. A good example of adaptive radiation was when mammals adapted to new ecological niches following the extinction of dinosaurs. More recently, adaptive radiation likely occurred when humans first left Africa, quickly expanding across Asia and perhaps diversifying into different species.

Natural selection, as proposed by Darwin, remains an important concept in evolution. In regard to human evolution, for example, it is likely that it was

BOX 3.3   On the Notion of Ape-Human Hybrids

In recent years we have become aware of an increasing number of hybrid animals – both in the wild and in captivity. A common example of a hybrid animal is a mule, the hybrid offspring of a horse and a donkey. A horse-zebra hybrid is a zorse. Another example is the lion-tiger hybrid, known as a liger or tion.

There are no verifiable instances of ape-human hybrids, but apparently not for lack of trying. A Russian scientist in the 1920s apparently made multiple attempts at impregnating female chimpanzees with human sperm, presumably through artificial inseminations. No pregnancies were documented. He also apparently had plans to impregnate "volunteer" human females with ape sperm, but it is likely this never progressed beyond the planning stage.

It wasn't only Russians who were interested in ape-human hybrids, however. Russell Tuttle (2014) reports, for example, that besides the Russian attempts, American, Dutch, German, and French scientists have all either encouraged or planned ape-human hybridization in the past, though the ethical and legal dilemmas posed by such mating would presumably prevent such experiments and attempts from occurring in the twenty-first century. But one never knows. Perhaps in the eyes of some it would make for good television.

In the 1970s, a chimpanzee that looked to some people to have some human-like features was promoted as an ape-human hybrid, known as Oliver. DNA tests confirmed, however, that Oliver was 100 percent chimpanzee. Many believe that such hybrids, known to believers as *humanzees* or *chumans*, exist, but there is no verified evidence of their existence.

In experiments, one researcher showed that human sperm could penetrate the egg of a gibbon, and presumably other apes, but not monkeys. The ability of human sperm to penetrate the egg of an ape is a long way from hybridization, however. It is possible that during the initial million years or so after the split of humans and chimpanzees from a common ancestor about seven million years ago there may have been some hybrids, but it is very unlikely there have been any since.

---

natural selection that led to significant loss of body hair and changes in skin color (discussed in Chapter 4). Even Darwin recognized that there were likely other selection processes as well, such as **sexual selection**, which essentially means personal mate selection. For instance, a variety of male animals create colorful displays and performances to entice the female of their species to have sex (consider the mating displays of fireflies or the architectural feats of the bower bird). Similar kinds of selection among humans, with mate selection having nothing to do with the potential for increased survival of the species, have likely been going on for millions of years. These mate choices may have been based on

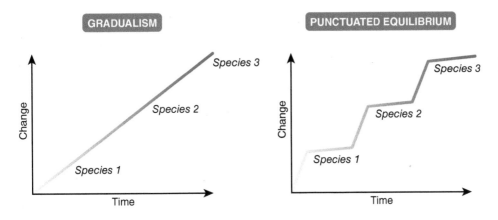

Figure 3.6
**PUNCTUATED
EQUILIBRIUM VERSUS
GRADUALISM.**
Two models of how
species evolve.
Gradualism is a slow
and steady evolutionary
process. Punctuated
equilibrium is mostly a
slow and steady process
interrupted by short
periods of significant
change.

physical features, such as facial symmetry, or cultural factors, such as the ability to heal others.

Gradualism and **punctuated equilibrium** are models pertaining to the speed at which evolution occurs. Gradualism suggests a slow, steady change, with a new species eventually emerging. Punctuated equilibrium suggests slow, steady change occasionally interrupted by short periods of significant change. Punctuated equilibrium explains lack of transitional forms in the fossil record. Both models can be used to explain human evolution at various times. The transition from *Australopithecus* to *Homo habilis* more than two million years ago, for example, has left few fossils that can be described as transitional, so punctuated equilibrium best explains the transition. On the other hand, *Homo erectus* fossils show evidence of well over one million years of fairly slow, gradual change. These transitions are further discussed in Chapter 4.

There are many working definitions of **species**. The definition used throughout this book is that a species is a population of individuals that can mate and produce fertile offspring in the wild. For instance, a horse and a donkey are different species. When they mate, they can produce a mule, but mules are almost always sterile.

Hybrids are known to occur in controlled conditions and increasingly in the wild as natural habitats decrease. In regard to human evolution, when anthropologists refer to different species of humans, the assumption is usually that they would not have been able to mate and produce fertile offspring. In recent years, DNA research is increasingly showing that different kinds of humans, such as Neandertals and modern humans, were indeed mating and producing fertile offspring, leading to the conclusion they were the same species.

Speciation is the process by which new species emerge. It can happen in numerous ways. Sometimes new species emerge from geographic isolation. It was likely the development of the Congo River in Africa 1.5 to 2 million years ago that separated

Figure 3.7

BRAIDED STREAM MODEL OF HUMAN EVOLUTION.

This drawing broadly conceptualizes the braided stream model of evolution as it applies to humans. The DNA strand represents the roots of the human lineages, the footprints represent bipedalism, the hand represents dexterity, the brain cognitive development, and the "abc" the development of language and symbolic thought.

Credit: Katherine Cook

the common ancestor of chimpanzees and bonobos, leading to those two separate species. Sometimes it is a single population of a species that evolves into another species, coexisting for a time with its ancestral species before the ancestral species becomes extinct. This is probably the more common situation in human biological evolution.

The evolution of species is rarely a simple, linear event with all populations of a species evolving into another; nor is it always useful to consider evolution as a tree with branches representing new species. In some cases, such as with humans, it may be best to consider evolution through a model known as the "braided stream" analogy, in which various populations of a species sometimes branch off and evolve independently for a time before re-joining the main part of the stream again, bringing new evolutionary changes with them.

The concept of **extinction** is important. Sometimes an entire species evolves into another species and the originating species therefore becomes extinct. In other cases, a species becomes extinct for other reasons, such as environmental change to which it cannot successfully adapt since it does not have enough variability from which to choose. This is likely what happened in the case of the dinosaurs 65 million years ago and in the extinction of mammoths and mastodons about 10,000 years ago. In other situations, species may become extinct because of the introduction of a new species that outcompetes them for resources. Extinction is normal, but occasionally extinction rates are extremely high. It is commonly accepted that there have been five periods of **mass extinctions**, when approximately half the species on earth became extinct. All previous mass extinctions occurred before the emergence of humans. Many believe we are on the verge of a sixth mass extinction, except this time humans appear to be the cause. Some biologists claim the rate of animal extinctions occurring in recent and contemporary times is 1,000 or more times above normal.

Terms that have appeared in recent years include **genome, genomics, epigenetics**, and **genetic engineering**. Genetics tends to refer to the study of individual genes and their role in inheritance. Genome refers to the entire genetic makeup of an individual or species, including all its DNA and genes, and genomics is the study of genomes. Epigenetics is a new area of research. We realize that genes are the primary way traits are inherited but that other factors, such as chemical reactions due to life events or stressors, may also have a role. Epigenetics thus refers to the study of how factors other than DNA or genes may influence the occurrence of specific traits. In other words, epigenetics can change a phenotype without a corresponding change in the genotype. Recent research indicates there are a variety of reasons that can lead to this inheritance, including the environment, diet, age, and disease. For example, a pregnant mother's exposure to pollution can increase her child's susceptibility to asthma. Genetic engineering involves the deliberate alteration of genes, often to increase the productivity of plants and animals used by humans for food. Traditionally, food producers would select specific plants and animals to breed based on phenotypes, but in recent years they have begun to use genetic engineering, selecting for, and modifying, genotypes. Consideration is now also being given to the potential genetic engineering of humans and how that may transform humanity as we know it (Kirksey, 2021).

The combination of the basics of natural selection and genetics, developed in the nineteenth century, with our greater understanding of processes such as mutation, gene flow, genetic drift, and speciation are often termed the *modern evolutionary synthesis*, or the *extended evolutionary synthesis*. While some researchers consider modern evolutionary theory and extended evolutionary synthesis to be one and the same, others prefer to reserve the use of the term extended evolutionary synthesis to refer to a more comprehensive set of theoretical concepts and recent findings that include the potential role of epigenetics and of other such factors that may lead to evolution.

BOX 3.4

Extinction and De-extinction

Extinctions are part of the evolutionary process. It is likely that more than 99 percent of all species that have ever lived are now extinct. Sometimes, species evolve into one or more new species. Others simply dwindle until the population is too low to maintain a breeding population. Sometimes a species simply does not have the variability within its population to adapt to changing environments (i.e., not enough variability from which nature can choose favorable traits); the variability may be reduced by genetic drift, or the existence of a new species evolving in the region or coming from elsewhere may outcompete the existing species.

There have been several mass extinctions in the past, where many taxonomic groups have become extinct. The last mass extinction occurred about 65 million years ago, causing many animals to become extinct, including the terrestrial dinosaurs. The consensus opinion is that an asteroid hitting earth caused significant climate change, to which the dinosaurs could not effectively adapt (not having the requisite variability). Small mammals had coexisted with dinosaurs, and many mammals likely became extinct as well. However, with the removal of dinosaurs and changing environments, some species of small mammals were able to quickly evolve, adapting to changing environments and occupying new ecological niches. Ultimately, one of these mammal species led to primates.

The primate fossil record is filled with now-extinct forms. Even the human fossil record is replete with species that have become extinct over the last several million years. One of the challenges for paleoanthropologists is to figure out which of these early species are in the line that ultimately led to *Homo sapiens*.

De-extinction refers to the idea of bringing back extinct animals through a form of cloning. Some attempts have been made (e.g., unsuccessful attempts with a kind of extinct goat), but the technology is very close to making it feasible.

SUMMARY

This chapter has provided overviews of the nature of science, the history of evolutionary thought, and the key concepts and vocabulary of contemporary evolutionary theory. Mirroring the Learning Objectives stated in the chapter opening, the key points can be summed up as follows:

- Science is a framework with a specific set of principles and methods.

Some researchers have suggested it may be possible to bring back a Neandertal, although much more discussion revolves around bringing back mammoths. Complete genomes have been reconstructed for both Neandertals and mammoths.

The advances in technology are such that we are very close to having the ability to clone mammoths – perhaps already there – and Neandertals are not far off. The question, therefore, is not so much "Can we do it?" but "Should we do it?"

Reasons for bringing back extinct species are varied. Some suggest that bringing back now-extinct species would be an effective way of recovering natural habitats. Introducing such animals to environments is often known as "re-wilding." Conversely, opponents suggest that introducing previously extinct animals to environments could wreak havoc on contemporary ecosystems.

Bringing back a mammoth would require gestation in a similar kind of animal, probably an elephant. Bringing back a Neandertal would presumably require a female chimpanzee or human. The associated legal and ethical issues would be tremendous. One American geneticist has already put word out that he is looking for a woman volunteer.

It is usually ecology scientists who are front and center in the debate about bringing back mammoths, and anthropologists who are front and center on the question of Neandertal de-extinction. Some anthropologists are keen for what they might learn about human evolution and believe that they may learn new cultural and biological adaptive strategies from Neandertals. Many, however, believe that little can be learned from bringing back a Neandertal, suggesting that whatever we could learn from a Neandertal clone brought up and controlled by humans would have serious limitations and pose serious ethical concerns.

- Charles Darwin gets most of the credit for figuring out how evolution works, and deservedly so. However, he did not work in a vacuum, and he depended on the work of others who came before him. Alfred Wallace also had it figured out.
- Gregor Mendel pioneered genetics with his work on pea plants.
- Contemporary evolutionary theory has built on the work of Darwin and Mendel, and we now have a far greater understanding of the variety of ways in which populations evolve.

Review Questions

1. What are the principles and methods of science?

2. Who were the key people involved in the development of evolutionary theory before 1859, and what were their contributions?

3. What are the key elements of Darwin's theory of evolution by natural selection?

4. What was the contribution of Gregor Mendel to evolutionary theory?

5. What is the braided stream model of evolution?

6. What is the modern evolutionary synthesis, and how may it be distinguished from the extended evolutionary synthesis?

Discussion Questions

1. If you could go back in time, with whom would you rather have a conversation – Darwin, Wallace, or Mendel? Why?

2. What are some of the difficulties in predicting how any kind of plant or animal species will continue to evolve biologically?

3. What are the implications of genetic engineering for the future of humanity?

4. How can the emergence and spread of COVID-19 be explained using the concepts outlined in this chapter?

Visit **www.lensofanthropology.com** for the following additional resources:

| SELF-STUDY QUESTIONS | WEBLINKS | FURTHER READING |

4

HUMAN BIOLOGICAL EVOLUTION

LEARNING OBJECTIVES

In this chapter, students will learn:

- *the basic methods, concepts, and issues of paleoanthropology.*
- *about the nature of the human fossil record.*
- *ideas about why the bipedal adaptation may have emerged.*
- *about the biological changes that occurred to allow efficient bipedalism.*
- *the genus and species of fossil humans over the last several million years.*
- *about the general trends in human biological evolution.*
- *that race is not a valid biological category.*

Anthropologists agree the birthplace of humans is Africa.
#WeAreTheScatterlingsOfAfrica #JohnnyClegg

INTRODUCTION

There are many thousands of human fossil remains dating back some several million years. They reveal that human evolution was not simply a case of one species evolving into another until *Homo sapiens* was reached. There is evidence that multiple genera of humans and at least a dozen species of humans have lived in the past, often coexisting at the same time in the same regions.

This chapter highlights the study of human biological evolution. It includes consideration of methods, why and how humans emerged several million years ago, and overviews of human evolution, including the major taxonomic groups and trends.

PALEOANTHROPOLOGY - METHODS, CONCEPTS, AND ISSUES

Finding Sites

There are several ways of finding paleoanthropological sites. Researchers often return to the same general area year after year. There is good reason for this: the chances of researchers finding human remains in the same area where they have been found before is usually much better than finding them in areas where nobody has looked.

Rarely do paleoanthropologists dig blindly. Typically, they search in areas where there are sediments exposed from the time period of interest. For example, a paleoanthropologist interested in early humans from approximately two million years ago would look in an area where the sediments that were initially deposited two million years ago had been covered over and have now been re-exposed. Thus, a paleoanthropologist merely has to walk over the ground surface to see bones that were initially deposited two million years ago. Every year, rains tend to remove fine layers of sediments and thus expose new remains. Paleoanthropologists also want to search in areas that have good preservation of organic remains. When humans are found, they are usually only a very small percentage of the total assemblage of bones identified.

A very high proportion of all human remains over one million years old, and all of them over 1.8 million years old, have been found in Africa, mostly in the **Great Rift Valley** of East Africa (including parts of Ethiopia, Kenya, and Tanzania) and in the country of South Africa. Most of the finds in the Great Rift Valley have been discovered through the search process just described.

Sediments of the time period of interest are not always exposed, however, or sometimes the preservation quality is not very good, but fortunately there is another way of finding early humans, which involves looking in caves. Indeed, many of the

Figure 4.1
LIFE HISTORY OF A PALEOANTHRO-POLOGICAL SITE.
This drawing represents the life history of a paleoanthropological site beginning with the death of an individual, the body's burial by natural processes, the body eventually being re-exposed by natural processes such as rain, and, finally, its discovery and excavation by anthropologists.
Credit: Katherine Cook

discoveries of early humans in South Africa have been in caves. Because they are in caves, bones are often protected from the normal elements that lead to decomposition. It is important to understand, however, that, although there are exceptions, most people did not live in caves (see Box 5.5).

There are many important paleoanthropological cave sites outside Africa as well. These include, but are certainly not limited to, Zhoukoudian in China (where 40 *Homo erectus* individuals were found), Shanidar Cave in Iraq (a purported Neandertal burial site), and Liang Bua on the island of Flores in Indonesia (where the remains of *Homo floresiensis*, popularly known as the "hobbits," were discovered).

The Fossil Record

The word **fossil** is applied very loosely in anthropology. For some, it means the organic remains that have turned into stone, or left an impression in stone, but this is not the case in anthropology. In anthropology, *fossil* is used to describe any preserved early human remains, no matter their condition. Thus, when one speaks of human fossil remains, it simply means they exist, and they may be in an extremely soft or fragile state.

The human **fossil record** has multiple meanings. In one sense the fossil record may be taken to mean the interpretation of human evolution, based on the data of the collected remains. In another sense, the fossil record may be taken to simply mean the assemblage of human remains collected.

The assemblage of human remains constituting the fossil record is substantial. Nearly two decades ago, popular writer Bill Bryson (2003) quoted well-known paleontologist Dr. Ian Tattersall as saying that the entire collection of early human remains could fit in the back of a pickup truck. Many fossils have been discovered since, adding thousands of bones and bone fragments to the assemblage. However, there are still unlikely to be enough to fill a dump truck, and the total assemblage probably represents hundreds, rather than thousands, of individuals.

Preservation and Taphonomy

Not all human remains preserve equally well. Teeth tend to preserve the best, followed by the bones of the skull. The mandible (lower jaw) is the thickest and most dense bone of the skeleton and thus

KR Cook

Figure 4.2
OLORGESAILIE,
KENYA.

A *Homo erectus* site in the Great Rift Valley of Africa. The elevated walkway passes over hundreds of stone tools that have been exposed hundreds of thousands of years after their initial deposition.

Credit: Barry D. Kass/Images of Anthropology

preserves the best of all the bones; it is sometimes found with the teeth intact. The bones beneath the head, commonly known as the post-cranial bones, are preserved less often. Except for the mandible, bones are rarely discovered complete. They are typically highly fragmented, and pieces of bone are frequently missing. This is usually due to natural processes acting on the bone in the years since the individual died.

To make proper inferences from human remains, paleoanthropologists are familiar with **taphonomy**, the study of what happens to organic remains after death. It is through taphonomy that anthropologists can identify the natural and cultural processes that may have acted upon the assemblage. For example, somebody familiar with taphonomy should be able to determine if bone breakage, marking on bones, the distance between bones, and what bones are present were due to specific kinds of natural or cultural causes.

The preservation of human remains before deliberate burial (probably about 50,000 years ago) is rare. In order for the remains to be preserved, the natural processes of scavenging and decomposition had to be impeded. Many animals, including humans, left on the ground surface will be eaten by other animals. Even without large-scale scavengers like wild dogs or hyenas, it is only rarely that preservation occurs. Decomposition essentially occurs when millions of microorganisms and insects eat the body. Millions of these microorganisms are already inside humans while they are living, but the human body, while alive, prevents them from leaving those areas where they are most useful, such as the digestive tract. In addition, animals such as flies may lay eggs on the corpse, which then hatch as maggots and eat the body.

In order for preservation to occur, then, the remains almost always have to be removed quite quickly from the ground surface. This removes them from scavengers and also usually puts them in an environment not conducive to decomposition by microorganisms, such as in areas with low oxygen, like below the ground surface. In order for early human remains to be preserved they typically had to be covered very soon after death, perhaps by sinking into the soft sediments near the shore of a lake, being covered by sediments being washed down a slope, or by slipping and falling into a subsurface cave system.

Osteology

The study of the human skeleton is known as human osteology. At a minimum, paleoanthropologists must be able to identify very small bone fragments as human, as opposed to other kinds of animal remains. The human skeleton is illustrated in Figure 4.3, and Table 4.1 provides some basic information on the human skeleton. In addition to identifying bones as human, anthropologists also should be able to determine the age of the individual when they died. They can do this primarily by looking at how many adult teeth have erupted and the degree of bone fusion (e.g., the ends of long bones fuse to the shafts and several of the bones of the skull fuse at certain times). Anthropologists also attempt to determine if the individual was male or female. They do this primarily by examining characteristic features of the pelvis (Figure 4.4).

Issues and Debates in Paleoanthropology

Prominent debates among researchers involve the assignment of a particular genus or species to a particular assemblage of human remains. This is often framed as the "Lumpers vs. Splitters" debate. Lumpers tend to assume there is considerable variability within genera and species and therefore have relatively few taxonomic categories for the several million years of human evolution. Splitters, on the other hand, tend to assume there is relatively little variability within genera and species and therefore recognize many different genera and species over the millions of years. An extreme version of lumping, for example, would consider that over the past four million years there have only been two genera of humans – *Australopithecus* and *Homo*. Looking at the same assemblage, a splitter may see more genera (e.g., *Australopithecus, Paranthropus, Kenyanthropus, Homo*) and many more species of the genus *Homo* than suggested by lumpers.

Another prominent debate involves making the links between various populations. Most anthropologists accept that one population of a species of *Australopithecus*

TABLE 4.1

Basic Osteology

Total Number of Bones in Adult Human Skeleton		206
Total Number of Teeth in Adult Human Skeleton		32
Human Dental Formula		2 incisors, 1 canine, 2 premolars, 3 molars
Key Bones, Bone Groups, and Features	Skull	Bones of the cranium and mandible
	Cranium	Skull, minus the mandible
	Mandible	Lower jaw
	Maxilla	Upper jaw
	Bones of the arm	Humerus, radius, ulna
	Bones of the wrist	Carpals
	Bones of the hand	Metacarpals
	Hip bones	Pelvic bones
	Leg bones	Femur, patella, tibia, fibula
	Ankle bones	Tarsals
	Foot bones	Metatarsals
	Fingers and toes	Phalanges
	Foramen magnum	Hole at the base of the skull through which the spinal column connects to the brain

(or maybe *Kenyanthropus*) evolved into the first members of the genus *Homo*. Early *Homo* was ancestral to *Homo erectus*, which was ancestral to *sapiens*, but the links between the many different species are debated. While there is general agreement that there have been many different species of hominins, which ones were something akin to evolutionary cousins (and eventually became extinct) and which ones are indeed ancestral to *Homo sapiens* is contested.

Problems arise in part because the taxonomic classification systems are essentially artificial constructs created by humans to make sense of their world. We

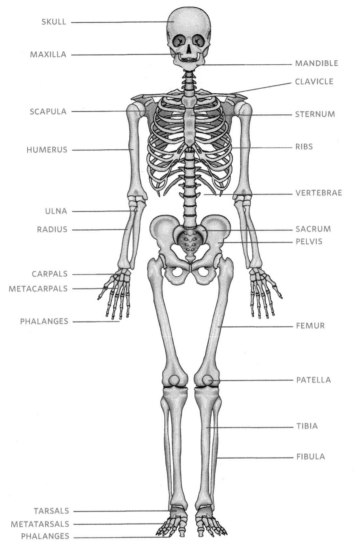

Figure 4.3
THE HUMAN SKELETON.
There are a total of 206 bones in the adult human skeleton.

SKULL

MAXILLA

MANDIBLE

CLAVICLE

SCAPULA

STERNUM

HUMERUS

RIBS

VERTEBRAE

ULNA

RADIUS

SACRUM

PELVIS

CARPALS

METACARPALS

PHALANGES

FEMUR

PATELLA

TIBIA

FIBULA

TARSALS

METATARSALS

PHALANGES

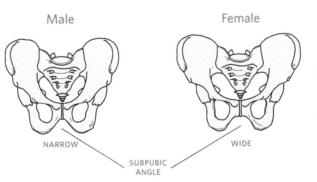

Male

Female

Figure 4.4
MALE AND FEMALE HUMAN PELVIS.
A female pelvis is typically wider, more basin-shaped, and has a wider subpubic angle than a male pelvis.

NARROW

WIDE

SUBPUBIC ANGLE

assign assemblages of human remains into specific species, for example, but we do not know if all those assignments reflect what we commonly accept as a definition of species (the ability to reproduce and produce fertile offspring). We often make guesses on the ability to reproduce and have fertile offspring based on skeletal features alone. This is problematic, however, since we do not know the variability with the populations. DNA studies are helping to clear up some of the debates, such as showing that many living people contain DNA at one time found only in Neandertal populations.

Another problem is that there is no consensus on what defines specific genera and species. There is no ultimate authority that decides upon the criteria for placing specimens in a specific genus or species. Thus, if paleoanthropologists are using different sets of criteria, different assignments are likely to occur.

Dating Techniques

There are several ways of determining how long it has been since a fossil human died or since an archaeological site was created. The three most widely used techniques are potassium argon dating, radiocarbon dating, and dating by association.

Potassium argon dating, often abbreviated K/Ar, is generally considered to be the best technique for determining the age of sites over 200,000 years old. The basic principle of potassium argon dating is that when volcanic sediment (such as ash or lava) is hot (as it is when being expelled from a volcano), it contains potassium, but no argon. As the sediments begin to cool, however, potassium begins to change into argon, and we know its rate of change. It is therefore simply a matter of sending a small sample of the sediments to a lab to measure how much potassium and argon are in a particle of volcanic sediment, and then the dates when the sediments began to cool are determined. Since the process of changing from potassium into argon is very slow, this is not a useful technique for sediments more recent than a few hundred thousand years. It is usually applied in paleoanthropology by dating the volcanic sediments below and above the biological or cultural human remains. The past several million years in East Africa was a time of intense volcanic activity. This area is where many of the early human remains and most important archaeological sites are found, so, fortunately, there are many volcanic layers that can be dated. Even when human remains are found on the ground surface it is usually still possible to use this technique by correlating volcanic layers on nearby hillsides.

For sites assumed to be less than 50,000 years old, the best technique for determining the antiquity of sites (assuming no written records exist) is generally considered to be **radiocarbon dating**, also known as carbon-14 or C-14 dating. The basic principle is that all living things contain carbon-14. At the instant of death, carbon-14 begins to decay at a known rate. Consequently, with lab analysis involving measuring the amount of carbon-14 left, the date at which the organism died can be inferred. Anything organic can be dated in this way, including human biological remains and also other kinds of plant and animal remains. The reason this method is not reliable beyond about 50,000 years is that after that time there is not enough carbon-14 left to measure.

Sometimes there are no volcanic sediments or organic remains to date, or they fall outside the range of potassium argon and radiocarbon dating, so paleoanthropologists use a technique known as **dating by association**. The basic premise of this technique is that if two things are found in the same stratigraphic layer, and the antiquity of one object is already known, then the other object is likely the same age. This technique is used widely in South Africa, where many of the most important early human remains have been discovered. For example, if human skeletal remains are discovered in the same deposit as bones of an extinct form of elephant that has been dated by potassium argon dating elsewhere, then the human remains are assumed to be the same age.

There are certainly many more techniques, but these three are by far the most common. When determining antiquity, it is normal practice to use as many different kinds of dating techniques as possible and also to use as many samples as feasible.

DEFINING HOMININS

There is no consensus definition of human or hominin. It is common, however, to consider hominins to include all members of the genus *Homo* and other taxa with evidence of bipedalism that have emerged since the split from the common ancestor of humans, chimpanzees, and bonobos about seven million years ago. The diversity of hominins that has existed over the past several million years is substantial. In a review, Wood and Boyle (2016) describe more than 30 species assigned to several genera of hominins. During most of the past several million years, multiple species of humans have coexisted.

The primary characteristic that distinguishes hominins from other hominoids (i.e., apes) is bipedalism. There are several other secondary characteristics that are used as well, although none as important as bipedalism. These secondary characteristics are often used in support of classifying fossils as hominins but are not enough to make the distinction on their own.

These secondary characteristics focus on teeth and features of the skull. The **dental arcade** of hominins is usually parabolic (i.e., the rows of teeth widen as they go back), whereas the dental arcade of apes is usually u-shaped (i.e., all the molars on each side of the arcade are the same distance apart), although there are

The Killer Ape Hypothesis

The killer ape hypothesis is based on the notion that aggression and violence are the driving forces of human evolution and remain at the core of our being. Fans of the science fiction movie *2001: A Space Odyssey* may recall that the opening of the movie depicts this scenario – one population of apes begins to become aggressive and violent, ultimately leading to the development of humans.

The killer ape hypothesis is used by some to rationalize violence, particularly male violence, based on the idea that humans are essentially genetically programmed to be aggressive and violent. There is, however, no research that shows such behavior to be genetically determined. Dr. Robert Sapolsky (2017), who is in a unique position as both a neurologist and a primatologist, describes the important role of both biology and culture for humans:

> What is most consequential is childhood, the time when cultures inculcate individuals into further

propagating their culture. In that regard, probably the most important fact about genetics and culture is the delayed maturation of the frontal cortex – the genetic programming for the young frontal cortex to be freer from genes than other brain regions, to be sculpted instead by environment, to sop up cultural norms.... It doesn't take a particularly fancy brain to learn how to motorically, say, throw a punch. But it takes a fancy, environmentally malleable frontal cortex to learn culture-specific rules about when it's okay to throw punches. (pp. 326–7)

Anthropologist Dr. Agustin Fuentes (2017) provides a similar view, stating that "violence is an option for humans, not an obligation" (p. 286).

Support for the killer ape hypothesis is often based on observation of aggression and violence in other living primates, especially among male chimpanzees. Chimpanzees in the wild are usually quite social and demonstrate considerable friendly behavior within their communities,

certainly exceptions. Hominins tend to have smaller canines than apes, and apes have **diastemas** to accommodate the large canines on the opposite jaw, which are rare in hominins. Hominins also tend to have thicker tooth enamel and reduced **prognathism.**

BECOMING BIPEDAL

One of the most significant areas of interest in paleoanthropology is in understanding the events surrounding the evolution of bipedalism. Why did bipedalism occur? What biological changes accommodated bipedalism? What was the biological variability that was selected for?

but they are also known to be occasionally aggressive and violent toward those within their group and to neighboring communities. One group of chimpanzees is known to have tracked down and killed several male members of a neighboring community.

Opponents of the hypothesis often look to bonobo groups that are genetically just as close to us as chimpanzees but tend to be much less aggressive and violent. Some critics of the killer ape hypothesis also suggest that even though male aggression and violence has been witnessed among wild populations, those communities had been habituated by humans for some years, which may have altered the chimps' normal behavior.

Dr. Russell Tuttle (2014) explains how the popularization of the killer ape hypothesis changes with current events. He notes, for example, that it was following the horrors of World War I and World War II that the killer ape hypothesis first

reached prominence, and then again in the 1960s and 1970s when North Americans were involved in Vietnam. There has been a revival of interest in North America in recent years, perhaps correlating to the ongoing violent international conflicts the United States has been involved with in the early twenty-first century. The reasoning here is that accepting that humans are innately aggressive and violent may serve to rationalize lethal violence.

The idea that aggression and violence were the driving forces that led to becoming human has little support. Similarly, many anthropologists are skeptical that humans are innately aggressive. It would be a mistake to believe our ancestors lived in a perpetual state of bliss and harmony. We do see evidence of violence in ancient skeletons and wounds from battles as well as from weapons. In general, however, it appears that aggression and violence in humans started to increase in significant ways only within the last several thousand years.

The transition to bipedalism wasn't all good. One of the negative consequences of walking on two legs is that it likely would have made our ancestors more vulnerable. They would have been easier to see by predators, for example. Moreover, moving on two limbs, rather than four, would have presumably made our ancestors slower. Walking on two legs puts enormous stress on the skeleton, caused by the entire weight of the human body being supported by two limbs instead of four. Just ask a middle-aged or older adult: sore backs, sore hips, sore legs, sore knees, and sore feet are often caused by decades of supporting weight on two legs.

Yet despite the negative impacts of bipedalism, our ancestors did make the transition. There are a few important things to consider: (1) there had to be the necessary variability within the skeleton in order for the transition to be made; (2) there must have been some advantages that outweighed the disadvantages; and

(3) just because it worked for our ancestors does not mean that an entire species became bipedal.

It was bipedalism that set us on a separate evolutionary track, but that does not mean that other populations would have necessarily done the same, even if given similar variability and circumstances. Consider, for example, that other populations of our last common ancestor with chimpanzees continued to evolve as well without being bipedal. While bipedalism worked for our ancestors, it was not necessarily the only – or necessarily the best – option. Perhaps the variability in the skeletons that make bipedalism possible did not exist before several million years ago, or perhaps it did, but the advantages in the skeleton did not outweigh the disadvantages. We may never know.

Why Bipedalism?

There are multiple explanations for why humans became bipedal and for the changes that occurred to facilitate bipedalism (Table 4.2). Until the 1970s it was widely thought that bipedalism, a significant increase in brain size (and corresponding intelligence), and tool use evolved together. If one was the trigger, it was probably the increase in brain size. A common line of thought was that a larger brain led to the intelligence required for making and using stone tools. Making and using stone tools was facilitated by standing and walking upright, which freed the hands. This made sense to many people who like to believe that the principal thing that distinguishes humans from other primates is our intelligence. Discoveries in the 1970s indicating that bipedalism occurred at least a million years before significant increases in brain size and evidence for tool use caught many by surprise but were quickly accepted.

Throughout the 1970s and 1980s it was widely thought that bipedalism was probably an effective way to adapt to a **savannah**-grassland kind of environment. Many paleoenvironmental reconstructions associated with finds of *Australopithecus* and early members of the genus *Homo* supported this idea. Consequently, hypotheses to explain the transition to bipedalism were commonly based on the assumption that it was an effective adaptation to living in savannah-grassland environments. Examples include the hypothesis that by being bipedal, people were taller and thus exposed less of their body to direct sun and had better exposure to air currents above the grasses. Others suggested that being taller in the savannah-grasslands allowed people to look over the grasses, significantly increasing the areas in which people could search for food and predators. Some suggested that to be effective in the savannah-grasslands, our ancestors may have started scavenging animals killed by others; in this scenario, it is suggested that the primary advantage of bipedalism is that it provides greater endurance than using four limbs. Using four limbs may be

TABLE 4.2
Bipedalism – Why and How?

Explanations for Becoming Bipedal	Skeletal Changes Accommodating Bipedalism
Carrying model (food, children, rocks, sticks)	Repositioning of foramen magnum
	Changes to vertebral column (adding curves)
Effective heat management (heat dissipation)	Changes to the pelvis (widening, basin-shaped, stabilizing weight distribution)
Greater endurance (energy efficiency)	Lengthening of the femur (increased stride length)
Increasing height (for vision, more food, and display)	Modification to knee (allowing full knee extension and locking in place)
Walking in trees (e.g., like orangutans)	Angling of femur inward
	Changes in the foot (e.g., arch, realignment of big toe)

faster, but it requires much more energy to power four limbs than two, and people likely had to walk long distances to scavenge.

Over the past few decades, there has been increasing research on both the early humans and the environments in which they lived. Evidence suggests that rather than the first humans living in an open savannah-grassland environment, they were likely living in more of a mixed environment, perhaps with patches of open woodlands, forests, grasslands, and savannahs. The hypotheses that are based on the assumption that bipedalism was an adaptation to the open savannah-grasslands now accordingly receive little attention.

For many, the best explanations for the transition to bipedalism are based on the assumption that the primary benefit was that it frees up the hands. Not needing the forelimbs when moving around means that humans can carry things while standing and moving. In this vein, some believe that it may have been tools that were being carried. This would include tools yet to be discovered but could include sticks and sharp rocks near impossible to identify millions of years later. It may have been useful, for example, for our early ancestors to be carrying rocks and sticks to dig edible roots out of the ground, to scare off other scavengers from animal carcasses, or for protection.

Others suggest the principal advantage was to carry food. Our ancestors that were perhaps able to carry food could, for example, take advantage of a particularly productive kind of food that was in an otherwise dangerous area (such as a place frequented by lions or dangerous snakes) by gathering the food quickly and carrying it to a tree or other area of safety. Carrying food may also have been associated with food sharing. Some have hypothesized that it was the ability of males to provision females and children with food that was the driver of bipedalism, leading also to monogamy. This hypothesis is certainly not without criticism, however, as we are aware that monogamy is not the favored form of relationship in most cultures of the world (see Chapter 11). Others still have suggested that the primary advantage of free hands was the ability to carry babies. Presumably being able to carry babies (rather than have them simply hold on to their mothers while moving, as other primate babies do) was good for the babies and thus the population. Paleoanthropologist Dr. Meave Leakey (2020) is among those who believe the principal advantage of bipedalism was in the freeing of the hands:

> Becoming bipedal is the pivotal event that enabled further evolutionary changes that set humans apart. No longer bound to the physical demands of walking quadrapedally, the forelimbs were free for other tasks, and this enabled the development of a dexterous hand over time. Manual dexterity vastly improved the efficiency with which an individual could gather food, enabling even a young infant to easily put berries and other hard-to-reach foods in its mouth – but, crucially, it also allowed for further developments later on, such as the manufacture of stone-tool kits and the ability to carry and store foraged food. (p. 107)

Some others have suggested that the major advantage of bipedalism was for display. There are two aspects of this. For some, the primary advantage of bipedalism was that standing and moving upright was an advantage in threatening situations. Many animals, including other primates, make themselves look larger as a show of dominance or aggression. Thus, those who could do this on a regular basis became dominant, leading to more sex and more babies who also carried the trait. Another view of the display hypothesis reasons that those who stood upright exposed their genitalia more, likewise leading to more sex and more babies who carried the trait. An interesting, but not likely, scenario.

Another hypothesis suggests that bipedalism occurred neither as an adaptation to living in savannah-grasslands nor as an adaptation to spending significant amounts of time on the ground in a mixed environment. This hypothesis, commonly known as the walking-in-trees hypothesis, suggests that bipedalism evolved as a more efficient way of moving around in trees. It is suggested that our ancestors may have moved like orangutans, who often walk along tree limbs in a bipedal way but using their arms to hold onto branches above. This hypothesis was not given much serious consideration until a 2009 report in a special issue of the journal *Science* that *Ardipithecus ramidus*, which many accept as an early human, had opposable big toes, which would presumably have been an advantage for grasping onto tree limbs while walking.

How to Become Efficiently Bipedal

There are several variations that were selected for to support bipedalism. To be efficiently bipedal, the **foramen magnum** has to be positioned centrally at the base of the skull. This is so the head is balanced properly on the vertebral column (spine). In other primates, the foramen magnum is positioned further back in the skull since the vertebral column enters the head from the rear rather than from directly below. Curvatures in the vertebral column are another change that likely occurred to facilitate bipedalism. Curves (an S-curve) are generally thought to have been selected for to facilitate the distribution of weight of the upper body and pelvis when upright. Similarly, a broader pelvis was likely selected for to deal with the weight of the upper body. In the lower body, skeletal changes that are widely thought to correlate with bipedalism include angling of the femurs inward from the pelvis to the foot, once again to facilitate the distribution of weight; modifications to the femur and knees to allow a fuller extension; and structural changes in the foot. One change in the foot is the development of an arch, which is thought to help absorb weight as well as add propulsion to walking. It was once widely thought that the loss of opposability in the big toe was related to bipedalism, but this has been called into question since the discovery of opposable toes in *Ardipithecus ramidus*.

THE FIRST HOMININS

Sahelanthropus, Orrorin, and Ardipithecus

There is some uncertainty about which fossil assemblages represent the first hominins. The three primary candidates are genera *Sahelanthropus*, *Orrorin*, and *Ardipithecus*. *Sahelanthropus* was discovered in Chad, while *Orrorin* and *Ardipithecus* were discovered in the Great Rift Valley of East Africa. Collectively, they date from between seven and four million years ago. There is no consensus that any of these are absolutely hominins, but based on the current assemblages of fossils, these are the best contenders.

Sahelanthropus dates between seven and five million years ago. Hominin-like features include small teeth, no diastema, and the position of the foramen magnum indicating bipedalism. Ape-like characteristics, on the other hand, include a relatively small brain, a u-shaped dental arcade, and thin enamel on the teeth.

Orrorin dates to approximately six million years ago. Hominin-like features include a femur indicative of bipedalism, a relatively large body, and small teeth with thick enamel. Ape-like characteristics include canine teeth that are ape-like in size and shape.

Ardipithecus, which dates between 5.8 and 4.4 million years ago, has two species, and it is *Ardipithecus ramidus* that provides most evidence of being a hominin. Several features of both the cranial and post-cranial skeleton suggest bipedalism, and it has small canine teeth. One of the most interesting aspects of *Ardipithecus ramidus* is that it had opposable big toes, the only hominin or potential hominin in the human lineage with such a feature.

Australopithecus, Paranthropus, and Kenyanthropus

There is no ambiguity about the genus *Australopithecus*. There is consensus that it is a hominin. Some generalizations about the genus include the view that its antiquity ranged from at least 4.2 million to about 1 million years ago. It was restricted to Africa, where there were several species with average bodies ranging in size from about 65 pounds and 3.5 feet tall to about 100 pounds and about 5 feet tall. At least some populations probably opportunistically hunted or scavenged. One population of one species of *Australopithecus* likely evolved into the first species of *Homo*, but which one remains debatable. The primary contenders are *afarensis* and *africanus*.

Three species of early hominins exhibit features that are relatively rugged, often termed robust. For lumpers, they are simply different species of *Australopithecus*. For splitters, they represent a different genus – *Paranthropus*. These species are clearly not ancestral to the genus *Homo*. They became extinct about one million years ago.

There is another assemblage that splitters term *Kenyanthropus*. It is quite *Homo*-like, and some believe that it, rather than *Australopithecus*, may be ancestral to *Homo*. Lumpers consider it to be *Australopithecus*. The evidence consists of only one skull, and although, based on physical characteristics, it does look like a potential *Homo* ancestor, it is difficult for many to support this claim based on such meager evidence.

THE GENUS *HOMO*

The Emergence and Early Varieties of Homo to One Million Years Ago

Out of one population of *Australopithecus*, or perhaps *Kenyanthropus*, evolved the first members of the genus *Homo*. This probably occurred between 2.5 and 3 million years ago. Some recent discoveries in Africa of bone and teeth dating to 2.8 million

TABLE 4.3

The Hominins (Splitter's View)

Genus/Species	Region	Approximate Dates (Range of Antiquity)
THE EARLY CONTENDERS		
Sahelanthropus	Africa	7–5 million years ago
Orrorin	Africa	6 million years ago
Ardipithecus	Africa	5.8–4.4 million years ago
THE AUSTRALOPITHECINES		
anamensis	Africa	4.2–3.9 million years ago
afarensis	Africa	3.9–2.9 million years ago
bahrelghazali	Africa	3.5–3 million years ago
africanus	Africa	3.5–2 million years ago
deyiremeda	Africa	3.4 million years ago
garhi	Africa	2.5 million years ago
sediba	Africa	2–1.8 million years ago
PARANTHROPUS		
aethiopecus	Africa	3–2 million years ago
robustus	Africa	2–1 million years ago
boisei	Africa	2–1 million years ago
KENYANTHROPUS		
platyops	Africa	3.5 million years ago
EARLY *HOMO*		
habilis	Africa	2.5–1.4 million years ago
rudolphensis	Africa	2.5–1.4 million years ago
MIDDLE *HOMO*		
ergaster	Africa	2 million–500,000 years ago
erectus	Africa, Asia, Europe	1.7 million–200,000 years ago
georgicus	Asia	1.7 million years ago
antecessor	Europe	1.2 million–700,00 years ago
heidelbergensis	Africa, Asia, Europe	700,000–200,000 years ago
naledi	Africa	335,000–236,000 years ago
LATE *HOMO*		
Neandertal	Asia, Europe	400,000–40,000 years ago
floresiensis	Asia	100,000–60,000 years ago
luzonensis	Asia	67,000–50,000 years ago
Denisovans	Asia	60,000 years ago
sapiens (modern)	Africa	300,000 years ago–present
	Asia	60,000 years ago–present
	Europe	40,000 years ago–present
	Americas	20,000 years ago–present

BOX 4.2

On the Discovery and Interpretations of *Homo naledi*

In 2013, anthropologist Dr. Lee Berger was made aware of some potentially early human remains found in a deep and very difficult-to-access cave chamber in South Africa. The cave system, known as Rising Star, is within a large area in South Africa designated as the Cradle of Humankind, a World Heritage Site (based on it having been the location of many significant discoveries of early humans).

A former student of Berger's, Pedro Boshoff, was working for Berger searching for possible early human sites in the region. Boshoff enlisted the aid of two amateur cave explorers, Rick Hunter and Steven Tucker, to help him scout possibilities, and it was these two amateurs who discovered what is now recognized as the largest accumulation of early human remains in Africa. Hunter and Tucker were able to squeeze through very small areas and make vertical drops to a chamber where they took photos of potential human remains. Berger describes the first time he looked at the images, during a late-night visit from Pedro and Steven at his home:

> Steven quickly flipped open his laptop and brought a picture up on the screen. There, upon a piece of dirty map paper, was a hominin mandible.

A cloth measuring tape next to it gave scale. It wasn't human; that much was clear. The teeth were in the wrong proportions to be from any recent population. A second image showed more bones, all seemingly hominin. A third showed a rounded broken white outline in the dirt floor of the cave. It was the right shape to be the cross section of a skull – a tiny skull.... The pictures, Steven explained, were from a well-explored and well-mapped series of connected cave systems.... Steven and Rick had climbed a large underground rockfall called the Dragon's Back. When they reached the top, they found a jagged slot that narrowed to a pinch point 18 cm wide.... Steven went in first, and then Rick followed. The squeeze was tight, and vertical, and almost 12 meters to the bottom.... A small drop of a couple of meters to the bottom, and they found themselves in a chamber. On the floor they found bones, many bones, lying loose there on the floor. (Berger & Hawks, 2017, pp. 110–11)

Subsequent research at the site has resulted in the recovery of more than 1,300 hominin remains, representing at least 15 individuals ranging in age from infants to an older adult. Many more bones and teeth, perhaps in the thousands, are yet to be

years ago have been tentatively categorized as *Homo*, but fossil evidence of *Homo* prior to 2.5 million years ago is not incontestable.

Compared to *Australopithecus*, *Homo* is usually characterized as having a larger brain, smaller teeth, and a less prognathic face. Whereas the brain size of *Australopithecus* ranged between about 350 and 500 cubic centimeters (cc), the brain size of the first species of *Homo* ranged between about 500 and 800 cc.

recovered. Collection has been done primarily by six women scientists with experience in caving and anthropological excavation who could maneuver through the very narrow passages. These women, sometimes referred to as underground astronauts, are Hannah Morris, Alia Gurtov, Marina Elliott, Elen Feuerriegel, Becca Peixotto, and K. Lindsay (Eaves) Hunter.

The remains were so different from any known hominins they were assigned a new species name – *Homo naledi*. Based on the mosaic of traits that were evident, it was initially thought they may have been as old as two million years, but they have now been dated to be between 335,000 and 236,000 years old. Berger estimates their adult heights to have been between 4.5 and 5 feet and their weights ranged from 90 to 120 pounds.

The sheer number of bones and teeth and the number of individuals represented make this one of the most important finds of early humans anywhere. It is not likely that *Homo naledi* were ancestral to *Homo sapiens*; rather they coexisted with other populations that were in a more direct line to modern humans. Beyond adding in a very significant way to our understanding of

the biological diversity of early human populations, the finds at Rising Star are also intriguing for understanding human thought, particularly as it relates to belief systems, more than 200,000 years ago. Berger and others believe that the presence of individuals in the chamber may reflect the deliberate disposition of the bodies there. Other than six bones of a bird and some rodent teeth, no other bones of animals were found in the chamber, which suggests that it was unlikely that the human remains found their way into the chamber naturally. They have ruled out the notion that entering the chamber would have been easier in the past, so we are left with the idea that more than 200,000 years ago people were deliberately placing their dead deep in this cave system, carrying bodies substantial distances with vertical climbs through very narrow passages. This, some believe, may be the first clear evidence of intentional disposal of the dead and – because it would have been so difficult to accomplish – may have elements of ritual associated with it. Also, if people were traveling through this cave system, they likely would have needed fire for light. Claims of deliberate disposition, ritual, and the use of fire may be borne out, but they are all currently subject to debate.

These earliest members of *Homo* are referred to by lumpers as *Homo habilis*. Splitters see enough variability to suggest another species as well – *Homo rudolfensis*. The range of antiquity for these early *Homo* is from about 2.5 to 1.4 million years ago, and like their predecessors, they have only been found in Africa. As outlined in Chapter 5, these early species of *Homo* are the first undisputed makers of stone tools, which is probably associated with meat eating. The lack of forms with a mosaic of

Figure 4.5
RECONSTRUCTED HEAD OF *HOMO ERECTUS*.

This reconstruction is based on fossil evidence of skulls and knowledge of primate anatomy. *Homo erectus* lived from about two million to 200,000 years ago.
Credit: Giorgio Rossi/ Shutterstock

Australopithecus and early *Homo* features suggests that the evolution into *Homo* was through punctuated equilibrium.

Another variety of human, referred to as *Homo erectus* by the lumpers (and *Homo erectus* and *Homo ergaster* by the splitters), emerged in Africa by about 2 million years ago and appeared in eastern Asia by 1.7 million years ago. There are also some interesting finds from the Republic of Georgia, dating to about 1.7 million years ago, which some refer to as *Homo erectus*, but which splitters call *Homo georgicus*. *Homo erectus* was larger both in body size and brain size than its predecessors. The average height was likely over five feet tall, but some likely exceeded six feet. The brain size averaged about 1,000 cc (compared to about 700 cc for the earlier species of *Homo* and the modern average of 1,350 cc). As described in Chapter 5, *Homo erectus* is associated with many cultural developments, including full-scale hunting, the control of fire, and cooking.

Varieties of *Homo* over the Last Million Years

Adaptive radiation of the genus *Homo* has continued over the past million years. Some finds from Spain, classified as *Homo antecessor*, may be ancestral to more recent species, but the evidence is too meager to have confidence. *Homo heidelbergensis* appears about 700,000 years ago, overlapping in time with *Homo erectus*, and may have evolved from *antecessor*.

The transition from *Homo erectus* to *Homo sapiens* is probably more a case of gradualism rather than punctuated equilibrium. In Africa, Asia, and Europe there are individuals that appear to have a mosaic of features of both *Homo erectus* and *Homo sapiens*. Some anthropologists prefer to describe these individuals as "archaic *Homo sapiens*," and they first appear about 800,000 years ago. Some prefer to describe some or all of these specimens as *Homo heidelbergensis*.

Neandertals likely evolved from a population of *Homo heidelbergensis* in Europe about 400,000 years ago. Their core territories appear to be focused in Europe, but they also inhabited the Middle East until about 50,000 years ago and Asia as far east as Siberia until about 30,000 years ago.

There have been more than a hundred years of debates about whether Neandertals are *Homo sapiens* or a separate species, *Homo neanderthalensis*. Recent DNA research indicates that modern *Homo sapiens* living in Europe and Asia at the same time as Neandertals were able to mate and produce fertile offspring with them, lending support to the notion that Neandertals were simply a subpopulation of *Homo sapiens*.

Modern-looking *Homo sapiens* evolved in Africa about 300,000 years ago. There are indications that populations of modern *Homo sapiens* moved into regions of Asia and Europe at least a few times before 50,000 years ago, sharing the lands and resources with others. DNA research indicates that the last significant wave of modern *Homo sapiens* moved out of Africa about 60,000 years ago and overwhelmed the pre-existing populations in Asia and Europe, both genetically and culturally.

There have been a few recent discoveries that have challenged conventional thinking about human evolution over the last 50,000 years. One was the discovery of a previously unknown variety of human known as the Denisovans, so named from the Denisova site in Siberia. Based only on DNA extracted from a finger and a tooth, researchers have identified a previously unknown kind of hominin living in the region between about 50,000 and 30,000 years ago, coexisting with Neandertals and modern *Homo sapiens*.

Another find that has challenged conventional thinking is the discovery of about a dozen very small individuals on Flores Island in Indonesia. These specimens are classified as *Homo floresiensis* and are commonly referred to as "hobbits." They stood about three feet in height and had a cranial capacity of about 400 cc – smaller than most australopithecines – and a brain size roughly that of a chimpanzee. (They did not, however, as far as we can tell, have pointy ears or hairy feet like the hobbits of Middle Earth fame.) Some believe they may simply have been suffering from a disease that causes extreme dwarfism, but most accept they are a newly defined species of human that lived from about 100,000 to 50,000 years ago. Evidence of *Homo erectus* on the island dating to about 700,000 years ago suggests they may be ancestral.

The discovery of *Homo floresiensis* challenges conventional thought in multiple ways. First, a general trend in human evolution, with few exceptions, has been for species to get larger through time, both in body and brain size. If *Homo floresiensis* is indeed descended from *Homo erectus*, it is an interesting and important means of illustrating the variety of ways people can adapt. Second, there has been a widely held assumption that developments in culture parallel developments in the brain, including increases in brain size. Cultural evidence associated with *Homo floresiensis* suggests they were hunting and cooking large animals, including a kind of elephant and Komodo dragons, which are deadly lizards that grow up to 10 feet in length and commonly weigh more than 100 pounds. Since the cranial capacity of *Homo floresiensis* averages only about 400 cc, this poses a serious challenge to the notion

BOX 4.3 **Neandertals**

Neandertals (also spelled Neanderthals) are a very well-known variety of human that occupied much of Europe, the Middle East, and parts of Asia as far east as Siberia. The first evidence of Neandertals was discovered in 1856 in the Neander Valley in Germany, thus the name Neandertal.

Neandertals probably evolved from *Homo heidelbergensis* in Europe about 400,000 years ago. They lasted in the Middle East until about 50,000 years ago, when they probably could not effectively compete with modern humans living there. They remained dominant in Europe until about 40,000 years ago, when modern humans moved into the same areas. There also, they probably could simply not compete with the modern humans effectively, and the remnant populations were pushed to fringe areas such as Gibraltar and Siberia.

Neandertals were typically rugged, with prominent brow ridges, large noses, and powerful limbs. Their average cranial capacity, at 1,450 cc, was about 100 cc larger than the average among modern populations. Their body type is often described as being short and stocky, similar to that of contemporary Inuit. As with Inuit, the Neandertal body type was probably an effective adaptation to the cold. It is estimated that when Neandertals were living in Europe, the temperature was an average of 10 degrees Celsius colder than present.

Neandertals were adept tool makers, had a diverse diet that included many kinds of animals, used spears, and controlled fire. Numerous Neandertal skeletons have been found in caves, leading most to infer that they were deliberately buried, though some archaeologists remain skeptical on this point. Neandertals probably hunted effectively with throwing spears, cooked their food, wore clothing (not tailored), and used speech as their primary means of communications. Many believe that for the last 100,000 years they could probably speak as well as modern humans. Studies of cave art in Spain, reported in 2016, indicate

that humans were only able to make tools when they achieved a brain size in the range of early *Homo* (i.e., closer to 700 cc).

The most recent species to be recognized by anthropologists in 2019 likewise comes from east Asia and has other similarities to *Homo floresiensis* as well, including an average height of less than five feet. This new species is *Homo luzonensis* and dates to between 67,000 and 50,000 years ago.

SUMMARY OF TRENDS IN HUMAN BIOLOGICAL EVOLUTION

There are several trends in human biological evolution over the past several million years. Our early ancestors became increasingly more proficient in bipedalism. While *Australopithecus* are properly described as being bipedal, for example, their skeleton

Neandertals were producing cave art at least 65,000 years ago, and some believe there is evidence of Neandertal jewelry, including pierced eagle talons, dating to about 100,000 years ago.

Recent research indicates that although Neandertals were genetically and culturally overwhelmed by modern humans, their DNA lives on. DNA researchers have determined that modern-looking *Homo sapiens*, probably arriving from Africa about 60,000 years ago, mated with Neandertals (a good example of gene flow), and some traces of Neandertal DNA remains in almost all people of European and Asian descent.

Figure 4.6 **NEANDERTAL IN A BUSINESS SUIT.**
While generally more rugged than contemporary humans, Neandertals are still clearly human in appearance. If one were living in a large, present-day city, they might go unnoticed.
Credit: Neandertal Museum

was not as fully adapted as later members of the genus *Homo*. With some exceptions, humans have become larger in body size over time, and our brains have also become larger, both in real size and in proportion to the rest of our bodies. The shape of our skulls has changed, most notably with the development of vertical foreheads, and the widest part of our skull is now near the top. Our faces are now less prognathic and our teeth have become smaller.

We don't know for sure, but it is commonly thought that we lost most of our body hair at least a few million years ago as a way to regulate our body temperature, either as a response to being active in open areas, such as savannah-grasslands with direct sunlight, or simply from spending considerable time being active on two legs, which creates body heat. The primary way of regulating our body temperature is by sweating, and thick body hair would have clogged the sweat glands. Thus, those with less hair were selected for. It isn't that people have lost their body

Figure 4.7
HOMO FLORESIENSIS SKULL.
Homo floresiensis is a very small human species, often described as "hobbits," that lived as recently as 60,000 years ago in Southeast Asia.
Credit: Nadine Ryan

hair, however; we still today have approximately the same number of hairs as apes. Rather than a reduction in the quantity of hair follicles, hominin hair simply got finer, lighter, and shorter.

Probably around the same time as the loss of body hair a million or more years ago, a darker pigmentation was selected for to protect the skin from the intense sun in Africa. Until recently it was widely assumed that as people moved into the northern regions of Europe and Asia, lighter skin was selected for. People need sunlight to create vitamin D in their bodies. Enough sunlight can penetrate dark skin in climates with lots of sun, but it becomes a problem in northern climates where there is less intense sunlight. A lighter skin is thus beneficial in these conditions. However, DNA research from 2018 on an individual dated to 10,000 years ago in Europe (known as Cheddar Man) indicates that he and others of that time period had dark skin.

THE CONCEPT OF RACE

Race is a term that is used widely in North America, but it is often misunderstood to be a natural or biologically based category. It is not. The category was invented, suggesting race is something akin to a subspecies, identified by a combination of physical and behavioral characteristics.

Tracing Ancestry through DNA

There is little doubt about the significance of DNA research in tracing ancestry. It has been enormously important in determining relationships of populations extending tens of thousands of years into the past. We now know modern-looking *Homo sapiens* (descended from African populations) mated with Neandertal populations (with European and Asian ancestry) more than 50,000 years ago. We know this because we can identify DNA that was distinctive to each of those populations. In fact, almost all modern populations with European and Asian ancestry continue to carry some of that Neandertal DNA.

DNA research is also significant in studying the ancestral relationships of the Indigenous peoples of North America. Research supports an Asian connection 18,000 years or more in the past, and there are multiple cases where Indigenous peoples have been shown to have direct ancestry with human remains thousands of years old in their territories. One research study has linked living descendants with individuals who lived in the same territory more than 5,000 years ago.

While academic-based DNA testing has been important for understanding ancestral relationships, the kinds of DNA testing marketed to the general population is a different story. As described by anthropologist Dr. Jonathan Marks (2017), "genetics has become more than just a scientific authority ... it has also become a cash cow ... where the goal of the advancement of knowledge is accompanied by the goal for profits." Marks calls this kind of genomic research that is marketed to the public to trace their own ancestry "recreational genomic ancestry testing," and while it is "legitimate, honest, and ethical," it is also basically a "fabrication of meaning" of the DNA tests for people. In other words, the recreational DNA testing companies are providing a connection that may or may not exist, and often their claims of being able to provide a link to specific populations is meaningless.

Many Native American populations have been specifically targeted by DNA testing companies. Dr. Kim Tallbear (2013) conducted a comprehensive study of several companies that did just that, taking advantage of some people seeking support of their self-identification as Native American or more specifically hoping to prove biological links to specific tribes for a number of reasons, including economic. Tallbear explains that the entire process is flawed for multiple reasons, including the fact that there are no genetic markers for specific tribes.

Many have tried to validate the concept by using a variety of physical qualities to separate people around the globe, but all attempts have failed. There simply are no physical criteria that can separate humans into different so-called races. Anthropologists who study human biological variability know that human variation is continuous; it does not cluster in racial categories. While DNA research is increasingly being used to determine ancestral relationships, and some genetic companies make claims that they can determine affiliation with specific races or ethnic groups, these claims should be viewed with skepticism (see Box 4.4). Chapter 8 covers more on the concept of race.

SUMMARY

Mirroring the Learning Objectives stated in the chapter opening, the key points can be summed up as follows:

- There are multiple important aspects of paleoanthropology to understand. Researchers often revisit the same localities to look for sites. Principal dating techniques include potassium argon, radiocarbon, and dating by association. Paleoanthropologists need to have a good understanding of osteology and taphonomy, and there are often debates about how best to classify various specimens.
- There are several reasons why bipedalism may have evolved, and several skeletal changes occurred to make it workable. Most anthropologists think the primary reason bipedalism occurred was because it freed up the hands to carry things while moving. Biological changes that occurred include repositioning of the foramen magnum, adding curves to the spinal column, widening the pelvis, and changes in the foot.
- There have been many varieties of hominins over the past several million years. Often there were multiple species coexisting at the same time. The precise ancestral line leading to modern humans is not known. It appears likely, however, that the first members of the genus *Homo* emerged from a population of one *Australopithecus* species. From the earliest members of the genus *Homo* multiple other varieties of humans evolved.
- There are exceptions, but the general trends in human biological evolution include becoming more efficiently bipedal, becoming larger, increasing brain size, and reducing prognathism.
- Race is not a valid biological concept.

Review Questions

1. What are the basic methods, concepts, and issues in paleoanthropology?

2. What are the principal explanations for becoming bipedal and the skeletal changes that accommodated it?

3. What are the widely recognized genera and species of humans?

4. Why is race not a valid biological category?

Discussion Questions

1. How can basic evolutionary concepts such as mutation, gene flow, genetic drift, and punctuated equilibrium be applied to human biological evolution?

2. What are some of the advantages and disadvantages of having no fixed criteria for assigning specimens to a specific genus or species?

3. What are possible explanations of the relatively small size of *Homo floresiensis* and *Homo luzonensis*? Were they likely descended from *Homo erectus*, or some other species? Can their small size be attributed to island dwarfism, or something else?

Visit **www.lensofanthropology.com** for the following additional resources:

| SELF-STUDY QUESTIONS | WEBLINKS | FURTHER READING |

5

CULTURAL DIVERSITY FROM THREE MILLION TO 20,000 YEARS AGO

LEARNING OBJECTIVES

In this chapter, students will learn:

- *about the nature of the archaeological record and its key components.*
- *about problems of archaeological visibility, bias, and loaded language.*
- *about speculations on human culture prior to two million years ago.*
- *the principal cultural periods and developments between two million and 20,000 years ago.*
- *when people expanded to new areas around the globe.*

INTRODUCTION

Chapter 4 outlined the last several million years of human biological evolution. This chapter focuses on the corresponding elements of culture associated with humans, from their first undisputed physical manifestation about 2.5 million years ago to 20,000 years ago. So, while Chapter 4 essentially provided an overview of evolution from the perspective of biological anthropology, this chapter provides an overview from the perspective of archaeology. It is important to understand the nature of archaeology and potential problems associated with it, so the chapter begins with this coverage.

THE ARCHAEOLOGICAL RECORD

As with the fossil record, there is no consensus on precisely what is meant by the term **archaeological record**. At a minimum, it is taken to mean the actual physical remains of human activities that have been recorded by archaeologists. Some expand this definition to include, in addition to the recorded remains, all the records associated with archaeological investigations in the field and remains (e.g., catalogs, maps, photographs, reports on excavations). Others use the term *archaeological record* to refer to the basic facts about the past that are based on the physical remains of human activity.

The primary database of archaeology is the physical remains of human activities. These may include, but are certainly not restricted to, human biological remains. The other major kinds of material remains investigated by archaeologists are **archaeological sites**, **artifacts**, **features**, **ecofacts**, and **cultural landscapes**.

An archaeological site can be broadly defined as any location with physical evidence of past human activity. In practical terms, archaeologists often limit what will be described as an archaeological site based on a minimum age or number of artifacts. Major kinds of sites recorded by archaeologists, especially in the time

Figure 5.1
PROJECTILE POINTS.
Stone tools, such as these projectile points, are a common kind of artifact found throughout the world. People have been making stone tools for about three million years, although stone points usually date to within the last 100,000 years.
Credit: Nadine Ryan

before 20,000 years ago, include **base camps**, **habitation sites**, **pictographs**, and **resource processing sites**.

Base camps are generally recognized by the presence of artifacts and ecofacts, often in specific patterns that can be identified as a feature. Some of the earliest archaeological sites, for example, appear to be base camps where people were carrying out butchering activities, reflected in discrete accumulations of lithic tools and butchered bones. In very early times, such as before about 500,000 years ago, it is difficult to ascertain if these camps where processing was occurring were also used for habitation.

The term *habitation site* is based on the inference that people were living at the site, at least on a temporary basis. As will be described in more detail in Chapter 6, people did not begin to live in permanent settlements until about 10,000 years ago, but it is very likely that for at least several hundred thousand years they would spend at least several days and perhaps weeks or months in the same camp.

Rock art includes pictographs, which are paintings on immovable rock surfaces, such as boulders, cliff faces, or cave walls. Another kind of rock art is **petroglyphs**, which are engravings made on rock surfaces. Rock art begins appearing in the archaeological record about 65,000 years ago.

Resource processing sites are areas where the physical remains indicate that people were harvesting resources (e.g., hunting, gathering, scavenging) and/or processing them, including butchering. The term is also used to describe areas where people were obtaining raw materials such as stone for artifact manufacture (often referred to as a quarry), and/or where they made artifacts from the stone.

An artifact may be broadly defined as any object that has been manufactured or modified by people, or that shows evidence of being used by people. Many archaeologists, especially in North America, also make the distinction that it must be portable. As with the definition of an archaeological site, in practice archaeologists often use a narrower definition. Some archaeologists, for example, may choose to catalog only an intended tool as an artifact and not the waste flakes removed from the original cobble of stone. Other archaeologists may choose to catalog the waste flakes as artifacts as well.

A feature is defined as a nonportable entity that has clearly been created by humans. Common examples include **hearths**, **lithic scatters**, **middens**, and shelter or house structures. A fire hearth indicates a discrete, contained fire. It does not necessarily have to have a ring of stone around it, but a fire hearth is usually about the size of a campfire. Evidence of fire is common in archaeological sites, but it is often difficult to be certain that fires were cultural rather than natural. Lithic scatters, meaning the accumulation of waste flakes created and left behind from the manufacture of lithic (stone) tools, are common. Middens are discrete accumulations of trash. It is apparent that people have never liked to live among their trash, so once people started to stay in one place for several days, they separated their trash from their living space. This is convenient for archaeologists since it is easier to identify a midden than widely scattered trash. Shelters or house structures are often identified by depressions in the ground surface, which people created as a sort of foundation or to level the floor. They may also be identified by a pattern in the sediments, indicating that posts from wooden poles were once there (known in archaeology as post-holes), or other alterations to the ground surface (including pathways, ditches, and sediments brought from elsewhere for flooring).

Ecofacts include plant and animal remains. Archaeologists are interested in ecofacts for two primary reasons: they are used to make inferences about (1) **paleoenvironments** and (2) diet. Plants and animals in archaeological sites, even if they occur naturally, provide indications of the kind of weather and climate to which people were adapting. Of course, plants and animals also are important for determining what kinds of food people were eating. If ecofacts are to be used for reconstructing diet, however, it is important that they be in a good cultural context, such as a midden, or show use of modification by people, such as butchering marks on bone or evidence of cooking.

Plant remains in archaeological sites are commonly referred to as botanical remains or floral remains, and they include seeds, nuts, pollen, **phytoliths**, and wood. Since they are organic, plant remains do not tend to preserve well, and therefore they are often rare or absent in archaeological sites. Charred wood, however, is often found where people had fires. Burning removes nutrients from wood, so the microorganisms that contribute to decay tend to leave it alone. Plant remains, where they exist in sites older than 20,000 years, are often only visible microscopically.

Animal remains, also known as **faunal remains** in archaeology, may include any part of an animal, including its bone, teeth, shell, hide, hair, fur, nails, claws, and internal soft tissue. Bones are easier to identify in archaeological sites than plant remains. This is at least partially due to the fact that bone preserves better than plant tissue, and archaeologists must thus be careful to recognize this bias when making inferences about diet (e.g., fewer plants at a site do not necessarily indicate few plants in the diet).

PROBLEMS OF ARCHAEOLOGICAL VISIBILITY, BIAS, AND LOADED LANGUAGE

When examining the archaeological record of human culture, one has to consider that it is vastly incomplete. Many aspects of human culture have what archaeologists describe as low archaeological visibility, meaning they are difficult to identify archaeologically. Archaeologists tend to focus on tangible (or material) aspects of culture: things that can be handled and photographed, such as tools, food, and structures. Reconstructing intangible aspects of culture is more difficult, requiring that one draw more inferences from the tangible. It is relatively easy, for example, for archaeologists to identify and draw inferences about technology and diet from stone tools and food remains. Using the same kinds of physical remains to draw inferences about social systems and what people were thinking about is more difficult. Archaeologists do it, but there are necessarily more inferences involved in getting from physical remains recognized as trash to making interpretations about belief systems.

Other things to consider include the fact that, in general, the further back in time one goes, the less visible evidence of culture will be. This is due to multiple reasons, including the fact that (1) the older the site, the more likely it is to be covered up; (2) the older the site, the less likely it is that organic remains will be preserved; (3) the further one goes back in time, the fewer the number of humans there were to leave physical evidence behind; (4) the further one goes back in time, the fewer the kinds of physical evidence of culture there were (first only tools, then shelters, etc.). Also, until about 10,000 years ago, most human groups were fairly mobile, moving within territories and peripheral areas but not likely settling for months at a time in the same place, where trash could accumulate (making it more visible).

Archaeologists also recognize that most archaeological sites have already been destroyed by both natural and cultural processes. Archaeologists appreciate that the earth is a very dynamic system, forever changing the landscape and often destroying or burying archaeological sites through a wide variety of processes, sometimes catastrophic (such as by glaciations, landslides, and tsunamis) and sometimes more gradual (such as through erosion). Archaeological sites near water are particularly susceptible since sea levels have fluctuated widely in the past (only stabilizing in their current position about 5,000 years ago), lakes are often temporary, and rivers and streams often change course over time.

Humans, as well, lead to the loss of archaeological visibility. Many human activities lead to the loss of archaeological sites, both known and unknown. Every time landscapes are altered, there is a good chance archaeological sites are being destroyed or buried. Some archaeological sites remain intact but are no longer visible due to modifications on the surface.

There are other kinds of biases to consider as well. Archaeologists recognize that there is a very strong bias toward inorganic materials, such as stone and ceramic artifacts, simply because they preserve better than organic materials. There is also a bias toward things recognized as trash and sites that were deliberately abandoned. Although there are exceptions, the overwhelming majority of artifacts recovered were recognized as trash by the people who left them. Similarly, most archaeological sites were deliberately abandoned.

Like many disciplines, archaeology has suffered from a male bias, especially insofar as the history of the discipline has tended to be dominated by male archaeologists focusing on activities long thought of as primarily male activities, such as hunting. Fortunately, this male bias has diminished in recent decades, especially due to the significant increase in women becoming archaeologists.

Another important kind of bias to consider is that of place. Especially in regard to the time period before 20,000 years ago, there has been a strong bias toward archaeological research in East Africa and Europe. Much of what we know of the archaeology of early humans, for example, comes from the Great Rift Valley in Africa. This is understandable since so much research has been done there. However, it should be appreciated that there may be many other areas where early humans were active, but these areas simply have not been examined. There is also a strong Eurocentric bias in archaeology. This is not totally surprising, since archaeology itself developed in Europe. We should be aware that although much of the focus in archaeology is on Europe, this is at least partially due to the interest in Europe by Europeans. It is common, for example, to accept that many of the great cultural achievements – such as cave art, deliberate human burial, ceramic technology, spear throwing, and more – developed in Europe. They may have, but we should not lose sight of the fact that not all areas of the world have received as much attention by archaeologists as Europe has, especially during the period before 20,000 years ago.

Archaeologists are increasingly becoming aware of the language they use, some of which has negative impacts on peoples of particular regions. On a broad scale, many archaeologists cringe when they see terms such as "exodus" or "migration" in popular media because it leaves some with the impression that the location people were leaving was an inferior place. Media frequently refers, for example, to an "exodus" from Africa at various times in the human past, when it may be more appropriate to consider those events an expansion of territory. Archaeologists working in North America are also aware that some of the language they use continues to support colonial perspectives, including notions that Indigenous peoples were primitive, and the words they choose often serve to disassociate Indigenous peoples from places, things, and resources important to them. As agents of decolonization, some archaeologists are now choosing their words more carefully in order to negate those words that suggest primitive peoples and words that disassociate. For example, some archaeologists now try to avoid using the word "prehistory," since it elevates the time period after Europeans arrived, and instead use the word "archaic" to describe the same period of time. Some archaeologists are increasingly using "ancestors" to replace "skeletal remains" and "belongings" to replace "artifacts." Rather than describe a location of activities in the past as a "site," some archaeologists are using more meaningful words such as "village" or "gathering place." Even the term "evolution" remains problematic. Anthropologists understand that "evolution" simply means change over time, but for many others it carries notions of a linear trajectory and is based in notions of colonialism. Despite the recent movement of decolonization in which many archaeologists are involved, however, the traditional vocabulary remains the norm in the early decades of the twenty-first century.

SPECULATING ON HUMAN CULTURE PRIOR TO TWO MILLION YEARS AGO

Inferences about human cultures from the past two million years are typically supported fairly well, based, as they are, on substantial evidence provided by artifacts, sites, features, and ecofacts; it is around the two-million-year mark that a significant amount of this physical evidence begins to appear in the archaeological record. Of course, the further one goes back in time, the less physical evidence one has to rely on, and the more speculation tends to increase.

Physical evidence of culture prior to two million years ago does exist, but it is not plentiful, and it is primarily restricted to evidence of stone tool manufacture and use. There is good evidence in Africa of stone tool use extending back in time to 2.6 million years ago, and, though there isn't a lot, there is also evidence that some stone tools may have been made as far back as 3.3 million years ago. There are also occurrences of marks on animal bones in Africa between 2.5 and 3.4 million years ago that many archaeologists accept as evidence of stone tool use – cuts on the bones would have been made in the process of butchery. Beyond stone tools, early humans likely had wooden digging sticks and may have used parts of plants and animals to create containers, but these have not survived.

In the absence of an abundance of physical evidence, archaeologists can only speculate on what human cultures were like prior to two million years ago. Besides the few occurrences of stone tools and butchered animal bones, we can also use paleoenvironmental evidence to suggest the kinds of environments early humans were living in, and we can use analogies with nonhuman primates and contemporary humans to further help in our reconstructions. Of course, we realize that any speculation on human cultures based on analogies with nonhuman primates and more recent humans is fraught with difficulties, including that none of the groups are of the same species.

Recognizing the difficulties and uncertainties of speculation, it is still reasonable to assume that there were considerable differences among human cultures prior to two million years ago. In the period between four and two million years ago, for example, there were likely at least a dozen different human species that were occupying different environments. In the period between three and two million years ago, it is likely that most were living in open woodland or savannah-grassland environments.

In this period between three and two million years ago, it is likely that people were mostly eating plant foods. Despite finding butchered animal bones, most humans likely depended mostly, or even entirely, on plants for sustenance. It is possible some groups included insects in their diet (Lesnick, 2018), and some scavenging may have occurred, though there is much debate about whether early humans ever scavenged for animal carcasses. Stone tools may have been used to sharpen sticks to help dig plant foods, and early humans may also have carried sticks to ward off predators. Hunting of small animals may have occurred among some groups, but if it did, it was probably opportunistic and not an essential part of the group's dietary strategy.

Humans living between two and three million years ago were likely living in groups of about 20 to 30, with fluid membership. They would have been aware of numerous other groups in their immediate region and been on mostly friendly terms with them. People would have had family members and friends living in neighboring groups. Either the females or males would have likely left the group they were born into to join another group in their teen years. Everyone in the group likely had roughly equal status.

Each group probably had a core territory of several square miles, but they would have occasionally ranged outside of this to meet with others. They had no permanent camps to which they returned each day, and beyond childcare, there would have been little division of labor between the sexes. It is also reasonable to assume that humans living between two and three million years ago had substantial culture, such as tool use and foraging strategies, and that much of it was taught through language. The language capabilities of humans at this time were rudimentary, but they would have been enough to pass on knowledge and practices, values, and group norms.

OVERVIEW OF CULTURAL CHANGE TO 20,000 YEARS AGO

Principal Cultural Periods

Archaeologists use frameworks when referring to prehistory. The most common for the period before 20,000 years ago are outlined in Table 5.1. Essentially, **Paleolithic** means "Old Stone Age," with the prefix *Paleo* meaning ancient or old, and *lithic* meaning stone. **Lower Paleolithic** is used widely to describe the peoples and cultures associated with *Homo habilis* and *Homo erectus* (lumper's view); **Middle Paleolithic** is often equated with the people and cultures associated with archaic *Homo sapiens*, *Homo heidelbergensis*, and Neandertals; and **Upper Paleolithic** (rarely used outside Europe) primarily refers to the culture and peoples who replaced Neandertals in Europe beginning about 40,000 years ago. The range of antiquity is approximate and varies among archaeologists and regions. Table 5.2 provides an overview of the principal cultural developments in each period.

TABLE 5.1

Principal Cultural Periods 3 Million to 20,000 Years Ago

Period	Antiquity	Region
Lower Paleolithic	c. 3 million–500,000 years ago	Africa, Asia, Europe
Middle Paleolithic	c. 500,000–40,000 years ago	Africa, West Asia, Europe
Upper Paleolithic	c. 40,000–12,000 years ago	Mostly Europe

TABLE 5.2

Principal Cultural Developments Prior to 20,000 Years Ago

Period	Cultural Developments
Lower Paleolithic	First undisputed evidence of culture, in the form of stone tools, was likely created and used by *Homo habilis*. Flakes of stone were chipped off one end of a cobble; the now sharp cobble was used as a tool, as were sometimes the flakes themselves. This technology is often referred to as Oldowan. *Homo erectus* made more complex tools, characterized by the Acheulean hand ax. *Homo erectus* likely controlled fire, had base camps, and had a division of labor. Hunting and meat eating was probably opportunistic or small-scale among early *Homo*, but a major part of the subsistence strategy among *Homo erectus*.
Middle Paleolithic	Continued changes in lithic technology and evidence of finely crafted spears by 400,000 years ago. Some suggestion of deliberate burials, art, and jewelry, but the evidence is debatable. Peoples extended territories into northern latitudes.
Upper Paleolithic	Continued advances in technology. Undisputed evidence of deliberate human burials and art. Invention of atlatl (spear-thrower).

Figure 5.2
HOW TO MAKE STONE TOOLS.

This drawing represents the principal methods of making stone tools. The earliest stone tools were probably only made by hardhammer percussion. Eventually, probably during Middle Paleolithic, softhammer percussion was added in order to make better tools. During the Middle and Upper Paleolithic, people added indirect and precision (or pressure) flaking to further increase the efficiency.

Credit: Katherine Cook

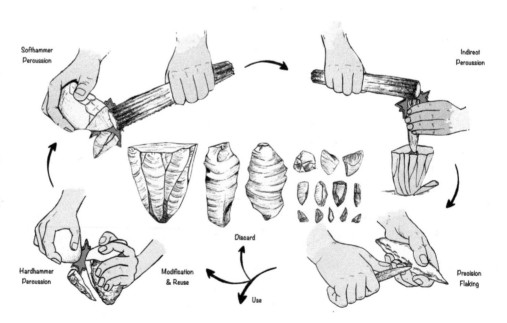

Softhammer Percussion

Indirect Percussion

Hardhammer Percussion

Modification & Reuse

Discard

Use

Precision Flaking

BOX 5.1 **Food Matters: Reconstructing Diet**

Archaeologists have several ways of reconstructing the diet of people living in prehistoric times. The major categories of remains that archaeologists use are (1) plant and animal remains in cultural context, (2) human skeletal remains, (3) **coprolites**, (4) human soft tissue, and (5) residue on artifacts.

Plant and animal remains in good cultural context are one of the most common ways to study diet. However, archaeologists need to be confident that the remains represent food and not just naturally occurring plants and animals or those used for other cultural reasons. Some of the ways cultural context is inferred include if the remains show evidence of cooking, if they are in a discrete midden associated with other refuse, if they are in a distinct fire hearth, or if the animal bones show evidence of butchery with a tool.

An analysis of the bones and teeth of human skeletons provides good indications of major food groups. An analysis of specific kinds of carbon isotopes, for example, can indicate a diet based on different kinds of grasses, shrubs, or fruits. Isotope analysis can also indicate the amount of marine versus terrestrial resources in the diet, and the amount of meat in the diet is reflected in trace elements. A diet rich in meat, for example, will usually include high levels of copper and zinc.

Coprolites, the word used by archaeologists to describe preserved human feces, are an excellent way of determining diet. Depending on the diet and health of the person, coprolites may appear as cylinders, pellets, or pads. Some reports suggest that when remains are reconstituted for analysis, the smell sometimes comes back. Archaeologists examine the coprolites, trying to identify partially digested food fragments such as seeds and small bones. The oldest reported human coprolites date to about one million years ago in Africa and 250,000 years ago in Asia, but whether these are human is in some doubt, and there are no reports of them being studied. The largest recorded coprolite came from a Viking; it measured 23 cm in length and indicated previous meals of meat and bread. The man also had intestinal worms.

Soft tissue and residue on artifacts are other ways of determining diet. When archaeologists find preserved bodies, they are often able to determine diet by examining stomach contents, but since soft tissue does not preserve well, using this method is not common. Residue analysis includes a chemical analysis of food remains in or on artifacts. Residue in pottery, for example, provides an indication of the kind of food or liquid stored in the pot. An analysis of blood residue on a projectile point or butchery tool can indicate the animal on which it was used.

Subsistence and Diet

There are many ways to reconstruct prehistoric diets. These include finding plant and animal remains in good cultural context, examining residue left on artifacts, and examining isotopes in human skeletons. These are covered more fully in Box 5.1.

The earliest members of the genus *Homo* likely depended primarily on plant foods but incorporated more meat in their diet than their australopithecine ancestors. Beginning about 2.5 million years ago, animal bones and stone tools indicate that meat eating became important. Accumulations of animal bones with evidence of butchery by stone tools appear, and many stone tools, presumably used for butchery, appear as well. These bones and tools are usually associated with *Homo habilis*. The significance of meat in the diet of early humans is debatable, but it is fairly clear that for at least some, meat became an important resource.

One of the biggest debates in the study of early human culture focuses on whether our ancestors transitioned to hunting via a period of scavenging. Many believe that scavenging animals killed by lions or other predators on the savannah would have made some sense. Presumably, early humans would not have been able to scare the large predators off a fresh kill, but they could have scared off scavengers such as wild dogs, hyenas, or vultures at least temporarily using rocks and sticks, which would have allowed some humans to move to the fresh kill and remove meat. There is relatively little evidence, however, for the scavenging hypothesis. Some support is offered in the identification of large animal bones smashed in a particular way with a rock to extract marrow. Perhaps once the scavengers had left a kill, humans moved in to extract the marrow that scavengers could not.

Others think it unlikely that humans ever went through a scavenging phase. The reasoning is that while scavenging may occur among contemporary human foragers, it is rare and opportunistic rather than a planned strategy. Those who dispute a scavenging phase also point to the knowledge that nonhuman primates that eat meat hunt rather than scavenge.

While it is widely accepted that early *Homo* incorporated some meat in their diet, the significance of meat and the adoption of hunting strategies clearly increased with the emergence of *Homo erectus*. The driving force may be linked with the use of fire for cooking, although this idea is controversial (see Box 5.2).

There is considerable evidence that hunting, at least for the past few hundred thousand years, included big game such as mammoths and mastodons. This would undoubtedly have required group cooperation, such as several people ambushing and targeting one specific animal. It is likely that hunting big game often occurred when the animals were most vulnerable, such as when crossing water. Hunting big game would have likely been a dangerous activity in early times, especially before the invention of throwing spears about half a million years ago. Prior to this, hunting even small and medium-size game likely depended on thrusting spears (i.e., held in the hands while thrusting into the animal). Some have suggested that hunting may have occurred by chasing animals until they died from exhaustion. This technique is called persistence hunting and essentially means that a small group

BOX 5.2 **Food Matters: Was Cooking the Driving Force of Human Evolution?**

Anthropologists are often interested in the driving force of human evolution. One popular hypothesis is that the driving force was cooking. The hypothesis is valid but not widely accepted. It makes sense to many, but skepticism remains because of the lack of evidence in the archaeological record, including the lack of evidence for the control of fire close to two million years ago.

The person most often associated with the hypothesis that cooking was the driving force of human evolution is Dr. Richard Wrangham, who outlined the hypothesis in his book *Catching Fire: How Cooking Made Us Human*, published in 2009. Wrangham summarizes his idea as follows:

> I believe the transformative moment that gave rise to the genus *Homo*, one of the great transformations in the history of life, stemmed from the control of fire and the advent of cooked meals. Cooking increased the value of our food. It changed our bodies, our brains, our use of time, and our social lives. It made us consumers of external energy and thereby created an organism with a new relationship to nature, dependent on fuel. (p. 2)

Wrangham suggests that the initial change that made us human was increased meat eating about 2.5 million years ago (associated with *Homo habilis*), followed by cooking about 1.8 million years ago (associated with *Homo erectus*). Here, Wrangham describes the value of cooking:

> Cooked food does many familiar things. It makes our food safer, creates rich and delicious tastes, and reduces spoilage. Heating can allow us to open, cut, or mash tough foods. But none of these advantages is as important as a little-appreciated aspect: cooking increases the amount of energy our bodies obtain from our food.

> The extra energy gave the first cooks biological advantages. They survived and reproduced better than before. Their genes spread. Their bodies responded by biologically adapting to cooked food, shaped by natural selection to take maximum advantage of the new diet. There were changes in anatomy, physiology, ecology, life history, psychology, and society. Fossil evidence indicates that this dependence arose not just some tens of thousands of years ago, or even a few hundred thousand, but right back at the beginning of our time on Earth, at the start of human evolution, by the habiline that become *Homo erectus*. (pp. 13–14)

A little further on, Wrangham elaborates on the benefits of cooked food:

> In humans, because we have adapted to cooked food, its spontaneous advantages are complemented by evolutionary benefits. The evolutionary benefits stem from the fact that digestion is a costly process that can account for a high proportion of an individual's energy budget – often as much as locomotion does. After our ancestors started eating cooked food every day, natural selection favored those with small guts, because they were able to digest their food well, but at a lower cost than before. The result was increased energetic efficiency. (p. 40)

of people would simply chase a selected animal, perhaps for days, until the animal died from exhaustion. This makes sense to some since, while most game animals are quite quick over short distances, they usually cannot maintain their speed over long distances. Bipedalism in humans, on the other hand, leads to extended endurance. People may not be as quick as some animals over short distances, but they can outlast them over long distances.

Social Systems

Inferences about the number and organization of people living during the Paleolithic is based on (1) analogy with nonhuman primates and human foragers of recent and contemporary times, and (2) archaeological evidence. Analogy with nonhuman primates and human foragers suggests that a group size of approximately 25 to 30 people was common. Membership would likely have been fluid (meaning that people could come and go), and groups would almost certainly have been exogamous (meaning that upon reaching mating age, members would find partners from outside the group). Because of exogamy, no group lived in isolation. People from various bands would have known and interacted with neighboring bands.

There is little evidence of social stratification in the Paleolithic, suggesting groups were mostly egalitarian. This is assumed by the fairly equitable distribution of resources within habitation sites. Stratification likely started in the Upper Paleolithic.

There has likely been a division of labor based on sex for close to two million years. This is based on analogy with nonhuman primates that hunt and with human foragers. As mentioned in Chapter 2, when nonhuman primates hunt, it is mostly the males that are involved and meat is often shared, including with females. It is a similar situation when human foragers hunt. It is primarily a male activity, and the meat is shared. This is not to imply that males were more important than females in subsistence activities. For example, what we have learned from human foragers is that while meat obtained by men is shared, so are the plant foods collected by women, and it is often the plant foods that are more important for daily nutrition.

It should be recognized that this explanation of the division of labor is based primarily on inference; archaeological remains provide no incontrovertible evidence for it. However, besides the analogies from nonhuman primates and contemporary human foragers, it makes sense to many that women were not usually involved in hunting because it would have been difficult to hunt when pregnant or caring for a child. The division of labor also likely created the need for a home base, a place where both males and females could return to at the end of the day. We recognize that a two-gender binary is overly simplistic and, as in the present, there were probably multiple genders. Although tasks may have been distinguished as being either primarily male or female, such distinctions were unlikely to be absolute.

Controlling Fire

One of the most important achievements in the development of human culture was the ability to control fire. It is difficult to overestimate the importance of fire in human evolution. It provides warmth, light, and protection, enhances diet, and provides a focus for social interaction.

Anthropologists are uncertain about when or why people started controlling fire, but archaeological evidence suggests that it has been common for at least tens of thousands of years, probably hundreds of thousands, and perhaps almost two million years. Evidence comes from a variety of sources including remnants of fires themselves, such as charred wood or ash, as well as from things that have been heated, including bone, stone, and clay. Based on observations of savannah chimpanzees' behavior around fires, Drs. Jill Pruetz and Nicole Herzog (2017) suggest that humans up to several million years ago also may have exhibited a comfort level around fire, at least initially being able to conceptualize fire, predict its movement, and understand its benefits, especially in regard to food.

Problems for archaeologists include determining whether the remnants of fire they observe are natural or cultural. Similarly, determining whether a bone or stone in an archaeological deposit has been heated by a natural or culturally controlled fire is problematic. Even if heated bones are found in a cave with other cultural remains, there is often some uncertainty over whether the heated bones were cooked by people or whether perhaps some other animal scavenged a burned bone and brought it to the cave.

The ability to control fire was significant. It enabled more kinds of foods to be eaten and increased the nutritional value of some others. By providing light, it increased the practices that required or were enabled by light, and by providing heat, it enabled the expansion into territories otherwise too cold. It also afforded protection from most animals and provided a focus for social interaction. Eventually, it also enabled advances in technology, such as heating rocks to enable better fracturing qualities, making ceramics, and strengthening wooden spears and other artifacts. In recent times, controlled fire has also been used in subsistence activities, such as lighting fires behind herd animals to drive them, and deliberately burning vegetation to speed the release of nutrients back into the soil or to provide a different kind of vegetation regrowth preferred by animals used in the diet.

Although some anthropologists believe humans started controlling fire almost two million years ago (see Box 5.2), most anthropologists are comfortable with claims ranging from a few hundred to several hundred thousand years ago. This comfort level usually comes from the quantity of evidence in cultural sites; even though some uncertainty remains, it is not so much a question of whether there is

Figure 5.3

FIRE.

Humans have probably been controlling fire for hundreds of thousands of years.

Credit: Datourdumonde Photography/Shutterstock

evidence of fire, but rather whether the evidence is of natural or culturally controlled fire. Most archaeologists agree that the evidence of widespread control of fire by about 40,000 years ago is indisputable.

When people began to start fires is unknown, but it probably dates to the Upper Paleolithic. The most common technique of starting fires in the distant past was likely by generating heat through the consistent friction of one stick against a stationary piece of wood, surrounded by some flammables such as dried botanical remains. However, the archaeological visibility of these items is very low, since they are unlikely to be preserved, and if they were, they may be difficult to recognize. Another way of starting fire was by creating sparks by hitting certain types of rocks together. Archaeologists have found these kinds of rocks in European Upper Paleolithic sites, showing evidence of repeated striking in the same place, something akin to an Upper Paleolithic lighter.

Evolution of Technology

The evolution of technology during Paleolithic times is profound. In particular, the differences in the level of sophistication of lithic technologies is astounding. Research on the average amount of cutting edge produced from a single pound of stone, for example, shows that the earliest members of the genus *Homo* (*habilis/rudolfensis*) were able to create an average of 2 inches of cutting edge per pound of stone. *Homo erectus* was able to create an average of 8 inches of cutting edge. People living during the Middle Paleolithic manufactured an average of 40 inches of cutting edge per pound of stone, and those of the Upper Paleolithic (at least those in Europe) an astounding 120 inches of cutting edge per pound.

Two of the earliest known types of stone tools are the **Oldowan** and **Acheulean** tools. Oldowan tools are usually associated with *Homo habilis* and were typically made from a cobblestone with a few to several flakes struck off one side of one end, creating what is known as a unifacial tool (or unifacial chopper). These would have been quite effective for many purposes, including butchering animals, sharpening sticks, and perhaps even digging roots and cutting plants. In addition to the cobble itself with flakes removed, some of the larger flakes were also used, presumably for cutting.

In 2015, researchers reported discovering in Africa what may be the oldest known human tools, dating to 3.3 million years ago. Predating the Oldowan by several hundred thousand years, these apparent stone cobble and flake tools are different enough from the Oldowan that the researchers suggested a new name for this tool-making industry or tradition: **Lomekwian**. Dating them to 3.3 million years is important insofar as they may provide the first evidence of tools created by *Australopithecus* or *Kenyanthropus*, or push back the origin of *Homo* to this time. Whether the finds truly represent a 3.3-million-year-old discovery of human tools remains debatable, however, and it may take some years before they are widely accepted or rejected.

Acheulean tools are associated with *Homo erectus* and were typically bifacial, meaning flakes were taken off both sides. There are several recognizable kinds of Acheulean tools, but none receive as much interest as the Acheulean hand ax. Thousands of hand axes have been recorded but their function remains an enigma. Explanations range from the hand ax merely representing a core left behind after flakes for artifacts had been removed, to their being multifunctional tools, throwing weapons, and ways to express sexual fitness or goodwill. These notions are explained more fully in Box 5.3.

Anthropologists are not certain when projectiles may have first been used, but it is widely assumed that spear technology was in practice by at least several hundred thousand years ago. One of the problems, of course, is that spears are unlikely to be preserved; it seems probable to many that the first spears were sharpened sticks that were thrust rather than thrown. The first undisputed throwing spears are dated to 400,000 years ago and are associated with *Homo heidelbergensis* in Germany. Attaching stone points to the ends of spears likely started about 100,000 years ago.

The first evidence of an **atlatl** (spear-thrower) appears in Europe about 30,000 years ago. An atlatl is essentially an extension of the arm. It requires shorter spears (commonly known as darts) and results in better distance, accuracy, and velocity than by throwing by hand alone. Typically, an atlatl was made of a piece of wood about one meter long, with a stopper at one end. A dart would be placed on the atlatl, with one end resting against the stopper. The thrower would move the atlatl much like a tennis racket in an overhand "serving" motion, releasing the dart (held between the fingers and thumb).

There is some suggestion that bow and arrow technology may have emerged as early as 65,000 years ago in Africa, but there is no consensus that the evidence is good enough to make the claim. There only exists one small point from that time period of the size that may have been used on an arrow. Most

BOX 5.3

The Acheulean Hand Ax – Tool, Core, or Sexual Object?

The Acheulean hand ax (Figure 5.4) is an enigma. The axes are associated with *Homo erectus* and start to appear in the archaeological record about 1.7 million years ago, continuing in their basic form for more than one million years. Many are often found at the same site, such as at Olorgesailie (Figure 4.2). They are typically about the size of a human hand, teardrop shaped with flakes taken off both sides and usually the entire surface, pointed at one end, sharp around the entire circumference, and weighted near the base where they are also usually the thickest. Many people have commented on the aesthetic quality of the hand axes and consider them works of art.

Their function remains unknown, but there are many ideas, some rather strange. Many archaeologists suggest they may be multifunctional, to the extent that some refer to them as "Swiss Army rocks." In this view, they could variously be used for cutting, piercing, and scraping, for example. Others suggest the hand axes were merely the cores left behind after all the desired flakes had been removed. Some archaeologists believe they were thrown at animals as a hunting technique. Experiments with throwing them like a discus indicated this is certainly a possibility and would explain the weight distribution, shape, and cutting surface all around.

Where it gets a bit strange is with the idea that the hand axes were created by men to influence women, leading to a kind of sexual selection process, with the women presumably favoring the men who made the best nonfunctional hand axes – a difficult hypothesis to test. An equally difficult hypothesis to test is that the hand axes were created as a show of trustworthiness or goodwill to others.

Most archaeologists reject the "sexy hand ax" and "trustworthy" hypotheses and are more inclined to believe that they had a specific technological function, that they were some kind of tool or weapon. Despite their name, they were probably rarely used as hand axes, although they were likely, at least in some contexts, used as hand-held cutting tools. In a recent article, Wynn and Gowlett (2018) suggest that the hand ax design results from neither strict functionality nor aesthetics, but a combination of both.

Figure 5.4 ACHEULEAN HAND AXES.
A characteristic tool of *Homo erectus*. Their precise function is unknown, but it is unlikely that they were commonly held in the hand and used like an ax.
Credit: Album/Alamy

Figure 5.5
SPEAR-THROWER IN USE.
Spear-throwers, also known as atlatls, originated tens of thousands of years ago and are known to have been used around the globe, especially for hunting big game animals.
Credit: The Trustees of the British Museum

archaeologists are more comfortable with bows and arrows emerging much more recently (no earlier than the Upper Paleolithic, and probably even more recent than that).

Art and Ideology

Art is an area of considerable interest in archaeology, especially the study of its origins. Some believe that engravings on shells dating to about 500,000 years ago are evidence of art, but there is no consensus that they in fact represent art. The engravings were discovered in 2014 by a researcher analyzing shells collected from a *Homo erectus* site excavated in Asia about 100 years earlier. There is some question about the association of the shells with human remains, the dates the engravings were made, and whether the engravings are cultural or perhaps due to some taphonomic process. It is not uncommon to see reports of archaeologists making claims to evidence of art 100,000 years or more ago in Africa, but those claims are also subject to debate.

The earliest undisputed evidence of art, found in caves in Spain, dates to at least 65,000 years ago. Art did not start to become common and widespread until 40,000 years ago, however, which is when it became more evident in Europe as well as in Indonesia and Australia.

Art in the archaeological record, especially **cave art** or **rock art**, is often linked with religion and ritual. Cave art and rock art before 20,000 years ago is usually associated with shamanism, although it is not necessarily restricted to only shamans creating the art. Examples include paintings meant to manipulate supernatural powers, such as to ensure the ongoing fertility of animals or ensure success in hunting. This is supported by the fact that many of the paintings depict animals that were routinely hunted, or that were pregnant or had spears in them, and fertility symbols. As well, most of the paintings were done in the most remote parts of

BOX 5.4

Upper Paleolithic Figurines – Not Just Erotica

Approximately 200 small human-shaped figurines, mostly female, have been recovered from archaeological sites in Europe and Asia that date to the Upper Paleolithic. They are constructed from stone, bone, ivory, and clay (Figure 5.6), and are commonly referred to as Venus figurines.

In both scholarly and mainstream media, the focus is often on the sexual characteristics of these figurines, such as explicit genitalia and large breasts,

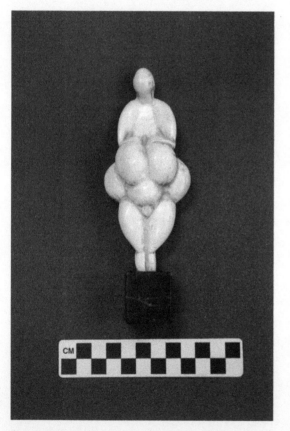

Figure 5.6 UPPER PALEOLITHIC FIGURINE.
Figurines such as this are usually described as art. Hundreds of figurines, made from stone, clay, and ivory, have been recovered from archaeological sites in Europe.
Credit: Nadine Ryan

and they are usually described in the context of being sexual objects. They are generally assumed to have something to do with sexuality, fertility, or gender. Some refer to them as goddess figures.

The notion that these figurines are best considered in the context of sexuality is critically examined by archaeologists Dr. April Nowell and Dr. Melanie Chang (2014) in an article titled "Science, the Media, and Interpretations of Upper Paleolithic Figurines." They challenge the common assumptions, for example, that they were made by men for men, that they are something like prehistoric Barbie dolls, that their function was to educate or titillate men, or that they may have been some kind of trophy commemorating acts of violence against women. The authors also challenge the assumption that most are representing women specifically, showing that some represent men and others depict animals.

Nowell and Chang suggest that alternative contexts and hypotheses be examined, including how the figurines may have been created and used to maintain alliances, how they may have been used in rituals, or how they may have served as some kind of charm or totem. Alternatively, they could be considered in the context of art, including as self-portraits. Instead of focusing primarily on sexuality, the authors suggest the figurines should be studied in the same way as other artifacts from the period are, with the examination of material, technology, skill, modification, decoration, and reuse. They further suggest that the study of the figurines in the context of sexuality and the follow-up media reports say more about the archaeologists and media than they do about the figurines or life in the Paleolithic.

caves, and the images are often superimposed on each other. It is apparent to most anthropologists that at least in some cases, it was the process of painting, rather than the product, that was most important.

Artifacts commonly referred to as Venus figurines are often categorized as art. They are associated with the European Upper Paleolithic, and often described within the context of erotica (see Box 5.4).

Ideology is perhaps the most difficult aspect of prehistoric human culture to reconstruct. Art is one way to reconstruct ideology, and the treatment of the dead, primarily in the form of burials, is another. At a minimum, burials are usually taken as reverence for the dead. Often, they are used to infer a belief in an afterlife. The idea that they represent a belief in an afterlife is especially supported if there is evidence of an associated ritual or objects such as food or artifacts buried with the individual.

There is considerable debate among archaeologists about when people first began burying their dead. As noted in Chapter 4, many believe that *Homo naledi* deliberately disposed of their dead in cave chambers more than 200,000 years ago, but this interpretation remains contentious. Some archaeologists believe they have evidence of deliberate burial from several hundred thousand years ago in Europe, but this interpretation is also debated. Many archaeologists believe Neandertals started burying their dead by at least 50,000 years ago. The evidence that Neandertals deliberately dug pits into which they placed dead bodies remains contentious, however, and some archaeologists only accept that it was modern *Homo sapiens* that began burying their dead, about 30,000 years ago.

There is some suggestion that Neandertals may have carried a tune, and even played flutes, but the evidence of this is controversial. Many anthropologists accept that Neandertals and other varieties of humans dating as early as 100,000 years ago had the same language capabilities as modern humans, including speech. Whether they could sing, however, is disputable. A hollowed-out and broken mammal bone with puncture holes that some believe was a Neandertal flute was discovered in Europe. The bone has two complete holes and parts of what may be one or two others, and their placement resembles that of a modern flute. Some have even made reproductions of this so-called Neandertal flute and played music on them. However, some archaeologists believe the holes were likely made by an animal and the fact that their placement resembles that of a flute is probably coincidental. The bones also show evidence of gnawing by animals on both ends, suggesting to some that it was made naturally. Those who accept the bone as evidence of a flute suggest the animal gnawing came after its use as a musical instrument.

Deconstructing Cave Men and Cave Women

Popular images and stereotypes of people living in the Paleolithic are problematic, particularly those imbued with a strong gender bias or the idea that people lived in caves and were stupid and unkempt.

Living in Caves

Living in caves would not have been a common occurrence in the Paleolithic. Certainly, some caves were used for habitation, but it would be a mistake to think that most people living in the distant past lived in caves. Living in caves doesn't make sense for an number of reasons, including the fact that (1) they are cold, dark, damp, and take a lot of wood to heat; (2) other, often large and dangerous, animals, such as bears, like caves; (3) because there is often only a single entrance, living in caves increases vulnerability for attack from other people or animals; (4) caves are often difficult to access; and (5) caves are often away from water.

Caves are likely overemphasized because they have high archaeological visibility. Because they are protected, they usually have better preservation of organic remains, as well as evidence of art and ritual. We are more likely to find evidence of humans and human activity inside a cave than out.

Being Stupid and Unkempt

Paleolithic peoples are often depicted as lacking language and being fairly stupid and unkempt. Research indicates that the language of peoples living perhaps as long as 100,000 years ago was likely as complex as our own. The idea that they were stupid is ridiculous. Only a small percentage of people today would likely be able to make a stone tool like *Homo habilis* was doing more than two million years ago without instruction. Being unkempt is also problematic. In the article "Bad Hair Days in the Paleolithic: Modern Reconstructions of the Cave Man," Dr. Judith

EXPANDING TERRITORIES

There are a few things that most archaeologists agree upon. One is that the birthplace of humans and human culture was Africa. Another is that once humans had culture, it facilitated a kind of cultural adaptive radiation around the globe (Map 5.1).

It is clear that by almost two million years ago, early humans (probably *Homo erectus/ergaster*) had spread throughout much of Africa and the southern latitudes of Asia. By one million years ago, they had expanded westward into Europe. Eventually, they started inhabiting northern areas. We know, for example, that they were in Siberia by at least 60,000 years ago.

Berman (1999) provides all kinds of evidence, including the depiction of humans in prehistoric art, that people were well-groomed. Berman writes,

> [T]he shaggy, grunting Cave Man, who fights dinosaurs, talks "rock," and woos prehistoric-bikini-clad Cave Women with a club, is firmly in place, and it is easy to see why ... [with] over 150 Cave Man films, animated cartoons, and television shows.... These filmed images are supported and reified by other popular media. (p. 289)

Although we have never seen Paleolithic humans in the flesh, we recognize them immediately in illustrations, art, cartoons, and museum displays. The familiar iconography of the "Cave Man" often depicts our early human ancestors with longish, unkempt hair. However, this conventionalized image is not congruent with available archaeological data on the appearance of Upper Paleolithic humans.

Gender Bias

There is considerable gender bias in popular images of Paleolithic peoples. Men are typically portrayed as the leaders and the providers, while women are largely featured in supporting roles and as sexual objects. An interesting popular article on this is "Paleohooters," written by Allen Abel (1997). The article includes many quotes from anthropologist Dr. Melanie Wiber, who responds to the idea that men were more important even in australopithecine times: "It's pandering to what we *want* to think our ancestors were doing. It's giving antiquity to what we do *now*" (p. 16). Wiber suggests that people have been programmed to believe that "females were secondary to the evolution of *Homo sapiens*, that light skin equals progress, and that woman exists to be domesticated and eroticized, even when she is a million-year-old, knuckle-walking, termite-eating ape" (p. 16). Anthropologists know that there is no reason to believe men had any more important roles than women in the past.

People had been in Asia for more than one million years. From there they started colonizing southern areas. They reached Australia, likely using watercraft, by at least 65,000 years ago. People may have ventured into the Americas from Asia before 20,000 years ago, but all such claims are still controversial.

The spread of humans during the Paleolithic was clearly enhanced by culture. Unlike all other animals, humans were not dependent on biology for survival. They could colonize colder environments, for example, because they had fire and, probably for at least a few hundred thousand years, clothing. Paleolithic culture also allowed people to utilize a larger range of food; combined with enhanced communication within and between groups, this made territorial expansion feasible.

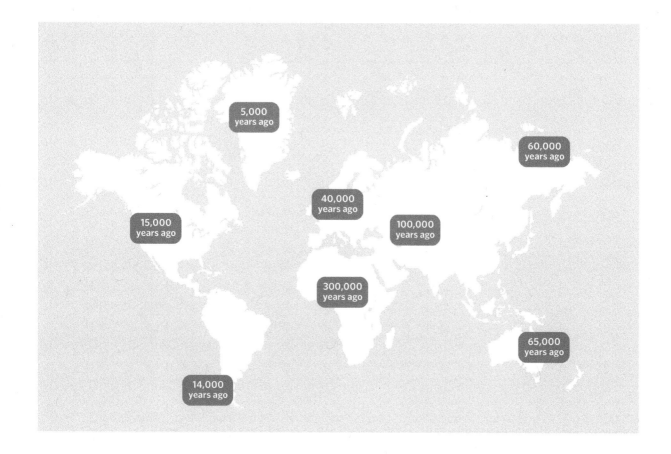

Map 5.1
This map illustrates the spread of modern *Homo sapiens*.

SUMMARY

This chapter introduces students to the archaeology of the period before 20,000 years ago. Mirroring the Learning Objectives stated in the chapter opening, the key points can be summed up as follows:

- The archaeological record of the period before 20,000 years ago consists primarily of sites, artifacts, ecofacts, features, and cultural landscapes.
- The archaeological record is incomplete and biased.
- The first evidence of culture dates to 2.5 million years ago; hunting and meat eating was evident throughout the entire Paleolithic; people have probably been controlling fire for at least a few hundred thousand years; organization in groups of about 30 was probably normal; and undisputed evidence of ideology and art begins about 40,000 years ago.
- Migrations and territorial expansions, including to northern environments and Australia, occurred during the Paleolithic.

Review Questions

1. What does the term *archaeological record* mean?

2. What are some biases in archaeology to be aware of?

3. What were the principal cultural developments in the Lower, Middle, and Upper Paleolithic, respectively?

4. What is the timing for human expansion into Asia, Europe, Australia, and the Americas?

Discussion Questions

1. How, using a holistic perspective, may changes in subsistence, diet, social systems, technology, ideology, and art be linked?

2. How, using a holistic perspective, may changes in culture identified in this chapter be linked with changes in biology (described in Chapter 4)?

Visit **www.lensofanthropology.com** for the following additional resources:

| SELF-STUDY QUESTIONS | WEBLINKS | FURTHER READING |

6

CULTURAL DIVERSITY FROM 20,000 TO 5,000 YEARS AGO

LEARNING OBJECTIVES

In this chapter, students will learn:

- *the principal cultural periods in North America and Europe.*
- *prevailing ideas about and evidence for North American prehistory.*
- *why and when people started developing new subsistence strategies.*
- *when new kinds of social and political systems emerged.*
- *how archaeologists reconstruct subsistence, settlement, and social systems.*
- *when and where civilizations and writing emerged.*
- *the conditions under which new technologies, such as pottery, were used.*

There were many changes in cultures around the world between 20,000 and 5,000 years ago, increasing cultural diversity.

#MultipleWaysOfAdapting

INTRODUCTION

Many significant events and changes in cultures around the world occurred in the period from 20,000 to 5,000 years ago. These include the settling of the Americas, as well as the emergence of new subsistence strategies, settlement patterns, technologies, and strategies of organization (social and political systems) around the world.

BOX 6.1   **Problems with "Paleo"**

There is a lot of interest in things "paleo" these days. It is apparently a good marketing gimmick, feeding on people's sense of the "good old days," taken to the extreme. Thus, we hear of the paleo lifestyle, the paleo diet, paleo exercise, paleo sleeping patterns, paleo medicine, paleo this, and paleo that. Some describe the interest as paleo nostalgia. Anthropologists sometimes refer to it as paleo fantasy.

There are problems with all things "paleo." These include a misunderstanding of what humans were like in the past, both biologically and culturally, and how we have evolved since. Another problem is that the past is misused to support assumptions, often incorrect, about some kind of natural state of humans, including how we should eat, sleep, have sex, and exercise. Many assume that if it is older, it must be better, but this isn't always true.

The very popular **paleo diet** provides a good example. The first edition of *The Paleo Diet: Lose Weight* *and Get Healthy by Eating the Foods You Were Designed to Eat* by Loren Cordain was published in 2002, and by 2011 had already sold more than 200,000 copies. Cordain (2011) describes the diet as

the diet to which our species is genetically adapted. This is the diet of our hunter-gatherer ancestors, the foods consumed by every human being on the planet until a mere 333 human generations ago, or about ten thousand years ago. Our ancestors' diets were uncomplicated by agriculture, animal husbandry, technology, and processed foods. Then, as today, our health is optimized when we eat lean meats, seafood, and fresh fruits and vegetables at the expense of grains, dairy, refined sugars, refined oils and processed foods. (p. xi)

Most anthropologists find this basic premise to be faulty. For archaeologists, the notion that there was

---

It was during this time that **food production** replaced food **foraging** as a primary subsistence strategy for many people around the world. Many people today believe that the transition to food production and the changes in lifestyle that go with it – including diet, exercise, and sleep patterns – are not good for us. They harken back to the "days of old," which they generally perceive as better, and create diets and other fads to mimic life in paleo times. These fads are addressed in Box 6.1.

PRINCIPAL CULTURAL PERIODS

There are several descriptive and analytical cultural periods used by archaeologists and others when considering the time period from 20,000 to 5,000 years ago. Some of the more popular ones are outlined in Table 6.1.

The use of these terms to describe cultural periods is in no way universal, and the start and end dates are very approximate. Not all archaeologists use this terminology,

a common diet in prehistory is absurd, and research shows evidence of grains in the diet long before 10,000 years ago. The notion that we have not evolved biologically with changes in our diet is also problematic. Lactose tolerance (the ability to drink milk without ill effects), for example, has evolved in different ways (from different mutations) at least three times in various parts of the world over the past several thousand years. And before this, people were able to enjoy the many nutritional benefits of dairy by processing milk into cheese and yogurt (see Box 3.1).

Those who take the paleo diet a step further and eat only raw foods often do not fully understand life in the past. Cooking food is a cultural universal, and probably has been for at least 30,000 years. Most archaeologists accept that cooking has been common for at least several hundred thousand years, and some suggest it may have originated about two million years ago (see Box

5.2). Cooking tends to enhance the nutrition of most foods and makes digestion easier.

In *Paleofantasy: What Evolution Tells Us about Sex, Diet, and How We Live*, Dr. Marlene Zuk (2013) describes some of the problems with "paleo," including how people often think that things were better in the past, and how adherents to the paleo lifestyle often do not know what life was really like:

> To think of ourselves as misfits in our own time and of our own making flatly contradicts what we now understand about the way evolution works.... The paleofantasy is a fantasy in part because it supposes that we humans, or at least our protohuman forebears, were at some point perfectly adapted to our environments. (pp. 6-7)

Even assuming we could agree on a time to hark back to, there is the sticky issue of exactly what such an ancestral nirvana was like.

TABLE 6.1

Cultural Periods 20,000 to 5,000 Years Ago

Period	Antiquity
NORTH AMERICA	
PaleoIndian	14,000–9,000 years ago
Archaic	9,000–5,000 years ago
EUROPE	
Upper Paleolithic	40,000–12,000 years ago
Mesolithic	12,000–10,000 years ago
Neolithic	10,000–5,000 years ago

they are not applicable to all regions, and when they are used, the start and end dates may vary by locality.

In North America, PaleoIndian and Archaic are two widely used frameworks. PaleoIndian generally refers to the period in which people first arrived and settled in North America, through to the end of the time in which they were hunting large animals such as mammoths and mastodons, about 14,000 to 9,000 years ago. Some archaeologists extend the use of the term to include Central and South America as well, but it is never used outside the Americas. In this context, the term Archaic is used to describe the period from about 9,000 to 5,000 years ago in much, but not all, of North America, which is characterized by an ongoing foraging adaptation. For some, "Archaic" may carry connotations of primitiveness, but in this case that is not appropriate. People in this time period were culturally adept, had sophisticated technology, and in many instances were probably healthier and had more leisure time than many people in more recent times.

Popular terms for cultural periods that are most applicable to Europe, but also used elsewhere, include the Upper Paleolithic, Mesolithic, Neolithic, Bronze Age, and Iron Age. As mentioned in Chapter 5, *Paleolithic* roughly translates into Old Stone Age. The Upper Paleolithic is the most recent era, beginning about 40,000 years ago, correlating with the arrival of modern *Homo sapiens* in the area, and ending about 12,000 years ago. The **Mesolithic**, from about 12,000 to 10,000 years ago, roughly translates as the Middle Stone Age. It correlates with climatic change in Europe (warming and deglaciation), advances in stone-tool technology, and changes in diet, including the addition of more maritime resources. The **Neolithic**,

TABLE 6.2

Major Cultural Developments 20,000 to 5,000 Years Ago

Time Period	Cultural Developments
20,000–15,000 years ago	Domestication of dog occurs. People begin colonizing what is now known as the Americas. Continued expansion and increasing population growth.
15,000–10,000 years ago	Transition from foraging to food production in many parts of the world. Megafauna (mammoth and mastodon) hunting in North America.
10,000–5,000 years ago	More people adopt food production as primary subsistence strategy. Significant population growth around the world. New forms of social and political organization emerge. Civilization and writing emerge.

from about 10,000 to 5,000 years ago, translates as the New Stone Age. Rather than being based on technology, it correlates with a shift to farming in Europe.

Metalworking provides the basis for labeling the Bronze Age and Iron Age in Europe. Natural copper began to be smelted about 6,000 years ago, but it was not widely used. Copper is much softer than stone and thus does not make durable tools. Approximately 5,000 years ago, people determined that by adding tin to copper they could create bronze, which required much hotter fires but was much more durable. A few thousand years later, people were able to create sufficiently hot fires to extract iron from ore, leading to what is popularly known as the Iron Age.

Table 6.2 provides an overview of some of the major cultural developments during the period from 20,000 to 5,000 years ago.

ARCHAEOLOGY OF NORTH AMERICA FROM 20,000 TO 5,000 YEARS AGO

Most archaeologists place the initial settlement of the Americas as occurring between 20,000 and 15,000 years ago. Some archaeologists suggest that evidence at some sites indicates much earlier dates, but many archaeologists believe the very high standards of proof for claiming early archaeological sites in the Americas have not been met in these cases. According to these standards, (1) the evidence must be undisputedly cultural, and (2) the dating has to be undisputedly reliable.

The purported evidence at some sites includes chipped stones that resemble stone tools, but many archaeologists are not fully convinced that those stones were not chipped naturally. At other sites, there is no question that artifacts were made by humans, but there are questions about the dating. Since carbon-14 dating is widely recognized as the best dating technique for early sites in North America (at least up until about 40,000 years ago), if a site is not dated by this technique, then the dates are usually questioned. Even dates determined by carbon-14 are sometimes questioned as well, especially if they come from areas in proximity to sediments with carbon that could contaminate the results.

Timing and Routes

There is much interest in the question of when and by which route people first came to the Americas (Map 6.1). Conventional archaeological thought indicates that the ancestry of contemporary Indigenous peoples in the Americas lies in Asia. This is supported primarily by biological similarities, with support from archae-ology. It also makes sense in regard to the general patterns of prehistory, in which we have evidence of people in Siberia and other areas of northeast Asia for tens of thousands of years.

The route from Asia to the Americas was via **Beringia**, a large, ice-free area connecting northern Asia to northwest North America during the last ice age. There is no undisputed archaeological evidence of a human presence in the North American part of Beringia before about 14,000 years ago, but it still makes sense to most archaeologists that they were there. Explanations for the lack of evidence are linked with the ever-changing landscapes in the area, which have likely destroyed much of it. Also, the temporary nature of the settlements meant that little evidence was left, and there was probably a low population density, with a consequential low number of sites.

Prior to about 12,000 years ago, most of what is now Canada, as well as much of northern Asia and northern Europe, was under ice. Beringia was an exception. Beginning about 14,000 years ago, however, undisputed archaeological sites dating to this time appear in the archaeological record of the United States and along the outer reaches of the west coast of Canada. Somehow, the people who left these sites got around or through the glaciers covering Canada. The theory that has the most popular acceptance is that the people came down the coast of what is now Alaska and British Columbia in western Canada, using boats or walking along the coast-line. This is often known as the **coastal migration route**. No sites clearly dating before 14,000 years have yet been discovered in these areas, but paleoenvironmental research indicates that the coastal area was certainly inhabitable in the time period before 14,000 years ago. Archaeologists continue to look for sites along the coast

Beringia

A   Coastal
     Migration Route
B   Ice-Free
     Corridor Route
C   North
     Atlantic Route

but, as with Beringia, there is the problem of changing environments, including rising sea levels, and the issue of low archaeological visibility.

Another possible entry route from Beringia past the glaciers was through a corridor between the two large ice sheets covering most of Canada. During warming trends of the last ice age, the two glaciers separated, creating a corridor linking Beringia to the areas south. This is known as the **ice-free corridor** route.

A third possible entry route was from Europe, via the North Atlantic Ocean. This is often referred to as the Solutrean hypothesis, named after the peoples identified as being part of the Solutrean culture of Europe, about 20,000 years ago. This route would have necessitated boat travel across the North Atlantic and southward down the glacial environment of eastern Canada. Although this hypothesis remains popular and is supported by some archaeologists, most believe there is not enough evidence to support it. The evidence of initial entry via Beringia is much stronger. This evidence includes cultural similarities to people of northeast Asia, as reflected in technology, and biological similarities, including DNA.

Once people traveled south of the ice sheets, they quickly spread throughout the region, with sites dating to about 14,000 years ago appearing in South America. It is likely that most Indigenous peoples of the Americas have ancestral ties to populations in Asia. Many probably are descended from those who first made their way around or through the glaciers before 12,000 years ago. Others may have come in subsequent migrations. Archaeologists remain uncertain about how many distinct migrations there were from Asia before 5,000 years ago, but there were probably at least several.

Cultures between 14,000 and 5,000 Years Ago

Despite the significant alterations to the landscape, thousands of archaeological sites predating 5,000 years ago have been recorded in North America. Some of the most significant are described in Table 6.3. Their locations are shown on Map 6.2.

Although there are sites in North America that some archaeologists claim are older than 14,000 years, these claims are all contentious. Most archaeologists only accept an age of 14,000 years for the earliest reliable evidence of human occupation. Problems with finding sites older than 14,000 years are many. Since sea levels have risen since the last ice age, most coastal sites from that time would now be under water. Thousands of years of both natural processes and cultural activity have undoubtedly destroyed many thousands of sites, and the activities of the early migrants and settlers would undoubtedly have low archaeological visibility due to relatively low population densities and the temporary nature of their settlements.

Two recently discovered sites, commonly known as the Triquet Island and the Calvert Island sites, have been excavated and further support the notion of the coastal migration route. The Triquet Island site shows evidence of human occupation 14,000 years ago, and the Calvert Island site includes multiple sets of human footprints dating to 13,000 years ago. These two sites are close to each, just off the central coast of British Columbia in western Canada.

It is clear that initial migrants and settlers of the continent adapted fairly quickly to the new ecological niches of the Americas. It is apparent that upon arriving south

Site Name	Description
Bluefish Caves	Located in present-day Yukon Territory, Bluefish Caves is significant because it provides evidence of people in Beringia during the last ice age. The site contains artifacts made of stone and bone as well as butchered animal remains. Most archaeologists accept dates between 15,000 and 12,000 years ago, although some suggest dates in the range of 25,000 years ago.
Cactus Hill	This site is located in Virginia. It is widely accepted as being older than 12,000 years, but the precise antiquity is uncertain. Some suggest that the site may be as old as 17,000 to 19,000 years, but these dates are contested.
Gault	This site, located in Texas, was a Clovis camp and was likely first occupied more than 12,000 years ago. The assemblage includes more than one million artifacts, including several hundred thousand that are more than 9,000 years old. The site is particularly significant for showing the diversity of diet, which, contrary to popular belief, indicates that mammoths and other large game were only a small part of the Clovis diet. The site is also significant in providing what may be the oldest art in North America, in the form of more than 100 incised stones.
Meadowcroft	Meadowcroft Rockshelter, located in Pennsylvania, is widely considered to contain deposits that are at least 12,000 years old. Some suggest the deposits may be as old as 19,000 years, but these dates are contested.
Paisley Caves	Located in Oregon, the oldest deposits at this site contain human coprolites dating to between 14,000 and 13,000 years ago.
Topper	This site, located in South Carolina, is widely considered to be at least 12,000 years old. Based on what appear to some to be artifacts below the 12,000-year-old layer, some believe the site to be older. Whether the so-called artifacts are really artifacts or naturally broken rocks remains debatable.
Charlie Lake Cave	This site, located in northern British Columbia, dates to almost 11,000 years ago and would have been in the ice-free corridor. The oldest levels contain a fluted point, a stone bead, and bones of several kinds of animals, including bison. DNA analysis of the bison and general similarities between artifacts here and in earlier sites in Montana suggest movement through the corridor at this time went from south to north.
Daisy Cave	Located on San Miguel Island, about 25 miles off the coast of California, this site dates to almost 11,000 years ago. It is particularly significant insofar as it provides evidence of basketry and cordage, as well as circumstantial evidence of watercraft.
Kennewick	Located in the State of Washington, this is where, in 1996, the infamous Kennewick Man was discovered. Initially described as a Caucasian and subsequently dated to about 9,000 years ago, the remains created considerable debate, both about the initial classification as Caucasian and the following controversy about whether the remains should be turned over to local Native Americans or kept for study by scientists. DNA studies in 2015 confirmed the remains as Native American.

Map 6.2
SIGNIFICANT
ARCHAEOLOGICAL
SITES IN NORTH
AMERICA

Bluefish Caves

Kwäday Dän Ts'ìnchi

Charlie Lake Cave

Head-Smashed-In Buffalo Jump

Ozette

Kennewick

Paisley Caves

L'Anse aux Meadows

Meadowcroft

Cahokia

Cactus Hill

Daisy Cave

Mesa Verde

Topper

Gault

of the ice sheets, some populations – probably those arriving via the coastal migration route – continued a maritime adaptation, with settlement expanding along the Pacific coast all the way to the southern tip of South America. Much of the evidence from the earliest sites on or near the west coast of North America indicates a maritime adaptation, including significant amounts of seafood in settlers' diets.

While some populations maintained a maritime adaptation, others adapted to inland resources. These may have been people who moved eastward once they got past the ice sheets along the coast, or perhaps they arrived through the ice-free corridor, preadapted to terrestrial resources.

Populations probably expanded fairly quickly beginning about 14,000 years ago, with essentially unlimited resources for the fairly small groups of migrants and early settlers. However, the overall population of North America remained relatively low for the first few thousand years of habitation. From perhaps hundreds of migrants traveling in small groups before 14,000 years ago, the population likely expanded to thousands by 12,000 years ago, when there was a population explosion of sorts.

The population explosion was likely triggered by the invention, about 12,000 years ago, of a new kind of projectile point, known as a fluted point. A fluted point is distinguished by its concave base, created by removing a flute (or channel) flake from one or both sides of the point. Well-known variants of the fluted point are known as the Clovis and Folsom points, and the people who used them are known by the same names, respectively. Fluted points were no longer used after about 9,000 years ago.

There is considerable evidence that the points enabled people to effectively kill large animals such as mammoths and mastodons. It is not unusual to find bones of these animals in sites of this period, often showing evidence of butchery and in association with fluted points. There is some suggestion that overhunting may have caused the extinction of these and other large animals, but it is more commonly believed that they became extinct due to environmental change. In any case, both the points and the presence of mammoths and mastodons only appear in archaeological sites dating from 12,000 to 9,000 years ago.

The purpose of the flute is uncertain. Popular ideas include that (1) it may have led to more blood loss from the animal by creating a channel for the blood to flow along, or that (2) it may have facilitated easier and quicker hafting to spear points. It is likely that the larger fluted points were hafted to spears thrown unaided by hand or perhaps thrust into the animals once they were wounded. Smaller fluted points were probably hafted to spears and thrown with an atlatl.

Sites with fluted points appear across the unglaciated parts of North America (i.e., mostly in what is now the "lower 48" United States) almost simultaneously about 12,000 years ago, and they start appearing in what is now Canada shortly after deglaciation (i.e., between about 12,000 and 10,000 years ago). One scenario has one group inventing the point and then quickly expanding through the continent. Another, probably more accurate, scenario has many groups pre-existing in unglaciated parts of the continent, then quickly adopting the new technology and incorporating large animals (often known as megafauna) into their diets. The reason for the numerous archaeological sites that start appearing around 12,000 years ago is probably due to a combination of (1) greater archaeological visibility and (2) increased population. Megafauna hunting – particularly with large stone projectile points – has high archaeological visibility. Mammoth and mastodon bones, for example, are easily identifiable, as are the spear points used to kill them. People before 12,000 years ago were likely hunting smaller animals, which leave far less trace in the archaeological record, and using either smaller stone projectile points or simply sharpened bone or wood spears, which leave no trace at all.

It would be a mistake to believe that people necessarily depended on megafauna before 9,000 years ago. It was undoubtedly part of the diet of many, but we should always be aware of the bias of archaeological visibility. Research at the site of Gault in Texas, for example, indicates that Clovis people had an extremely varied diet that included, but was probably not dependent on, mammoth.

Overall, diet was varied throughout the continent before 5,000 years ago, and the populations can be characterized as foragers. As everywhere, people used what was locally available in regard to both plants and animals, including maritime resources. Some of the best evidence of dietary diversity comes from coprolites (preserved human feces). The Hinds Cave site in Texas, for example, includes about 2,000 human coprolites spanning 8,000 years. In one coprolite alone, analysis revealed evidence of antelope, rabbit, squirrel, rat, and eight kinds of plants.

As was common with foragers elsewhere, people likely lived in groups of a few dozen or more, were familiar with and interacted with populations around them, and had steady, significant population growth over time. It is likely that by 5,000 years ago, the population of North America was at least in the hundreds of thousands, and there were probably at least dozens of distinct ethnic groups, each with its own unique culture.

THE TRANSITION TO FOOD PRODUCTION

The transition to food production is certainly one of the most significant changes in culture in the period from 20,000 to 5,000 years ago. It is likely that before this, foragers around the world effectively managed their resources by, for example, doing some weeding before leaving a berry patch, or otherwise altering environments to make them more conducive to the cultivation of specific kinds of plant foods. Beginning about 15,000 years ago, however, the manipulation of plants and animals got serious.

Ways of obtaining food started changing significantly between about 15,000 and 12,000 years ago. Many people around the world became **food producers**, meaning they began to manipulate plants and animals to increase their productivity, creating surplus. The process of food production involves **domestication**, which means that plants and/or animals are under the control of humans. Some people started to domesticate animals, which led to the subsistence strategy known as **pastoralism**. Others started to domesticate plants, which led to **horticulture**.

Archaeological research suggests that domestication first developed in the Middle East and Asia but was relatively quickly adopted by populations living in Europe and Africa as well. It developed independently in regions of the Americas. Early plant domesticates included rice, wheat, potatoes, and maize (corn). Early animal domesticates included sheep, goats, and cattle.

Why Domestication and Food Production?

Initially, it was likely that these new domestic plants and animals supplemented a diet of primarily wild food. However, by about 10,000 years ago, many people had become dependent on these newly domesticated plants and animals, and, with few exceptions, there was no going back. Domestication increased the **carrying capacity** of the regions that humans filled. Going back to subsistence based primarily on foraging would not support the increased numbers of people. Eventually, by several thousand years ago, some **horticulturalists** took it a step further, which led to **agriculture**, sometimes known as intensive cultivation.

There is no consensus on why people started to become food producers. Historically, the domestication of plants and animals was envisioned by anthropologists as a great idea that made life easier, and most hypotheses to explain domestication began with notions of intelligence or keen observation. Now, however, it is clear that while the lives of some people got better, that certainly wasn't the case for all. Ethnographic research indicates that for most people, pastoralism and horticulture required more time spent on subsistence than a life based on foraging. Analysis of skeletal remains also supports the notion that for many, health suffered as a result of domestication. Poor nutrition and diseases are reflected in the skeletons of early pastoralists and horticulturalists.

There were certainly trade-offs. One clear advantage of domestication is that it produces a food surplus, which could be used as a hedge against poor hunting and gathering or turned into trade items. Domestication is also associated with reduced mobility, which some view as an advantage. Staying in one place for longer periods, as one does when depending on domestic plants and animals, also means people can accumulate more. Further domestication increases the carrying capacity of a region, meaning more people can live together. With large populations and more permanent settlements comes more internal conflict, inevitable social inequality, and the emergence of more formal political systems.

Explaining the emergence of food production is a matter of considerable debate in archaeology. Many archaeologists recognize that since many people had to work more and suffered from worse health as a result of food production, a better explanation than it simply being a good idea is required. After all, it isn't as if people in various parts of the world 15,000 to 12,000 years ago all woke up one morning and said, "We have too much leisure time and are too healthy, so let's change the way we get our food." Archaeologists examine the trade-offs.

What some researchers see as an advantage, others see as a disadvantage. Many in contemporary societies, for example, view being sedentary (living in permanent settlements) year round as an advantage. Yet many foragers place a higher value on mobility. Many suggest the food surplus is a hedge against poor crops or hunting. On the other hand, domestication also ties up the resource base in fewer species, making people more susceptible to disease and vulnerable in the case of a drought.

Most popular explanations for the emergence of food production are linked with changing environments. A basic idea is that changing environments may have reduced the carrying capacity of a region. Hypothetically, for example, an area that once maintained a human population of 1,000 may now, due to environmental change, only support a population of 900. The population has some choices. The people could perhaps become aggressive and raid food from neighboring groups; they could migrate elsewhere; they could have fewer children; or they could simply start domesticating plants and animals to get the carrying capacity back up to 1,000. The big problem, of course, is that once domestication begins and food surplus is created, the population keeps increasing, and the cycle of increasing food production to keep up with demand continues.

There are, of course, other explanations for the transition to food production that are not based on environmental change. For example, archaeologist Dr. Brian Hayden suggests that at least some foods may have been domesticated to increase wealth and status (see Box 6.2). Others suggest that some plants may have been domesticated to make alcohol (see Box 6.3). This last idea, which has received considerable interest in recent years, has led to many jokes, including one explaining that the reason permanent settlements are associated with domestication is so that waiters know where to find us.

Transitioning from subsistence relying absolutely on wild foods to subsistence relying on domestication need not be considered a drastic event that developed in a vacuum. Food production can certainly be considered revolutionary, but it wasn't as if the change from depending on wild sources to domestication required special knowledge. It is very likely that people were already managing their resources in numerous ways. Archaeologists working in some coastal areas, for example, have found archaeological evidence of people altering the landscape of beaches to make

BOX 6.2 **Food Matters: Why Did People Domesticate Plants and Animals?**

Why people started to domesticate and eventually depend on plants and animals is one of the principal areas of research in archaeology. Until the 1960s, it was widely assumed that domestication was a good thing, allowing access to more food with less risk and less work, and leading to better health. Archaeological questions tended to focus on where the first domestication occurred and how the idea spread, all assuming that it made life better. Archaeological and ethnographic research in the late twentieth century, however, convinced many that this assumption was not warranted; while some may have benefited from the transition to food production, many suffered from worse nutrition and less leisure time. The question then became why, considering its negative effects, did many groups around the world initiate the process of domestication, which led to pastoralism and horticulture?

Over the last few decades, popular explanations for the origins of domestication of both plants and animals have usually focused on ecological reasons. For example, many archaeologists reason that environmental changes reduced the number of people who could effectively forage in a region. Rather than reduce their population to match the now-reduced carrying capacity, move to another area, or resort to raiding others, people began to increasingly manage their resources to the point of domestication. The tendency to focus on ecological explanations may be tied to contemporary concerns about changing environmental conditions.

Other popular, but not as widely accepted, ideas are that plant domestication may have occurred to produce alcohol, or perhaps to gain status. The notion that plants and animals may have been domesticated as prestige items, or luxury foods, to gain status and social and political advantages is the subject of an article by archaeologist Dr. Brian Hayden (2003) called "Were Luxury Foods the First Domesticates? Ethnoarchaeological Perspectives from Southeast Asia." Hayden suggests that certain animals and plants (including rice) were domesticated as luxury foods for feasting. He reasons that "the primary force behind intensified subsistence production is not food shortage, but the desire to obtain social and political advantages – to obtain the most desirable mates, to create the most advantageous alliances, to wield the most political power" (p. 465).

In Hayden's view, domestication was initiated in societies where people were attempting to gain wealth and status, primarily by impressing others, and one way of accomplishing this was to have feasts with what he describes as luxury foods. Eventually these luxury foods became staples. For an analogy of how early domestic foods like rice became staples, Hayden provides the following examples:

> Chocolate, once reserved for Mesoamerican elites, is now the bane of overfed multitudes. Oversized, out-of-season fruits and vegetables which once only graced the tables of kings and nobles have become everyday fare. Fat-rich meats, which formerly were used only for special occasions or for the highest ranks of society, are now commonplace.... Wines and spirits that played crucial roles in feasts for elites ... have now become the profane intoxicants of households throughout the industrial world. In short, our eating habits today largely are the result of, and reflect, the luxury foods of the past. (pp. 458–9)

Food Matters: Was Alcohol a Driving Force of Human Evolution?

Probably not. But some people like to consider it.

In 2014, researchers suggested that the ability to process alcohol was a driving force of becoming human. Their research, they claimed, indicated a mutation occurring in an ancestral human population about 10 million years ago made possible the ability to process alcohol and, thus, eat fermenting fruit. This, they suggested, created a new food source – fermenting fruit left rotting on the ground. The mutation thus became favorable, was selected for, and spread. To effectively make use of this new food source, the researchers claim, the ancestral humans likely started spending increasingly more time on the ground, which in turn led to bipedalism. Thus, according to this very unlikely scenario, the ability to process alcohol led to humans. This is an interesting hypothesis, but only a real stretch of the imagination, perhaps under the influence of alcohol, would place it among the most realistic views of human evolution.

In *Uncorking the Past: The Quest for Wine, Beer, and Other Alcoholic Beverages*, archaeologist Dr. Patrick E. McGovern (2009) describes the allure of alcohol:

> Humans throughout history have been astounded by alcohol's effects, whether it is imbibed as a beverage or applied to the skin. The health benefits are obvious – alcohol relieves pain, stops infection, and seems to cure diseases. Its psychological and social benefits are equally apparent – alcohol eases the difficulties of everyday life, lubricates social exchanges, and contributes to a joy in being alive. (p. xi)

Most archaeologists believe that the origins of plant domestication are linked to food shortages, although some suggest it may have been more political (see Box 6.2). Some take another view: that the early domestic plants, such as wheat and barley, may have been cultivated to produce alcoholic beverages, particularly beer. It remains

clams more productive. Similarly, there are indications that people in the past may have deliberately set fire to areas to manipulate the kinds of plants and animals that would repopulate. People have been smart for a very long time. It stands to reason that before leaving an area where they were gathering wild plants, for example, they may have done some clearing of unwanted plant material. Making the transition to domesticating plants and animals was likely a case of just doing more of what they were already doing in regard to managing food resources. And maybe doing it a bit differently.

Although not directly related to food production, many people are interested in the domestication of dogs. Many of us consider dogs to be our best friends, but this relationship wasn't always so cozy. It wasn't that long ago that dogs were wolves – predators, not companions. But at some point between 40,000 and 20,000

a minor, but interesting, hypothesis – that the driving force of domestication was beer drinking.

Evidence for the production of beer begins appearing in the archaeological record about 10,000 years ago, predating wine by at least a thousand years. How or why people started making beer is uncertain, but it quickly caught on. Besides the fact that it made people feel good, other reasons that have been offered to explain the adoption of beer include that fermentation increased the nutritional value of wheat and barley, and that by killing the pathogens in water through the fermentation process, it was a safer and healthier option than drinking untreated water.

McGovern (2009) suggests a few other reasons:

> Alcoholic beverages have other advantages. Alcohol spurs the appetite, and in liquid form, it also satiates feelings of hunger. The process of fermentation enhances the protein, vitamin, and nutritional content of the natural product, adds flavor and aroma, and contributes to preservation. Fermented foods and beverages cook faster because complex molecules have been broken down, saving time and fuel. Finally, as we have learned from numerous medical studies, moderate consumption of alcohol lowers cardiovascular and cancer risks. (p. 7)

Some breweries, with the aid of archaeologists, have begun recreating (and marketing) paleo beers. The beers are typically based on the residue analysis of pots evidently used to store beer, which, upon identification of the elements, is used to create a recipe. In more recent times, once writing began, the ancient recipes were recorded, and those recipes can be used today. Thus, some lucky connoisseurs have been able to drink beers as they would have been 9,000 years ago in China, as well as in the ancient civilizations of Mesopotamia and Egypt.

years ago, dogs were domesticated. It is likely that wolves first began interacting with humans by hanging around hunter-gatherer camps, which featured scrumptious trash from discarded meals. Humans saw a benefit to keeping wolves, as they could serve as companions, guard dogs, and clean up the waste that otherwise might attract rodents and other scavengers. People cultivated this mutually beneficial relationship, breeding the friendliest while culling the aggressive wolves from the packs.

Identifying Subsistence Strategies in the Archaeological Record

There are many ways of identifying subsistence strategies archaeologically. Foragers, for example, are usually distinguished in part by a wide variety of wild plants and animals as food refuse. Since foragers usually live in small groups, create temporary

Figure 6.1

BEER AND CULTURAL EVOLUTION.

Beer was part of the diet of many people from at least 10,000 years ago. It had many benefits, and some suggest it was the driving force of cultural evolution, including the domestication of plants.

Credit: Anders Nilsen

settlements, and are egalitarian, evidence of these factors can also be used to support inferences of foraging.

Because foragers are relatively mobile, archaeologists are often interested in determining in which season a site was used. The primary way of doing this is to look for the presence of seasonal plants at the site. The presence of migratory animals also allows for inferences.

Indications of pastoralism and horticulture are made by examining plant and animal remains and correlating them with other cultural elements, such as the size of settlements and populations.

Pastoralists, for example, usually have little diversity in the animals used for food but retain a diversity in plants, which they often continue to gather. Horticulturalists, on the other hand, exhibit little diversity in plants but considerable diversity in animals. Since pastoralists and horticulturalists tend to live in larger and more permanent settlements, indications of this can also be used to make inferences about subsistence.

Of course, inferences of pastoralism and horticulture are aided by determining whether the plants and animals can be classified as "wild" or "domestic." Ways of making these determinations are covered in the next section.

Figure 6.2
PASTORALISTS.
Pastoralism began more than 10,000 years ago. Pictured here is a Turkana woman in Kenya, watering livestock from a waterhole dug in the sand.
Credit: John Warburton-Lee Photography/Alamy

Identifying Domestication in the Archaeological Record

Archaeologists have multiple ways of distinguishing between wild and domestic varieties of plants and animals. Table 6.4 lists some of the criteria for distinguishing wild from domestic varieties of the same species; there are several ways to do this, but few are visible in the archaeological record. The attribute most commonly used in archaeology is the size of the edible part of the plant. Frequently, as domestication continues, the size of the edible part continues to increase. In the earlier deposits in an archaeological site, for example, corncobs may be quite small, but as the deposits get younger, the size of the corncobs gets larger. This is a clear indication of domestication. Alternatively, if

TABLE 6.4

Criteria for Distinguishing Domestic versus Wild Plants and Animals

Domestic Plants	Domestic Animals
The part of the plant that people use is usually larger.	The animals tend to be smaller (at least in early stage of domestication).
The plant may have lost its mechanism for natural dispersal.	There is a tendency to find more complete skeletons in the faunal assemblage.
The part of the plant that people use may have become clustered.	There is likely to be a high percentage of young male animals in the assemblage.
There is often a genetic change.	
There may be a loss of dormancy.	There is likely to be a high percentage of old female animals in the assemblage.
The plants tend to ripen simultaneously.	
There is a tendency for less self-protection, such as thorns and toxins.	

corncobs from the same time period are small in one area and large in another area, it may indicate that they are wild in the first and domestic in the other.

The primary ways of identifying domestic animals are based on the size and completeness of skeletons, and the age and sex ratios of the butchered animals. At least in the early stages, domestic animals tend to be smaller than their wild counterparts. This may be due to the smaller animals being selected for domestication or nutritional deficiencies. Faunal assemblages with relatively complete skeletons are another way of identifying domestication. When hunting wild animals, some preliminary butchering is often done at the kill site, so the entire skeleton is not brought back to the settlement. Relatively high proportions of young males and old females in the assemblage are another indication of domestication. Young males are usually the most difficult to kill in the wild, but in captivity only one or a few need to be kept for breeding. Females are often kept for breeding and eaten only when their usefulness for breeding has passed.

Besides the plant and animal remains, domestication may also be identified by other kinds of evidence. Indications of plant domestication, for example, include the identification of garden plots and irrigation ditches. Indications of animal domestication may include large accumulations of animal dung close to a village, or evidence of fencing.

Food production is strongly correlated with settlement patterns. When people started domesticating animals, for example, there was a necessary reduction in mobility. Pastoralists are typically semi-sedentary, meaning they move only a few times each year. They may graze their animals in a valley for several months of a year and at a higher altitude for the rest. This results in much more permanent structures. Knowing that they will be living in the same house for at least a few months each year, for example, leads to people spending more time building structures than if they were foragers spending relatively little time in the same location. The surplus created by pastoralism also means that more people can live together, leading to larger settlements than can usually be maintained by foragers.

Similarly, horticulture is correlated with increasing sedentism. Horticulturalists typically will build a settlement close to fields they have planted and only move once the nutrients in the soil are depleted, which is usually after at least a few years. Consequently, like pastoralists, horticulturalists will put more time into building structures, knowing they will be there for some time. Also, as in pastoralism, the surplus of food leads to larger population and settlement size.

Eventually, as food surplus increased, new kinds of settlements emerged. Villages and towns increased in number and size. The surplus of food led to many people not being directly involved in food production, and even larger settlements, known as cities, emerged. A standard archaeological definition of a **city** includes an assumption that there were at least 5,000 residents.

One of the most significant technological adaptations of the period was the use of **pottery**, which can basically be defined as **ceramics** (baked clay) used to contain something. Evidence from the Upper Paleolithic in Europe indicates that people had developed ceramic technology about 30,000 years ago (e.g., some of the figurines from the period were made from ceramics), but it wasn't until much later that people used the technology for containers. There is some suggestion that pottery may have been made in China as long as 20,000 years ago, and a recent discovery dates its use in Japan to about 15,000 years ago. Pottery did not become common in Africa, Asia, and Europe, however, until about 10,000 years ago. It did not become common in the Americas for another few thousand years.

It is likely that people had the ability to make pottery before 10,000 years ago, but it did not make sense to use it. Pottery has both advantages and disadvantages. One advantage is that it is easy to make. It is simply a matter of mixing clay and water and then heating it. Another advantage is that, because clay and water are so common, nobody can control or monopolize the resource, and it is therefore inexpensive. Pottery also makes excellent containers, including for liquids.

Figure 6.3
EXAMPLES OF POTTERY.
Widespread pottery use is correlated with settling down. Pictured here are coiled, pinched, and thrown pots. Each of these methods, as well as others, was in use before 5,000 years ago.
Credit: Nadine Ryan

But there are also many disadvantages of using pottery, including that it is (1) bulky, (2) heavy, and (3) fragile – reasons that probably led to it not being common before 10,000 years ago. Foragers, being mobile, were unlikely to want the burden of carrying pots. It is likely that only when people started to settle down, at least temporarily, did using pottery make sense.

There were certainly other important inventions, innovations, and adaptations besides pottery during this period. The bow and arrow, for example, was likely widely used around the world by the latter part of this period. The wheel was likely invented for use in making pottery (i.e., the pottery wheel) about 5,500 years ago in Mesopotamia and was only adapted for transportation more recently. Metalworking began with smelting copper, leading eventually to bronze and iron working in more recent times. Beginning about 6,000 years ago, plows were used in farming. Technology was sufficient to build large monuments and cities.

CHANGES IN SOCIAL AND POLITICAL SYSTEMS

Changes in social and political systems are intricately linked with food production. Although some archaeologists believe that the significant changes in all aspects of culture during this period may have been triggered by social and political factors,

Figure 6.4
MAASAI VILLAGE, TANZANIA.
The Maasai are pastoralists in East Africa. Settlements of pastoralists and horticulturalists are usually larger and have more permanence than those of foragers. Indications of subsistence and social organization can often be found by examining the kinds of plant and animal remains found in the settlement, the number and sizes of the houses, and the distribution of remains within houses and the village.
Credit: Luis A. González

many believe the driving force for change was food production, which in turn led to changes in social and political systems.

A basic idea is that the change to food production, perhaps triggered by an initial reduction in carrying capacity, led to larger populations, which in turn required new kinds of social and political structures in order to be effective. Essentially, the reasoning is that increasingly large numbers of people require leadership to coordinate activities within groups as well as to create and maintain relations with other groups.

There is considerable evidence of the correlation of food production and social and political structures in the archaeological record. Pastoralists, for example, usually exhibit the beginnings of social stratification and leadership. They would typically be organized as a **tribe**, with some kind of leader (either official or unofficial), sometimes referred to as a **Big Man**. Tribes of pastoralists were usually divided into several villages with a total population of no more than a few thousand. Each village had its own Big Man, whose tasks included both maintaining order within the village and representing the village when interacting with others.

Horticulturalists are typically organized as either a tribe or a **chiefdom**. Small-scale horticulturalist groups, usually numbering a few to several thousand divided into several villages, typically organized as a tribe. Larger-scale horticultural groups, sometimes numbering in the tens of thousands, were organized as chiefdoms. Unlike tribes, chiefdoms were typically rigidly hierarchical, leadership was based on heredity, and there was taxation, which was paid in the form of goods or labor to the leaders.

Agriculture significantly increased the carrying capacity of a region, leading to even more people – and to the problems associated with larger populations and agricultural activities. Populations based on agriculture are typically organized as **states**, which, even shortly after their emergence, about 6,000 years ago, could number in the hundreds of thousands.

There are multiple ways of identifying social and political systems archaeologically. The subsistence strategy often provides one good kind of evidence, correlating foraging with **bands**, pastoralists with tribes, horticulturalists with chiefdoms, and agriculturalists with states. In the absence of writing, archaeologists also look for indications of social stratification through the size of houses and distribution of remains. Differences in the sizes of houses or the material used to construct them, for example, are often a good indication of social stratification. Similarly, if highly valued items are unevenly distributed within a site, it is probably an indication of stratification.

CIVILIZATIONS, WRITING, AND ART

Civilizations emerged a few hundred years before 5,000 years ago. The earliest is commonly referred to as the Sumerian civilization, and it comprised 13 different city-states in what is commonly known as **Mesopotamia** ("between the rivers") in ancient times. Today, it is Iraq. Many may have heard Iraq being referred to as the cradle of civilization, which indeed it was.

To qualify as a civilization, most archaeologists suggest that a society must have had at least most of the following characteristics: an agricultural base, state-level political organization, monumental architecture, at least one city, and writing. For most societies making the transition to civilization, writing was the last requirement to be filled.

The earliest widely accepted form of writing is traced to Mesopotamia a little more than 5,000 years ago. It is a script inscribed on clay tablets, often referred to as cuneiform. Decipherment of some of the earliest tablets suggests that its primary function was as a form of record keeping. Prior to 5,000 years ago, the only known civilization and system of writing is that from Mesopotamia. Other systems of writing and civilizations soon followed and are described in Chapter 7.

Art has been pervasive in cultures throughout the world for many thousands of years. As we get closer in time to the present, it becomes more visible. Rock art, for example, is known from around the world. One of the best-known pictograph

Figure 6.5
ROCK ART
AT KAKADU,
AUSTRALIA.
Kakadu contains
one of the greatest
concentrations of rock
art in the world.
Credit: Gillian Crowther

sites from this period is Kakadu National Park in Australia. With an estimated 15,000 sites, some as old as 20,000 years, Kakadu National Park has one the greatest concentrations of rock art in the world (see Figure 6.5).

SUMMARY

This chapter has highlighted some of the key developments in culture around the world in the period from 20,000 to 5,000 years ago. Mirroring the Learning Objectives stated in the chapter opening, the key points can be summed up as follows:

- The principal cultural periods in North America are the PaleoIndian and Archaic. The principal cultural periods during this time in Europe include the Upper Paleolithic, the Mesolithic, and the Neolithic.
- People probably first came to North America, via Asia, between 20,000 and 15,000 years ago. For a few thousand years they incorporated megafauna in their diets.
- Between about 15,000 and 12,000 years ago, many populations began to domesticate plants and animals, which they ultimately became dependent upon.

- Associated with the emergence of food production, populations increased, settlements got larger and more numerous, and social stratification emerged.
- Archaeologists have multiple ways of determining subsistence, settlement, and social and political systems.
- Civilizations and writing emerged a little more than 5,000 years ago.
- Pottery became widely used only after people start settling down.

Review Questions

1. What are the names and time ranges of the principal cultural periods in North America and Europe?

2. What were the major cultural developments occurring between 20,000 and 5,000 years ago?

3. What are the principal explanations for the emergence of food production?

4. How do archaeologists distinguish domestic plants and animals from wild ones?

5. How can archaeologists distinguish specific kinds of social and political systems?

Discussion Questions

1. How might changes in subsistence, technology, settlements, and social and political systems be linked? Think holistically.

2. Do you think the development of new subsistence strategies, settlement patterns, social and political systems, and civilizations were beneficial for most people at the time they occurred?

3. If you could live any place on earth at any time between 20,000 and 5,000 years ago, where and when would that be? Why?

4. Should the word "civilization" be considered a loaded term? Why or why not?

5. Should the transition to new forms of subsistence strategies be considered progressive? Why or why not?

6. Should the transition to new forms of social and political systems be considered progressive? Why or why not?

Visit **www.lensofanthropology.com** for the following additional resources:

SELF-STUDY QUESTIONS   WEBLINKS   FURTHER READING

7

ARCHAEOLOGY OF THE LAST 5,000 YEARS

LEARNING OBJECTIVES

In this chapter, students will learn:

- *the sequence of early civilizations around the world.*
- *explanations for the collapse of civilizations.*
- *North American prehistory from 5,000 years ago to 1500 CE.*
- *that diversity in subsistence and political systems continues.*
- *why archaeologists work in areas for which written records exist.*
- *how archaeologists evaluate explanations.*
- *how archaeologists are involved with sustainability.*
- *about historic period archaeology and archaeology of the contemporary world.*
- *about pseudoarchaeology and how archaeologists evaluate competing explanations.*

INTRODUCTION

Many significant changes occurred in the period from 2.5 million to 5,000 years ago. By 5,000 years ago, most people were dependent on domestic plants and animals, and several states, as well as at least one civilization, had emerged. Things did not slow down in regard to change, though.

This chapter provides an overview of ancient civilizations, including possible reasons for their collapse. This is followed by a section on population estimates, continued geographic expansion, and maintenance of diversity. Finally, the chapter provides an overview of the last 5,000 years of prehistory in North America, archaeology of the historic period, archaeology of the contemporary world, and pseudoarchaeology, including how archaeologists evaluate competing explanations.

ANCIENT CIVILIZATIONS

There is no agreement on precisely what is meant by **civilization**. At a minimum, most archaeologists would accept that a society characterized as a civilization must have all, or almost all, of the following: at least one city, monumental architecture, subsistence based on agriculture, a state level of political organization, and a system of writing. In archaeology, **city** is usually defined as a settlement having at least 5,000 residents. Monumental architecture may include buildings, but it also often features other large structures like pyramids or megaliths. It is usually the addition of a system of writing that leads archaeologists to describe a society as a civilization; the other characteristics are generally already in place and have been for a thousand years or more.

Civilizations are sometimes, but not always, equated with a state level of political organization. Often "civilization" is used to describe several distinct states in a region, with each one meeting the criteria of civilization. When one state dominates or exercises control over others, it is often referred to as an **empire**. Empires are often identified in the archaeological record by the commingling of cultural traditions and connecting road systems.

There were many civilizations in the ancient world. As mentioned in Chapter 6, the earliest known civilization was that of the Sumerians, in the region of modern-day Iraq. In reference to ancient times, this area is also known as **Mesopotamia**. It was the Sumerians who developed the world's earliest system of writing, known as cuneiform, about 5,100 years ago. Other major civilizations that followed in the region include the Akkadians, Babylonians, and Assyrians. In more recent times, the region was dominated by the Persian, Roman, and Islamic empires.

The world's second oldest civilization developed in ancient Egypt. It is generally accepted that the Egyptian civilization began 5,000 years ago with the development of hieroglyphics, an early writing system that was finally deciphered in the 1800s using the **Rosetta Stone**. It is in the earliest stages of the Egyptian civilization that the famous pyramids were constructed. In more recent times, the region was dominated by the Roman and Islamic empires.

Figure 7.1
ROSETTA STONE.
Discovered in 1799 in the town of Rosetta, Egypt, the stone repeats the same message in three different scripts, providing the key for deciphering Egyptian hieroglyphics.
Credit: The Trustees of the British Museum

The Minoans were the first European civilization, centered in and around the Mediterranean, especially on the island of Crete. The civilization rose about 4,000 years ago, and its collapse was likely due to a tsunami that caused significant damage to the ports controlled by the Minoans. Many believe that the story of **Atlantis** is based on the collapse of the Minoan civilization. Subsequent civilizations in the area include that of the Hittites, Mycenaeans, Greeks, and Etruscans. In more recent times, the area was dominated by the Macedonian, Roman, and Byzantine empires.

Asia has many well-known ancient civilizations. The earliest is commonly known as the Indus Valley, or Harappan, civilization. This civilization, centered in the area now known as Pakistan, emerged about 4,400 years ago and included well-planned cities of tens of thousands of people. It provides the first evidence of sewage infrastructure, with piping from inside houses joining to larger piping that carried waste out of the residential areas. In China, the Shang civilization emerged about 3,800 years ago. Subsequent civilizations in the region include the Zhou and Han. The

Figure 7.2
EGYPTIAN PYRAMID.
Pyramids symbolize the
early stages of Egyptian
civilization, known as the
Old Kingdom.
Credit: Nadine Ryan

well-known **Terracotta Army** comprises about 8,000 life-size warriors deposited in formation near the tomb of the first emperor of the area.

There were many well-known civilizations of Central and South America in ancient times. The first civilization in **Mesoamerica** was the Olmec, which emerged about 3,500 years ago. Subsequent civilizations in the area include the Maya, Teotihuacán, Toltec, Zapotec, and Aztec. In South America, the first civilization to emerge is commonly accepted to be the Chavín, about 2,500 years ago. Subsequent to this were the Moche, Tiwanaku, Nasca, and **Inka**. Because of their territorial expansion and domination, the Maya, Aztec, and Inka civilizations are often referred to as empires.

All civilizations eventually collapse. In some cases, the explanation may be fairly obvious, such as the Aztec and Inka civilizations, which were decimated by Spanish invaders in the 1500s. For many other civilizations, however, explanations are not so clear, and there is considerable debate. Table 7.1 lists some of the most common explanations for the collapse of civilizations, which may be categorized as ecological, social/political, and ideological.

Ecological explanations vary. Most involve some environmental occurrence that makes it impossible to effectively feed the population and maintain the civilization's infrastructure. For example, climatic change may lead to reduced food supplies, resulting in abandonment of cities. Similarly, epidemic diseases among crops are suggested by some as a cause of reduced food production. Many suggest overuse

TABLE 7.1

Explanations for the Collapse of Civilizations

Ecological	Ecological catastrophe · Climate change · Diseases to crops · Depletion of soil nutrients
Social/Political	Failure of trading networks · Internal conflict · Conflict with other groups
Ideological	Too many resources spent on religious activity

of the soil for farming also leads to reduced food production, which can result in landslides, adding further difficulties. There is some evidence that some large sites were abandoned due to overirrigation, which causes increased salinity of the soils, rendering them deficient for agriculture. Ecological catastrophe includes such things as earthquakes and tsunamis; the latter, for example, caused significant damage to Minoan sites, from which they could not recover.

Social and political explanations often invoke conflict as an important variable in the collapse of civilizations. Some archaeologists, for example, suggest that infighting can lead to too many resources being used to support war or other forms of conflict at the expense of other, perhaps more important, aspects of a civilization, such as maintaining enough food for all. Examples of this type of internal conflict include the continual fighting among the independent states of the Sumerians in Mesopotamia and the fighting among the various groups of Maya in Mesoamerica. Some archaeologists suggest civilizations collapse because of their failure to create or maintain trading alliances.

Religion has also been invoked as a cause of collapse. Some, for example, suggest that the collapse of the Maya may have been caused by religious leaders redirecting resources for religion and ritual at the expense of subsistence. One line of reasoning holds that leaders might blame a poor crop year on the failure of the people to devote enough resources, including time, to religion or ritual. Accordingly, the

Figure 7.3
TERRACOTTA WARRIORS.
8,000 life-size terracotta warriors guard the tomb of an emperor in China.
Credit: lapas77/Shutterstock

leaders demand more resources for religious purposes; this in turn leads to an even worse crop the following year, ultimately ending with the civilization's downfall.

POPULATION ESTIMATES, CONTINUED COLONIZATION, AND MAINTAINING DIVERSITY

Population Estimates and Growth

Estimating populations from prehistoric times is fraught with difficulties, but archaeologists are still able to make approximations. Principal difficulties include (1) many prehistoric sites for which no records exist or have been destroyed, and (2) estimates that require many assumptions and analogies.

The primary method of estimating populations is to examine what are assumed to be residential structures in a settlement. If the group is assumed to move at least seasonally, archaeologists need to be aware that various settlement sites may be occupied by one group at different times of the year. Some researchers suggest, based on a wide range of ethnographies, that population size can be calculated by estimating 10 square meters (12 square yards) of floor space for every person. Archaeologists are aware, however, that without an indication of many other variables, such as environment, subsistence strategy, and social systems, using this average is not reliable. In many locales, house size is correlated with changing climates, such as smaller

Figure 7.4
MACHU PICCHU, PERU.
Machu Picchu is an Inka settlement located high in the Andes Mountains.
Credit: Barry D. Kass/Images of Anthropology

BOX 7.1   **The Collapse of the Maya**

For about 2,000 years, the Mayan civilization dominated the region commonly referred to in anthropology as Mesoamerica (including what is now known as Mexico and other Central American countries). Then, approximately 1,000 years ago, construction of monumental architecture and the writing of inscriptions ceased, the population in urban areas declined, and eventually the cities were abandoned. This is often described as a "collapse," but this term is loaded. The population of the urban areas certainly declined, the Mayan influence over other states stopped, and the Maya lost their empire, but Mayan civilization continued, and the Maya continue to exist today. In fact, the Indigenous people of the Yucatán area of Mexico are mostly Mayan.

The so-called collapse about 1,000 years ago is an area of significant interest in archaeology, and many legitimate hypotheses have been put forward to explain this apparent event. The hypotheses can be grouped into the categories of ecological, social/political, and ideological.

Ecological hypotheses include (1) the depletion of nutrients in the soil necessary for farming due to overuse, (2) landslides caused by deforestation, (3) drought, and (4) diseased crops. Social and political hypotheses for the collapse include devoting too many resources to conflict rather than to subsistence. This includes allocating resources toward conflict between Mayan states and kingdoms as well as allocating resources to conflicts with others such as Toltec invaders. This category also includes the failure to create or maintain sufficient trading alliances with other groups. Ideological explanations include suggestions that the collapse was caused by allocating too many resources to religion at the expense of either subsistence or maintenance of social and political harmony.

There is no consensus among archaeologists about why the collapse occurred. Anthropologist Dr. Richard Wilk (1985) wrote an interesting article on the collapse called "The Ancient Maya and the Political Present," illustrating how ideas about the cause of the Mayan collapse are biased by current events, at least among American archaeologists. Wilk notes, for example, that in the 1960s, most explanations for the Mayan collapse focused on warfare, which correlates with the US's military involvement in Vietnam. Presumably, an American archaeologist hearing daily reports on the war in Vietnam would have been biased toward warfare as an explanation. It was in the 1970s that ecological explanations for the collapse became popular, which Wilk correlates with the emergence of the US environmental movement. In similar manner, Wilk correlates the popularity of ideological explanations with the rise of religious fundamentalism in the US.

houses in colder times because they are presumably easier to heat. It is also likely that people will have larger living spaces when they spend more time indoors, as they do in areas with considerable rain. Archaeologists are also aware that in the past, much like today, people with higher status tend to live in larger residences.

Archaeologists usually rely on ethnographic analogy when determining how many people occupied a site. If they determine foragers occupied a site, for example, they will use ethnographic data from foragers to make inferences about populations. If they determine pastoralists or horticulturalists occupied a site, they will use relevant ethnographic data collected from recent pastoralists and horticulturalists. Archaeologists are well aware of the problems with using ethnographic data, even if it is used for interpreting sites only a few thousand years old. Foragers, pastoralists, and horticulturalists are not simply living fossils of the past; over time, they have changed as well. And importantly, while foragers, pastoralists, and horticulturalists once occupied the prime habitats on earth, they are now largely in much more inhospitable environments, pushed out of the prime areas by agriculturalists and those interested in resource extraction.

There are many secondary methods of calculating prehistoric populations. Some archaeologists use ecological information to calculate the potential carrying capacity of an area, and then assume that numbers of people were at, or just below, that capacity. Some use cemetery information (e.g., counting graves), but this has many difficulties, including the identification of graves, determining contemporaneity, and assessing preservation. Some archaeologists estimate population based on the size of middens, number of fire hearths, and numbers of discarded cooking pots, but these too require many assumptions and are not reliable on their own.

Because of these difficulties, estimates of prehistoric populations, even when calculated by archaeologists or demographers specializing in the past, should be considered only very approximate. Estimates from the historic period (i.e., when there are written records) tend to be more reliable, but it would be a mistake to consider that just because they were written down, they are accurate. Estimates of the Indigenous populations of North America calculated by Europeans in the 1700s, for example, fluctuate widely and were usually significantly lower than the probable population.

Despite the problems, many archaeologists have made estimates of population size in the past. A review of estimates indicates that there were probably around 10 million people in the world about 10,000 years ago. By 5,000 years ago, the population had likely increased to about 100 million, and by 1,000 years ago, the population was probably close to 350 million. The current population is close to 8 billion and climbing.

Continued Colonization

Most, but not all, of the major regions of the earth had been colonized before 5,000 years ago. Notable areas that have only been occupied within the last 5,000 years include the eastern region of the Canadian Arctic and many islands and island groups of the Pacific Ocean.

Figure 7.5
STATUES OF RAPA NUI (EASTER ISLAND).
Over the last 1,000 years, several hundred of these monolithic statues, called moai, have been made and positioned on the island. The precise reason is unclear, but archaeologists rule out pseudoarchaeological explanations, such as the theory that they were placed there by bored aliens awaiting rescue from the mother ship.
Credit: Nadine Ryan

The eastern Arctic was the last part of the Americas to be inhabited by humans. Settlement began about 4,000 years ago and has included several different groups. Today's **Inuit** are descended from the Thule people, who migrated into the area from the west about 1,000 years ago, replacing the populations already there.

New Zealand and many of the Polynesian islands were among the last places on earth to be colonized. Despite its relatively large landmass and proximity to Australia, which has been occupied for about 65,000 years, New Zealand only appears to have been occupied for about the last 1,000 years. Some of the Polynesian islands such as Fiji, Tonga, Samoa, and the Cook Islands were likely settled between about 3,500 and 2,500 years ago. It was only about 1,000 years ago that Rapa Nui (Easter Island) and Hawaii were settled.

Maintaining Diversity in Subsistence and Political Systems

There have been several significant changes in subsistence over the past few million years. As described in previous chapters, the first major change in human subsistence occurred about two million years ago when people started incorporating significant amounts of meat into their diet. The second major change is associated with cooking, which increased the diversity of the plants and animals that could be eaten. The third major change in subsistence began between 15,000 and 12,000 years ago when people began domesticating plants and animals, creating a substantial food surplus. The surplus significantly increased several thousand years ago, when people began practicing intensive agriculture, which included the use of plows, harnessing animal power (including animals to pull plows), and intensive irrigation. The surplus was enough that it enabled a large proportion of the population not to be directly involved in the collection or production of food. It is this surplus created by agriculture that basically allowed the population to expand to hundreds of millions worldwide.

The next major shift in subsistence occurred only a couple of hundred years ago. This involved relying on **industrialism** for food production. This essentially means that mechanized equipment, in many cases powered by fossil fuels, is involved in farming. This development significantly increased the food surplus once again, and it is likely that most people in North America eat food primarily produced by industrialization (see Chapter 10 for further discussion about food and industrialization).

The basic sequence of subsistence strategies starts with foraging and then moves through pastoralism and horticulture to intensive agriculture and then industrialization. It would be a mistake, however, to think that these changes were necessarily progressive. It is important to remember that there are usually many ways of successfully adapting to environments and other people and situations. What works for some people does not necessarily work for others.

Similarly, the basic sequence of political systems starts with bands and moves through tribes and chiefdoms to states and empires. Archaeologists know that subsistence and political systems are closely correlated and that no one system is necessarily better than another.

Thus, although many people now depend on food produced from industrialized farms and live in state-level societies, it is important to understand that diversity in subsistence and social systems remains. Foraging, pastoralism, horticulture, and agriculture are not extinct strategies, nor are bands, tribes, and chiefdoms. Indeed, recent and contemporary examples of these systems are covered in the remaining chapters of this book.

In no way should groups that are not willing partners in industrialized states be considered inferior or primitive. In many ways, people who have resisted industrialization have more leisure time and live a healthier life.

THE LAST 5,000 YEARS IN NORTH AMERICA

Hundreds of thousands of archaeological sites date from the period between 5,000 years ago to the arrival of Europeans in what is now the United States and Canada beginning around 1500 **CE**. Many of these are recognized as having "world heritage" status and are included in the United Nations list of World Heritage Sites. Some of the most significant and interesting sites dating to the last 5,000 years in the United States and Canada are described in Table 7.2 and shown in Map 6.2.

TABLE 7.2

Significant Archaeological Sites in North America Less than 5,000 Years Old

Site Name	Description
Cahokia	Cahokia, located near St. Louis, Missouri, is a United Nations World Heritage Site. With an estimated population of about 20,000 people, it was probably the largest prehistoric settlement north of Mexico. The site is an excellent example of the mound-building peoples from the prehistoric period. Cahokia itself contains more than 100 distinct mounds. The largest is estimated to have covered 12 acres and to have been close to 100 feet in height.
Head-Smashed-In Buffalo Jump	This site, located in Alberta, is another United Nations World Heritage Site. Its cultural deposits are 10 meters thick, containing projectile points and buffalo bones. The site was used for several thousand years and includes drive lanes to direct buffalo over the cliff. The site provides considerable evidence of technological and social evolution related to the communal hunting of buffalo, which was fundamentally important to the Indigenous peoples of the North American plains and prairies.
Kwäday Dän Ts'ìnchi	Located in the Pacific Northwest, close to the intersecting borders of Alaska, British Columbia, and Yukon Territory, this site yielded an extremely well-preserved body of an Indigenous man that was exposed by a melting glacier. The local Indigenous groups and archaeologists undertook studies of the body, after which the remains were returned to the land with ritual.
L'Anse aux Meadows	A United Nations World Heritage Site located in Newfoundland, L'Anse aux Meadows is best known as a Viking settlement dating to about 1000 CE. It is the oldest reliably dated site created by Europeans on the continent, but it was probably only used for a few years. The Vikings evidently did not have a good relationship with the local Indigenous peoples in Newfoundland and abandoned their settlement. Viking settlements have been found elsewhere in Arctic Canada and Greenland dating to approximately the same time. There is evidence of Indigenous occupation of the site up to 6,000 years ago.
Mesa Verde	Mesa Verde National Park is yet another United Nations World Heritage Site, representative of peoples and cultures of the American Southwest from several hundred to more than 1,000 years ago. The park protects more than 4,000 archaeological sites, including 600 cliff dwellings constructed mostly with sandstone blocks and adobe mortar.
Ozette	Ozette is a village site, sometimes referred to as the "Pompeii of North America" due to the excellent preservation of remains caused by a mudslide that buried the village. The site, located on the coast of the State of Washington, consists of several large multifamily houses and tens of thousands of wood and bone artifacts not normally preserved.

BOX 7.2   **The Archaeology of Pandemics**

The past can help us understand how people coped with unthinkable tragedies, such as pandemics. The term **pandemic** is used to describe a disease or virus that spreads over large portions of the globe and affects large portions of the population.

Archaeologists take an interdisciplinary approach to understand pandemics, relying upon the knowledge and expertise of specialists (Antoine, 2008). Zooarchaeologists examine the remains of animals for presence of a disease; environmental archaeologists, geologists, and paleoecologists determine the environment that gave rise to a pandemic; archaeobotanists identify the types of food consumed during the pandemic; and bioarchaeologists, forensic anthropologists, and physical anthropologists estimate the sex, ethnicity, and age of skeletal remains and study the remains for the presence of disease and/or other causes of death. Depending on the time period, historical archaeologists might be involved to comb through archival records for references to the pandemic and medical practices and interventions used to combat the disease.

By drawing upon a variety of specialists' findings, including scientists outside of the field of archaeology, we have, for example, learned a great deal about the Great Plague of Marseille that broke out in France in 1720 and lasted into 1722 (Leonetti et al., 1997). Approximately 50,000 people died in Marseille alone, representing half of the city's population. The plague was introduced into the region by a ship arriving from Syria in May 1720.

In 1994, archaeologists excavated a mass grave containing approximately 200 people who had died in Marseille. Researchers were surprised to find that two of the victims had bronze pins in their toes, one of which had been placed under the victim's big toenail. Because victims had to be buried quickly to prevent further spread of the plague, there was a very real fear that some people might be accidentally buried alive. Archival research revealed that physicians used a variety of methods to verify that a patient was in fact deceased, included placing pins underneath patients' skin and, in this case, toenails.

The Great Plague of Marseille was caused by a bacterium known as *Yersinia pestis*, which can be transmitted from rats to humans. Researchers at the Max Planck Institute for the Science of Human History extracted DNA from the teeth of several victims and made two important findings (Bos et al., 2016). First, they found that this particular strain of the plague no longer exists in the world today. Second, they argue that this strain had mutated from the original strain that resulted in the Black Death, which killed millions of people in Europe 350 years before the Great Plague of Marseille.

The **COVID-19** pandemic will also likely leave material traces behind for future archaeologists to discover. Plastic gloves and non-reusable masks now litter the ground, which will eventually make their way into the archaeological record. Moreover, many recycling programs have been halted and countless bottles of bleach and bleach wipes have been consumed to prevent the spread of COVID-19 through surface transmission. All of this materiality will contribute to our exponentially growing problem of plastic waste, which pollutes our oceans and contaminates our drinking water.

Figure 7.6
STONEHENGE.
The area around Stonehenge has been occupied for about 8,000 years, although the core parts of the monolithic stone structures were positioned at various times between about 5,000 and 3,000 years ago.
Credit: Barry D. Kass/ Images of Anthropology

When Europeans first arrived, most Indigenous groups could likely trace their ancestry to populations living in the same area for the previous several thousand years or more. There were certainly exceptions, though. The Apache and Navajo migrated from the subarctic region of northern Canada to the American Southwest about 1,000 years ago, and as mentioned earlier in the chapter, the ancestors of contemporary Inuit in Canada replaced other groups in the region about 1,000 years ago.

The Indigenous peoples of North America had a phenomenal understanding of plants for food and medicines. Archaeological and ethnographic research indicates that before the arrival of Europeans, more than 1,500 species of plants were used for food and more than 2,500 were used as medicines.

ARCHAEOLOGY OF RECENT TIMES, EXCLUDING CIVILIZATIONS

Much of the focus on the past 5,000 years is on ancient civilizations, and for those in North America, on the Indigenous populations there. It is important to understand that this is partly because of the high archaeological visibility of civilizations, and North Americans often assume at least some ancestral connection with ancient civilizations.

Certainly, there have been some great cultural achievements associated with civilizations, but it isn't as if they were, or are, necessarily superior. Stonehenge, for example, and many other similar monuments requiring advanced engineering

Figure 7.7

EXCAVATING A HISTORIC SITE IN NORTH AMERICA.

Historic archaeology often provides important information that was never written down or is in contrast to biased written sources. This site was a Japanese settlement in western Canada, for which no written records exist.

Credit: Emma Kimm-Jones

skills were built by those living in chiefdoms. The settlement of some Pacific islands required seafaring skills at least equal to those of European explorers. We know they weren't simply blown off course accidentally, because they brought domestic plants and animals with them.

A major subfield of archaeology is historic archaeology, which essentially means that archaeologists are doing work in an area and focusing on a time period for which written records exist. Some non-anthropologists question why, when there is a written record, we would even bother, but archaeologists are aware that written records are usually incomplete and biased. The history of enslavement, for example, is written almost entirely from the perspective of owners of enslaved people. Similarly, the histories of Indigenous peoples and other minorities are usually written from the perspective of the majority, the oppressors. Archaeology helps provide balance and gives voice to the voiceless.

In North America, the transition from prehistory (before written records) to history (with written records) occurs whenever Europeans first entered an area. Thus, historic archaeology starts about 500 years ago on the east coast of the continent, but not until the late 1700s on the west coast of Canada. Major areas of interest in historic period archaeology in North America include the archaeology of colonialism, civil war archaeology, African American archaeology, and Asian American archaeology.

One example of this type of research is Dr. Kathryn E. Sampeck's (2019a, 2019b) study of the globalization of chocolate due to European colonization in Central America (see also Sampeck and Thayn, 2017). Sampeck has studied the transformation of chocolate from a culturally significant ingredient in pre-Columbian Guatemalan recipes to an immensely desirable commodity in our contemporary world. Cacao, which is used in the production of chocolate, is found on a tropical tree known as *Theobroma* and was domesticated by Mesoamericans. According to recipes referenced in Maya texts, cacao was first cultivated for use in beverages. What we today call "chocolate" appears in historical Mexican and Spanish records starting around 1580 CE. It was at this time that chocolate became the first highly sought-after agricultural commodity in Guatemala.

It is believed that cacao was first used around 2000 BCE and likely restricted to elites. Around 400 BCE, non-elites in southern Mesoamerica started consuming it as a beverage. For thousands of years across Central America and Mexico, cacao was thought to embody magical properties. It was used in fertility, marriage, and death ceremonies, sometimes corresponding to specific days of the year or times of the day. In the sixteenth-century CE, cacao seeds were planted in Guatemala at the same time as human conception rites took place, both of which were to take place under the light of a full moon. The fertility of the land and of human beings was therefore seen as intertwined. Cacao was also thought to have medicinal properties, used to heal the wounded or ill, and victims of human sacrifice were even led to their deaths wearing necklaces of cacao, the human body and cacao both seen as valuable offerings.

Europeans were introduced to cacao in 1502 CE when Christopher Columbus encountered traders on the Yucatán peninsula. As Spain colonized Guatemala in the 1500s, Indigenous groups would slowly lose control over the production of cacao. Spanish colonization and violence resulted in the establishment of cacao plantations, which relied upon the labor of enslaved peoples in Central America, South America, and parts of Asia and Africa. Chocolate would become a commonplace good in Europe by the late seventeenth to early eighteenth centuries. The demand for chocolate has not ceased, nor have the exploitative relationships involved in its production, as detailed in Box 14.2.

WORLD HERITAGE

In 1972, the United Nations Educational, Scientific, and Cultural Organization, commonly known as UNESCO, created the Convention Concerning the Protection of the World Cultural and Natural Heritage, which led to the well-known World

TABLE 7.3

Selected World Heritage Sites from the Last 5,000 Years

Region	Sites
United States	Mesa Verde, Cahokia, Chaco Canyon, Pueblo de Taos, Independence Hall, Monticello and the University of Virginia's Academical Village, the Statue of Liberty
Canada	L'Anse aux Meadows, Head-Smashed-In Buffalo Jump, Old Québec, Old Town Lunenburg, SGang Gwaay
Central America (including Mexico)	Tikal, Copán, Monte Albán, Chichén Itzá, Palenque, Teotihuacán, Uxmal
South America	Tiwanaku, Rapa Nui, Machu Picchu, lines and geoglyphs of Nasca
Europe	Acropolis of Athens, Pompeii, Stonehenge and Avebury
Asia	Angkor, Moenjo Daro, Terracotta Army, Bamiyan Valley
Africa	Pyramids at Giza, Great Zimbabwe
Australia	Kakadu

Heritage List. Sites included on the list have been determined to have outstanding universal value. There are currently more than 1,000 sites on the list, with at least several new ones being added each year. At least several hundred are in the "cultural" (as opposed to "natural") category.

A small sample of sites that fall within the time period of the last 5,000 years are included in Table 7.3.

The World Heritage List has many benefits such as popularizing the importance of heritage, providing education about archaeology, providing some assistance and protection for the sites, and stimulating archaeo-tourism. Although there are strict criteria for nomination and inclusion on the list, it is not without bias. Even a quick look at the entire list provides evidence of bias toward specific countries (such as those in Europe and Central America), and bias toward sites from the historic period, especially with links to Europe or colonialism. The fact that a site is not on the list does not necessarily mean that it is not significant. Some countries simply choose not to nominate sites for inclusion.

ARCHAEOLOGY OF THE CONTEMPORARY WORLD

In the early twenty-first century, many archaeologists have turned their archaeological lens toward the contemporary world. Areas of interest include contemporary household trash, nuclear waste, and trash in outer space. Archaeologists have developed new subfields, such as forensic archaeology and disaster archaeology, and have become advocates for the disenfranchised. They also like to contribute in practical ways to issues related to sustainability.

This trend toward shifting the focus of archaeology to contemporary society started in a serious way in the late twentieth century with the study of contemporary North American trash, commonly known as **garbology**. The guru in this regard was Dr. William Rathje (2002). He led archaeological projects focusing on trash at levels ranging from looking through individual household trash bins that were put out for collection to excavating landfill sites of large North American cities. Results in the late 1900s were surprising. They often showed a disconnect between what people said they did at the household level and what they actually did, as indicated by their trash. (See Chapter 8 for a discussion of real culture versus ideal culture.) Similarly, popular perceptions of landfills were different from reality. One of the most important results of these landfill archaeology projects was the discovery that organic remains do not decompose very quickly. It was not unusual to find decades-old heads of lettuce and hot dogs. Research at both the household level and in landfills also indicates that North Americans are extremely wasteful, discarding a significant percentage of food, and that despite efforts to reduce trash, the average amount of trash thrown away by individuals continues to increase.

Many archaeologists have remained interested in contemporary waste in the twenty-first century, and they sometimes work alongside others from environmental geography, environmental science, discard studies, and sustainability studies to help solve some of the problems of trash. Of course, the problems are many and include the sheer volume of trash, the danger of the chemicals and toxins within the trash, pollution, animals feeding on plastics, and the costs of packaging and recycling. Edward Humes (2012) provides an overview of some basic data associated with trash created by Americans. This includes the fact that Americans create 25 percent of the world's total waste; the average American creates more than seven pounds of trash per day; there is more money spent on waste management than on fire protection, parks and recreation, libraries, and schoolbooks combined; there are about 60 million water bottles discarded daily; the amount of plastic wrap discarded each year could shrink-wrap Texas; and there is enough wood put into landfills each year to heat 50 million homes for 20 years.

Figure 7.8
WASTE AUDIT ARCHAEOLOGY.

Archaeology students put what they learn of archaeological method and theory into practice, sorting through campus trash. The university uses the results to effectively reduce waste.

Credit: Robert J. Muckle

Some archaeologists on campuses across North America are now tackling the problems of trash in a practical way. At Capilano University in North Vancouver, Canada, for example, archaeology students are involved in regular waste audits on campus. Waste is sorted, and the resulting data are used to improve discard behaviors, reduce total waste, and increase campus sustainability.

Archaeologists have also been involved in studies of nuclear waste. The US government has consulted archaeologists since the 1980s on how to best mark nuclear waste sites so that people of the future will recognize the danger. Who better to consult than those who know what is likely to survive and be interpretable? Based on their experience and knowledge, archaeologists have made many recommendations including that (1) multiple symbols, pictures, and languages should be used; (2) structures should be made of natural materials such as earth or stone with no perceived value, for if there is a perceived value it will inevitably be looted; (3) large monoliths should ring the site, so that a pattern will be visible; and (4) subsurface markers, made of ceramic, should be included at various levels.

Archaeologists in the early twenty-first century are also interested in trash in outer space, including the tons of trash left on the moon and planets as well as the hundreds of thousands of pieces of space junk in orbit, also known as orbital debris. Archaeologists both remotely document the remains of human culture left on the moon and planets and advocate for their protection from space tourists, who will inevitably be making trips to the moon. Archaeologists also document and study

The Archaeology of Food Insecurity

Over the course of history, famines have threatened the very survival of the human species. Although food is abundant in today's world, a disproportionate number of people have access to it. An estimated 20 to 25 percent of the population of Africa suffers from undernourishment. Archaeologist Dr. Amanda Logan (Logan, 2016a, 2016b, 2020) wants to know how food insecurity emerged in African countries.

Building off 30 years of research conducted by archaeologist Dr. Ann Stahl in Banda, Ghana, Logan argues that food supply is shaped by political factors often external to a community. Though Banda experienced a catastrophic drought from 1450 to 1650 CE, the residents of Banda did not suffer from food insecurity. Yet a more recent and much shorter drought that took place between 1968 to 1972 resulted in a devastating famine. So, what had changed in Banda between the 1400s and now?

Logan argues that severe food shortages began in the wake of the Atlantic slave trade dating from the sixteenth to nineteenth centuries and the nineteenth-century slave trading internal to Africa. According to oral histories, a slave raider destroyed Banda's village and seized its food supply in the late nineteenth century, forcing Banda residents to flee to neighboring villages. British colonization made food even more difficult to procure, with the British exerting authority over resources for export and profit. Local farmers and agricultural producers lost control over their crops and food, resulting in famines that were once preventable through the pooling of community resources.

the many thousands of pieces of orbital debris from satellites and other objects in space, which pose serious hazards to space travel.

Back on Earth, archaeologists are in a race against time to save archaeological sites from destruction due to climate change. In Scotland, for instance, archaeologists have developed a successful community outreach program that trains volunteers to identify and document archaeological sites in danger due to coastal erosion (SCAPE, 2021). In North America and the Swiss Alps, archaeologists are rescuing artifacts from melting ice patches caused by global warming (Schlumbaum, 2010; Hare et al., 2012; Reckin, 2013).

Many of the artifacts found in ice patches are not usually recovered during archaeological excavation, as materials like textiles and wood quickly decompose in the soil. Artifacts like leather pants and bows made of wood have been encased in ice patches for nearly 6,000 years since they were used by humans. But now that ice patches are melting, archaeologists have a brief window of time to recover these delicate artifacts that will disintegrate within a few years of exposure to the elements. These artifacts are costly and difficult to access, as ice patches are located in high altitude climates and on steep, treacherous terrain.

Climate change archaeologists are not only trying to save the past from destruction but also to learn from it to make our world a better place. In California, a state known for its devastating fires, archaeologists are collaborating with Indigenous communities to understand how Native Californians managed land prior to colonization. This work requires an interdisciplinary team of researchers – dendro-ecologists, palynologists, zooarchaeologists, plant population geneticists, and oral historians, to name a few – who reconstruct the plants and animals that once made up California's ecosystem.

According to the findings of Lightfoot (2013), Native Californians actively managed their environment with fire, using it "to clear undergrowth; to control insect and pest infestations; to facilitate game hunting; to encourage plants to produce young, straight stems and other raw materials for cordage, baskets, and other household materials; and to enhance the diversity and quantity of economic plants and animals in their territories" (p. 210; see also Lightfoot et al., 2013). By working with both the Amah Mutsun Tribal Band and the Muwekma Ohlone Tribe, researchers hope that together they can re-institute Native Californians' traditional methods of managing land to prevent fires and to bring back animals and plants of the past for Indigenous use in the present.

Other ways archaeologists bring their lens and skills to the contemporary world are through the emergent subfields of forensic archaeology and disaster archaeology. Forensic archaeology is the application of archaeology in legal contexts, usually in regard to assisting criminal investigations. Forensic archaeologists offer skills of identifying, recovering, recording, and interpreting physical evidence of human activities. They also occasionally focus on human biological remains, but that focus is usually considered to fall under the umbrella of forensic anthropology.

Disaster archaeology emerged in the United States in the twenty-first century. It got its start with offers from archaeologists to help with the identification, recovery, and interpretation of human remains immediately after the 9/11 disaster in New York City, in which thousands were killed after airplanes hit the Twin Towers. Since then, disaster archaeology has taken hold, and archaeologists are often on scene shortly after disasters, both natural and cultural, to aid in identification and recovery.

PSEUDOARCHAEOLOGY

Archaeologists are often faced with deciding which explanation is best for something that occurred in the past. There are often multiple explanations for a particular phenomenon, such as the origin of food production or the collapse of civilizations, and archaeologists must choose for themselves which one is best. Likewise, there are often some rather bizarre explanations that include aliens and other strange phenomena that should be subjected to critical thinking.

Some of the criteria for evaluating explanations or hypotheses about the past are described in Table 7.4. Testability is paramount. If the hypothesis is not testable, then it should not be considered, especially within the framework of science or archaeology. This is where odd ideas involving extraterrestrials, supernatural forces, or other strange or bizarre explanations are discarded. We have no way of testing for the presence of an extraterrestrial or the supernatural. Archaeologists need empirical evidence – things that can be touched, weighed, drawn, photographed, and analyzed. People who endorse explanations involving aliens or supernatural phenomena often are characterized as practicing **pseudoarchaeology**. Popular notions about the Egyptian pyramids, statues of Rapa Nui, and Stonehenge, for example, having something to do with alien life forms, are pseudoscientific. They are not considered seriously since the hypotheses cannot be tested. In response to one such popular claim that aliens mated with ancient Egyptians, providing them knowledge to build the pyramids to specific criteria, well-known and highly respected scientist Dr. Carl Sagan (1979) stated:

> We are thinking beings. We are interested and excited in understanding how the world is put together. We seek out the extraordinary, and if you think of these claims, if only they were true, they would be amazingly interesting – that we have been visited by beings from elsewhere who not only have created our civilizations

TABLE 7.4

Criteria for Evaluating Explanations about the Past

Question	Action
Is the hypothesis testable?	If not, consider it no further.
Is the hypothesis compatible with our general understanding of the archaeological record?	If not, be cautious.
Which hypothesis provides the simplest explanation? (i.e., Occam's razor)	The simplest explanation is usually the best. The simplest is the one that requires the fewest assumptions.
Have all competing explanations been considered equally?	Do not accept one hypothesis by merely rejecting the others.

BOX 7.4 **Food Matters: An Archaeologist Eats a Shrew**

One of the most interesting – some may say disgusting – stories about archaeology in recent years involves a bit of experimental archaeology. It concerns an archaeologist eating a small animal, followed by him and another archaeologist then sifting through the feces of the eater to see if all that went in also came out.

The experiments are described in an article called "Human Digestive Effects on a Micromammalian Skeleton" (Crandall & Stahl, 1995). The objective was to be able to better interpret the assemblages of animal bones in archaeological sites, which is necessary to make proper interpretations of diet. The authors describe how they skinned, eviscerated, and cooked a shrew before one of them ate it without chewing.

After days of examining the feces of the archaeologists who ate the shrew, the results were in.

Of the 131 skeletal elements of the shrew that went in, only 28 were recovered from the feces. The authors are confident that the missing bones completely succumbed to the human digestive processes.

The research gained much publicity in 2013 when the archaeologists received an Ig Nobel Prize (a parody of the Nobel Prize), sponsored by the *Annals of Improbable Research*. As stated on the *Annals* webpage (www.improbable.com), the awards are to "honor achievements that make people LAUGH, and then THINK." In that regard, the award to the archaeologists is deserved. It may seem disgusting, but the work is significant. Archaeologists interested in diet need to know what is likely to be preserved and what isn't. Sometimes, apparently, you have to sift through your own fresh feces, or that of your colleague, for the good of archaeology.

for us, but mated with human beings. In my view it's much more likely to successfully mate with a petunia than an extraterrestrial.

When evaluating various hypotheses, it is important to consider how well they fit with our general understanding of the archaeological record. When considering the initial colonization of North America, for example, the route from Asia is usually considered the strongest hypothesis. This explanation best fits our basic understanding of world prehistory, insofar as archaeologists know people were already close by in Asia immediately before the first evidence in the Americas, and their subsistence and technology was similar.

Occam's razor is well known in science. Essentially, applying Occam's razor means that the simplest explanation is usually the best, with the simplest being the one that requires the fewest assumptions. Considering the initial colonization of the Americas again, the simplest explanation is that the ancestry of the initial migrants lay in Asia. The only assumption necessary to accept this hypothesis is that they expanded their territory. The idea that the migrants came from Europe, on the other hand, requires multiple significant assumptions, including the fact that the migrants had the ability to build boats able to navigate the often stormy North Atlantic; that they had the imperative to seek distant lands they probably knew nothing about; that they were able to travel by boat for very long distances in glacial environments; that they were able to adapt to resources substantially different from those in Europe; and that they left behind the significant artistic traditions common in Europe at the time the voyages are assumed to have occurred.

It is important that all explanations be considered equally, and that we not simply accept one explanation by elimination of the others. Accepting one hypothesis by eliminating the others is a common ruse of pseudoarchaeologists. A pseuodarchaeologist, for example, may generate four hypotheses, systematically reject the first three, and then proclaim the fourth explanation, one that involves aliens, as the only possible explanation. This is very bad science.

SUMMARY

This chapter has provided an overview of the archaeology of the last 5,000 years. Mirroring the Learning Objectives stated in the chapter opening, the key points can be summed up as follows:

- Ancient civilizations were many, and they all collapsed. There are multiple explanations for this.
- A significant recent development is industrialization, beginning only a couple of hundred years ago.
- Foraging, pastoralist, and horticultural groups still exist, as do the forms of political organization known as bands, tribes, and chiefdoms. They are not inferior or primitive. In many cases they provide a better quality of life.
- North American prehistory over the past 5,000 years was highly developed, and some groups practiced horticulture.
- Many archaeologists focus on the time period for which written records also exist, providing balance and new kinds of information.
- Some archaeologists focus on the contemporary world, using the methods and theories of archaeology.
- The World Heritage List is important, but biased.
- Pseudoarchaeological claims can be evaluated and debunked by using established scientific criteria.

Review Questions

1. What is the history of early civilizations around the world?

2. What are the principal explanations for the collapse of civilizations?

3. How do archaeologists make population estimates for settlements and regions, and what are the estimates for global populations in the past?

4. What are the principal cultural developments around the world over the last 5,000 years?

5. What are some examples of archaeology of the contemporary world?

6. What is the World Heritage List?

7. How do archaeologists evaluate competing explanations?

Discussion Questions

1. Why do you think there are so few prehistoric World Heritage Sites in North America? Do you think it is the result of Eurocentric bias, historic bias, lack of significant sites, lack of features with high archaeological visibility, or something else?

2. Assume North American civilization collapses and all cities are abandoned within the next 100 years. What kinds of evidence will archaeologists have of twenty-first-century North America 5,000 years from now, and what might their interpretations of this evidence be?

Visit **www.lensofanthropology.com** for the following additional resources:

SELF-STUDY QUESTIONS	WEBLINKS	FURTHER READING

8

STUDYING CULTURE

LEARNING OBJECTIVES

In this chapter, students will learn:

- *the anthropological definition of culture.*
- *the meaning of race and ethnicity and how the idea of race is misunderstood.*
- *the usefulness of cultural relativism in studying culture and the importance of identifying ethnocentrism.*
- *how to evaluate whether cultural practices are adaptive or maladaptive.*
- *the functions of culture.*
- *how raising children and cultural practices are connected.*
- *how anthropologists study culture in the field.*
- *the applicability of anthropological research to solving problems.*

Culture is the app for human behavior.
#StudyingContemporaryCulture

INTRODUCTION

When I (co-author Laura González) sat down at the long wooden table on my first night of fieldwork in Ciudad de Oaxaca, Mexico, I wanted to make a good impression on the men who would be my cultural informants over the next several months. Of course, as good hosts, they had prepared some drinks and snacks to share: tequila shots and crispy fried *chapulines*, or baby crickets. Suddenly, I was confronted with the kind of decision that ethnographers make as a routine part of fieldwork. Do I smile, accept their hospitality, and eat the bugs? Or do I risk disrespecting my hosts' generosity by refusing? The crickets weren't bad, actually – reminiscent of crispy chips with *chile* and lime – although I did have to pull a tiny leg out of my teeth. The tequila helped wash it down, and we shared an evening of introductions and laughter.

This book has introduced the perspective of anthropology as a kind of lens through which one sees the world. In particular, the lens shows that each society functions with an intricate web of knowledge, beliefs, and practices – ones that may be quite different from our own. The anthropological lens sees every culture as valid and complex – a magnificent puzzle that the anthropologist attempts to piece together. In my experience above, my hosts and I had two different cultural views about what is and is not food. Food choices are an essential part of culture, of what is understood as "normal" or "natural." As an anthropologist, I knew that no food choice is inherently "right," only familiar or unfamiliar. I therefore accepted their offering as a gesture of goodwill, hoping that my willingness to partake gave a positive first impression.

Similarly, culture itself can be thought of as a kind of lens, although no special skill or knowledge is needed to look through it. We're not born with one particular lens or cultural perspective. All humans have the capacity for culture. That is, we are born ready to learn culture – to build that lens – but nothing in our genes or biology determines *what* culture we learn.

All humans use their own cultural lenses to understand and interpret their surroundings. The cultural lens guides our behavior and interactions with others; for instance, whether we consider crickets to be food. The lens is mostly invisible; that is, we learn how to act and interact largely automatically within our own culture, without requiring a rulebook for behavior. For example, an infant born to a Ghanaian mother and father will learn Ghanaian culture. Equally, an infant born to an Italian mother and a Polish father living in Ghana will learn Ghanaian culture. She'll likely wear kente cloth and enjoy *fufu* (a Ghanaian staple food) in her school lunches. Of course, she will also be exposed to aspects of her parents' upbringing (and may also enjoy the occasional linguine or kielbasa), but she doesn't receive "Italian DNA" or

"Polish DNA" that translates directly into culture or language. In other words, she acquires her culture based on experience.

All human groups have culture (and some animal groups, too, although to a lesser extent). Sharing culture means people understand what goes on around them in approximately the same ways. They can read their surroundings as if social interaction were a text. Another helpful way to think about culture is to imagine members connected by threads of a web. These webs of meaning are invisible and usually unspoken. But the unwritten rules are there, guiding our behavior as we choose which threads to take along the way.

Since culture can't be measured, held in your hands, or shown on a map, how do anthropologists understand it? We learn about culture by getting in the thick of it. Ethnographers go and live among the people they aim to learn about and slowly, over time, come to understand their world.

Practicing cultural anthropology means that fieldworkers participate in people's lives at the same time that they are observing and analyzing behavior. We call the process of studying culture **ethnographic research**, and the written or visual product of that research an **ethnography**. Importantly, the lens of anthropology shapes how **ethnographers** approach their subjects and what questions they ask. In particular, ethnographers seek to understand the **emic** – or cultural insider's – view, as well as the **etic** – or outside observer's – view.

Every anthropologist has a story about how their interest in anthropology developed, which field sites attracted and hosted them, and what issues called to them over the course of their career. Box 8.1 introduces Dr. Julie Lesnik, an anthropologist who studies **entomophagy**, or insect eating, in human cultures.

THE CULTURE CONCEPT

A basic way to define culture is as the shared understandings that shape thought and guide behavior. That is to say, members of a culture share a set of beliefs, customs, values, and knowledge. These shared understandings allow us to act in ways that make sense to others.

Symbolic anthropologist Dr. Clifford Geertz (1973) emphasizes the idea that culture is a set of functions, and not simply a list of attributes or behaviors. He argues that culture works like a computer program, in that it has rules and instructions for behavior. This prescient definition, written before computers were an integral part of our everyday lives, makes sense now more than ever. In other words, cultural practices can be observed, but what anthropologists really want to get at is what causes those behaviors. Anthropologists want to know how culture functions like an "app"

BOX 8.1   # Food Matters: Practicing Anthropology, Eating Bugs

Dr. Julie Lesnik is a professor of anthropology at Wayne State University in Detroit, Michigan, who specializes in eating bugs. Let me rephrase that, lest you picture her with a lunch box full of grubs in the university cafeteria. Her field of specialization is called entomophagy, or insect eating, as it relates to the human diet both in the past and present.

The idea of eating insects may sound strange or "disgusting" to most North Americans, but it may be the wave of the future. Insects are an excellent source of nutrition. Over two billion people in more than 80 countries in the world eat a variety of insects regularly. Edible insects range widely; however, the most consumed species are beetles (and their larvae), caterpillars, bees, wasps, termites, and ants. In particular, Lesnik found that insects likely have been an important nutritional supplement for pregnant and lactating women throughout our foraging past and into the present. Mealworms, for example, have about as much protein, vitamins, and minerals as the same amount by weight of fish or chicken. And, from this author's personal experience, they taste good, like roasted almonds.

While people who live in forests, deserts, and jungles have easy access to insects through gathering, urban people can benefit from eating bugs as well. In urban areas, eating protein-rich culinary insects decreases the pressure to create more and faster factory-farmed meat. With nine billion people expected to inhabit the earth by 2050, food production will need to double. Farming insects is one way to ensure there will be adequate protein for more people in the same amount of space.

Julie Lesnik has studied insect eating in South Africa and East Africa, especially as it pertains to the development of the human diet. She knows that Western culture has not yet embraced the culinary potential of insects. An ethnocentric feeling of disgust (the "yuck" factor) prevents most people from seeking out this alternative, sustainable source of protein. Nonetheless, she finds that North Americans today are much more likely to accept insects in powdered form, such as cricket flour, for example. This alternate flour source for baking can turn a muffin into a protein powerhouse. Even the United Nations published a recent report (Van Huis et al., 2013) stating that insects will be an important food source for the future. Indeed, the benefits of entomophagy for the world's growing population are clear. Visit Julie Lesnik's website at Wayne State University for more information: https://www.entomoanthro.org/.

Figure 8.1 **JULIE LESNIK.**
Dr. Julie Lesnik (Wayne State University) studies the way humans have used insects as a source of nutrition throughout human evolution. She also promotes them as a sustainable food source for people today.
Credit: Julie Lesnik

to guide our behavior. At the same time, depending upon their framework, anthropologists are also interested in how biology (such as genetics) and environmental influences affect the thoughts and behaviors of people in society.

What Are the Parts of Culture?

Culture has three basic parts: what we think (cognition), what we do (behavior), and what we have (artifacts).

First, what we think: The values we learn from our parents and the symbols we understand in our environment are cognitive. This includes the information and understandings that allow us to relate to other members of our culture.

Second, what we do: Actions and interactions with others are behavioral. How a person eats, works, and plays are all products of this shared knowledge. Shared culture guides behaviors in ways that allow people to understand and act appropriately with each other.

Finally, what we have: The material products of our society are artifacts (portable items) and features (non-portable items). This includes things such as pottery and clothing (portable) and buildings and roads (non-portable). Artifacts and features are also referred to as material culture: the things that people make, alter, and use.

Four Characteristics of Culture

Culture has certain important features that make it different from behavior based on biological instincts or **personality** traits. Instincts are automatic and coded in people's DNA as part of the legacy of *Homo sapiens*. Examples of instincts are to run from danger or recoil from a burn. On the other hand, personality traits arise in individuals due to their unique development or experience. Culture differs from genetics or personality because it is learned, based on symbols, holistic, and shared.

Culture is *learned*. Humans are not born with knowledge of their culture. They learn it actively and acquire it passively from the people around them. The process of learning begins with an infant's interactions with primary caregivers and family. Then the process extends in childhood to friends, schoolteachers, the media, and other influences.

Culture is *based on symbols*. When we talk about culture as being based on symbols, it doesn't only refer to peace signs and emojis. Although these and other graphic representations of ideas are symbols, a **symbol** can be anything that stands for something else and carries meaning. Language is symbolic because we agree that words on a page and the sounds we make when we talk stand for meaning.

Culture is *holistic*, or integrated. Anthropologists approach the study of culture with the knowledge that all aspects of a society are linked. If one aspect is altered, then the others will be affected as well. For instance, in a colonial situation, the

dominant society may impose new religious practices on traditional ones. With the loss of familiar ritual, the rites performed to ensure a good harvest may be lost, farming practices may change, and even family life may be altered. This is why some compare anthropology to an orchestra: the fieldworker must listen to the strings, winds, horns, and percussion to understand the whole musical piece.

Culture is *shared*. Finally, because the idea of culture involves more than just one individual, culture must be shared. Otherwise, a personality feature that wasn't shared by others could be called a quirk or unique attribute. For example, if one person wears earrings made of tomatoes, it is a quirk. If that person carries some status and others join in, then it may become a cultural fad (albeit a very strange one). The shared nature of culture allows people to understand each other's behaviors.

Culture as Community

In this book, several different terms are used to talk about people and their cultures. Depending on the context, people may be referred to as a community, group, or society. When people share a geographical space, they are referred to here as a **community**. A community of people lives, works, and plays together. A community can also be virtual, since online communities can develop the same kinds of relationships that physical ones do. **Group** is a looser term, referring to people who share culture. Members of a group generally live in the same region. Finally, **society** is used somewhat interchangeably with the term group, to refer to a large number of people with social connections.

Within any culture, we can find many communities based on **identity markers**. Markers may include ethnicity, socio-economic class, religious beliefs, age, gender, and interests. These subgroups, or **subcultures**, are made up of people connected by similarities. Subcultures may reflect ethnic heritage, such as Mexican Americans, or they may denote common interests, such as members of a sports team, car club, or cosplay community.

A group may share many identity markers, such as the Hadza people of Tanzania. The Hadza have close-knit communities and remain primarily hunter-gatherers in the twenty-first century. They largely reject outsiders joining their social group and marry within it. Therefore, although of course there are individual differences among Hadza people, they share many of the same beliefs, values, and behaviors. Sometimes these groups are referred to as more **homogeneous** (*homo* = same) than others.

Groups that share few identity markers may be described as more **heterogeneous** (*hetero* = different). An example of people who share heterogeneous culture are those living in the United States. The US has been described as a "melting pot" or "tossed salad," referring to the mixture of people of different ancestries who are all residents. There are many languages, religious beliefs, values, and ethnicities, but all share a set of understandings (i.e., culture).

Cultures can be large or small; they can be concentrated or **diasporic** (spread across the world). A culture is not a fixed entity, especially when so many **emigrants** or refugees have left their birthplaces to live in another region, state, or country. In fact, as of 2020, about 3.5 percent of the world's population (approximately 272 million people) are international immigrants, with 59 million of those in North America (International Organization for Migration, 2019). Many factors contribute to immigration, both "push" factors (such as poverty, high unemployment, the effects of climate change, natural disasters, oppression, economic injustice, persecution, or genocide) and "pull" factors (such as reunion with family or loved ones, seeking a higher standard of living, or for educational and employment opportunities for self or family). A vast number of people are first- or second-generation citizens of a country to which their parents or grandparents have immigrated. Individuals whose families have left their home countries and now reside in a different one are often bicultural or multicultural. They share values and practices of the culture of their parents and the culture of where they currently live.

Even though the term *culture* is used in this book and others to refer to a set of understandings that guide people's behavior, it is important to recognize that culture is not static. It is always changing. Some changes may be slow, such as the goals of gender or racial equality. Other changes may be fast as lightning, such as the adoption of smartphones or social responses to the COVID-19 pandemic. Nonetheless, every society adapts and evolves. No group is fossilized as if their culture were a museum exhibit.

Learning Culture

Since culture is learned, people are not born with instincts about what to do to be a fully functioning member of that culture. They don't yet know what and how to eat, how to behave appropriately, or what is right and wrong. Because members of a group

Figure 8.2
WOMEN'S WHEELCHAIR BASKETBALL CHAMPIONSHIPS. Athletes in sports subcultures learn specialized vocabulary that identifies them as knowledgeable players. In addition to knowing the rules of their game, they also learn the important unwritten rules about how to behave on and off the court, essentially the culture of the sport. These German and Japanese athletes are competing in the under 25 division of the 2011 International Wheelchair Basketball Federation World Championship in Ontario, Canada.
Credit: WBC/Kevin Bogetti-Smith/CC BY-ND 2.0

Figure 8.3
BOY AND DAD PLAYING WITH DOLLHOUSE.
This North American father is supporting his son's imagination and development through play with toys meant for all genders.
Credit: Josh Davis/CC BY-ND 2.0

share culture, the knowledge and understandings that make up that culture must be passed on from member to member. Culture is transmitted from one generation to the next, from parents and other adults to children, through the process of **enculturation**.

People who have the most contact with infants and young children act as the primary transmitters of culture. While these are usually the children's mothers and fathers, they may also be their grandparents or other close adults. As the children get older, they come into contact with many other people outside their families. Their peers also play a role in the enculturation process. As kids play together, they learn from each other, practicing cultural roles they will later step into, such as dad, mom, warrior, healer, or teacher.

RACE AND ETHNICITY

You learned in Chapter 4 of this book that anthropologists agree with biologists and geneticists that biological **race** does not exist for the human species. Since all humans, no matter where they live on the globe, share about 99 percent of their DNA, there are no meaningful divisions in their biology. Skin color has nothing to do with our personalities, behaviors, or capabilities. Even if two people share skin tone, they may speak different languages, have different family histories, and come from entirely different places.

Figure 8.4
YOUNG DOMINICAN BASEBALL PLAYER.
This young baseball player is taking a break from practicing in San Pedro de Macoris, Dominican Republic.
Credit: Adam Jones/CC BY-SA 2.0

"But if two people stand next to each other, you can see they're different!" Sure, but those differences that we notice on the outside are biologically superficial, representing less than 0.1 percent of our DNA. For racial divisions to be real, all the members of the "White" group would have to share at least 0.25 percent of their DNA *and* it would have to be different from those in the "Asian," "Latinx," or "Black" groups. Biologically, we are much more alike than we are different.

Many people see high numbers of people of certain ethnicities as successful athletes in particular fields and wonder if there is a genetic basis for their talent. Why do Kenyan runners seem to win every marathon? Why are Russian gymnasts so talented?

The truth is that there certainly may be some genetics involved in their potential – just like any parents may pass down traits to their children – but this is different from saying they are genetically "wired" to run fast or jump high. Talents are individual and are nurtured over a lifetime. Most genetic differences are also individual – not spread through groups – so individuals have different physical characteristics that may give them advantages over others.

To examine this question, let's look at the number of players in US Major League Baseball (MLB) from the Dominican Republic, a small island in the Caribbean. The population of the island is only 3 percent of the US population. However, Dominicans make up 12 percent of MLB players.

The national sport in the Dominican Republic is baseball. The first toy in many a young boy's crib is a tiny bat. He has every opportunity to play from a very young age, idolizing the great Dominican national heroes of the sport. Perhaps his uncle, father, or cousin plays for a team and provides encouragement, equipment, and coaching. Family and friends bring him into their networks. He may even be pressured to quit school so he can pursue baseball full time (Ruck, 1999). As he grows, he understands that excelling in this sport can bring him praise, the potential for fame, and an impressive income. It may allow him to leave the island and see the world. He may be able to build houses and revitalize his community, as Boston Red Sox pitcher Pedro Martínez did for his Dominican town of Manoguayabo.

For this young athlete, the cultural environment is supportive while he commits himself to working hard to reach his goals. Now imagine this same scenario for a young boxer in Cuba or a Brazilian soccer player. Talent is individual, but talents can be valued, honed, and nurtured by an entire society.

History of the Race Idea

So how did the term *race* become such a common part of our vocabulary? The sixteenth century was a time of exploration for our species. In a bold move, the pope decreed that all lands west of Brazil would belong to Spain. Europeans came to the Americas (the "New World") to discover they were not the only people with fully developed cities. However, since the Europeans could not understand the language or ways of the Indigenous peoples living in the Americas, they considered them to be primitive and savage. The conquest and colonization process was deadly for Indigenous peoples, who were enslaved, stripped of their rights and culture, and died in massive numbers from illnesses for which they had no immunity.

Some 200 years later, the field of science was developing. The Swedish naturalist Dr. Carl Linnaeus (who is also known by the Latinized first name *Carolus*) set to the task of assigning Latin names for all observable species in nature, as was explained in Chapter 3. In 1758, he added humans to the great taxonomy of living things, after analyzing the reports of those who had sailed across the seas.

However, looking at Linnaeus's "varieties" of humans, it's clear that they were based on biased observations and skewed positively toward Europeans. Besides the bizarre *Homo sapiens ferus* (four-footed, mute, hairy wild people) and *Homo sapiens monstruosus* (a category for people thought to be very tall, very short, or differently bodied), the other four varieties that can be identified more clearly are described in Table 8.1.

For the next several hundred years, these categories – based on reports from explorers and those involved in the slave trade – came to represent a pseudoscientific basis for different biological races of humans. It laid the foundation for centuries of slavery and oppression of non-White peoples based on "God-given" traits that

TABLE 8.1

Linnaeus's Four "Varieties" of Humans

The four types of humans that Linnaeus added to the 1758 version of *Systema Naturae* were based on tales from explorers but created the racial "color" classifications that are still in use today.

Homo sapiens americanus	Red, choleric [angry], upright. Hair black ... obstinate, content free ... ruled by habit.
Homo sapiens europaeus	White, sanguine [cheerful], muscular. Hair yellow, brown, flowing ... gentle, acute, inventive ... ruled by custom or law.
Homo sapiens asiaticus	Pale-yellow, melancholy, stiff. Hair black ... severe, haughty ... ruled by belief [opinions].
Homo sapiens afer	Black, phlegmatic [sluggish], relaxed. Hair black, fizzled ... crafty, indolent, negligent ... ruled by caprice [impulse].

were unalterable. Racial classification played a role in the horrific extermination of people during the ethnic cleansing of World War II under the guise of **eugenics**, a pseudoscientific plan to "purify" the human race. More recently, it has led to forced abortion and sterilization, laws against marriage, and anti-immigration policies.

Fortunately, Darwin's theories of natural selection and, later, an understanding of genetics shed light on the process of evolution by natural selection. Scientists began to apply those principles to the study of human variation. Later, in the twentieth and now twenty-first centuries, most of the pseudoscientific conversations have dropped out of public **discourse**. Now that we are able to compare the entire genomes of people around the world, we know that people's DNA differs by no more than 0.14 percent anywhere in the world. It is abundantly clear that all of this racial typecasting was done to maintain the social superiority of certain groups of people to the detriment of others.

Of course, humans have differences! But although we may refer to differences among groups of people as "racial," clearly those differences are not biologically important. The differences attributed to race are only meaningful as a set of categories based on social and cultural experience. That is to say, the experience of racism, **prejudice**, and **discrimination** is real. This is a problem of social injustice, with no direct connection to human biology.

Prejudice and discrimination can actually lead to poorer health and higher mortality rates. One way we can understand how racism affects health is to look at statistics that compare different populations in the same country. In 2011, the US Centers for Disease Control (CDC) reported major health disparities based on

BOX 8.2

BOX 8.2   Talking About: Sports Team Names and Mascots in the Time of Black Lives Matter

The use of Indigenous stereotypes in sports has never been free of controversy, but the past decade has brought a greater awareness of its potential harm. In particular, the **Black Lives Matter** movement in 2020 brought discriminatory practices in all areas of society into sharp focus. While some sports teams had begun the process of exploring changes to their team names or mascots before then, others found themselves under sudden economic pressure from investors and sponsors to do so.

Why focus on sports teams? Social science organizations agree that Indigenous-related team names and mascots can perpetuate racist stereotypes of Indigenous peoples. These names reduce the ethnic identities of people with diverse histories of colonization and resistance to disparaging and inaccurate caricatures. In a 2014 study, researchers from American University Washington College of Law found that because American Indian and Alaskan Native (AI/AN) students attend schools with derogatory and demeaning team names and mascots,

> these team names and mascots can establish an unwelcome and hostile learning environment for AI/AN students. [The research] also reveals that the presence of AI/AN mascots directly results in lower self-esteem and mental health for AI/AN adolescents and young adults. And just as importantly, studies show that these mascots undermine the educational experience of all students, particularly those with little or no contact with indigenous and AI/AN people. (Stegman and Phillips, 2014, p. 1)

Although sports teams often claim that their mascots are not demeaning to Indigenous peoples, this and other research demonstrates that Indigenous youth are negatively affected. In addition, stereotypical representations contribute to the development of cultural prejudices across all sectors of society.

It's important to note that these names are demeaning whether or not that is the intention of the users. Fans often argue that because they love their teams, they also honor the symbols that represent them. Yet, it's hard to see how red-faced "Chief Wahoo," named the most demeaning caricature of Native American peoples in US sports (Tracy, 2013), honors the Indigenous peoples of the Ohio area where the Cleveland Indians play.

"race" and "ethnicity" (the CDC uses these terms in accordance with the US Census terminology). These included higher rates for Black and Hispanic people in preterm births, infant mortality, coronary heart disease, strokes, and obesity. Mortality rates are higher overall for non-White populations in the United States.

The devastating effects of the **COVID-19** pandemic revealed major disparities in death rates across "racial" or ethnic communities. At the time of this writing, Black, Indigenous, and Latinx patients in the US were "at least 2.7 times more likely

Finally, after decades of pressure, Chief Wahoo was formally retired by the Major League Baseball team in 2018.

The Washington Redskins of the National Football League (NFL) long held the title for the most controversial sports team name. All major English dictionaries define "Redskin" as an offensive term. After decades of protest from Indigenous organizations and a formal lawsuit in 2014, the Washington team lost the copyright to the Redskins name. Then, after the death of George Floyd and a surge of Black Lives Matter protests in 2020, major corporate sponsors including Nike and FedEx began to exert severe economic pressure on the team. By the end of July 2020, the team had announced that they would use the Washington Football Team as their temporary name, and in 2022 they adopted the name the Washington Commanders.

Although culture often changes very slowly, sometimes a moment stands out in time as a driving force of rapid change. The Black Lives Matter protests shone a light on discriminatory practices in all aspects of social life, including policing, schooling, the corporate world,

legal practices, and health care. It extended beyond examining the racist policies and practices around the lives of Black people to any community that has suffered the effects of institutional racism or discrimination. It extended to our national pastimes, where the voices of Indigenous peoples were heard more clearly than ever before.

Figure 8.5 **MAN HOLDING SIGN AT WASHINGTON REDSKINS PROTEST.**
In November 2014, marchers convened at TCF Bank Stadium in Minneapolis to protest the name of the visiting Washington football team.
Credit: Fibonacci Blue/CC BY-SA 2.0

to have died of COVID-19 than White Americans," and Pacific Islanders were 2.3 times more likely to have died (Egbert and Liao, 2020). Indigenous communities on reservations in the US have been particularly devastated by the virus, with many contributing factors such as lack of running water, underlying health conditions, and under-equipped health services.

In Canada, Indigenous peoples (First Nations, Inuit, and Métis) also suffer health disparities compared to the non-Indigenous population. A review by the

National Collaborating Centre for Aboriginal Health (2012) provides evidence that due to educational, geographic, and economic barriers, the universal health care system that works well for most of the Canadian population does not reach Indigenous populations with the same success. Indigenous people suffer from poorer "maternal, fetal and infant health; child health; certain communicable and noncommunicable diseases; mental health and wellness; violence, abuse and injury; and environmental health" (p. 4). Although infection and death rates from COVID-19 have been lower overall in Canada compared to the US, communities with more than 25 percent ethno-cultural (non-White) minorities have had the highest death rates.

Today, First Nations, Inuit, and Métis peoples suffer the kinds of social, economic, and health disparities felt most by African Americans, Latinx, and Native Americans in the United States. These groups all share lower rates of employment, higher rates of mental and physical illness, and rates of incarceration that are hugely disproportionate to their numbers in the population. All these results show evidence for continued systemic racism in both Canada and the US. There are many ways in which biological health is tied to our history, economics, and social and cultural environment. Where discrimination exists, it affects all areas of people's wellness and, ultimately, survival.

Ethnicity, like race, is a complex subject, but it may be closer to what we mean when we mistakenly use the term "race." It, too, does not have a biological basis. Ethnic origin most often has to do with a person's ties to their culture, language, and shared history. Therefore, ethnicity can have some connections to one's biological heritage, but not necessarily and not always.

To help define ethnicity, anthropologists offer these points to consider as guideposts:

- Ethnicity is a way of classifying people based on common histories, cultural patterns, social ties, language use, and symbolic shared identities (Fuentes, 2012).
- Ethnicity is created by historical processes that incorporate distinct groups into a single political structure under conditions of inequality (Schultz & Lavenda, 2009).
- Ethnicity is fluid and flexible, and it can be utilized to pursue particular goals, either of the group itself or of a dominant group.
- Ethnicity is not a unit of biology or genetics.

People identify with their ethnic heritage as a member of a particular group of people. However, society tends to group people into larger, often meaningless, groupings. The US Census, for example, counts people of very different origins as members of the same category even when their cultural practices, vocabularies, and national or regional histories are extremely different. For instance, far too many populations are lumped together under the umbrella term "Asian" for it to be useful as a category. Asian peoples include people from India, Sri Lanka, Nepal, Vietnam, Cambodia, Thailand, Japan, and China, not to mention all of the native populations within those countries. But by the terms of the US Census, millions of people with different histories, languages, customs, and belief systems are arbitrarily and artificially placed in the same category.

Nonetheless, because ethnicity is the core of a meaningful group identity – including religious beliefs and shared history – it can be a galvanizing force for individual **agency** and collective action. Ethnicity is also used as a path to political and cultural **resistance**. This occurs on all levels, from small, local protests to organized global movements, both peaceful and violent.

History has seen racial and ethnic-based movements for civil rights and human rights around the world. In the 1960s, the civil rights movement drew attention to the great inequities suffered by African Americans in the United States. Today, the Black Lives Matter movement again focuses on the injustices that continue to be suffered by African American individuals and communities.

Figure 8.6
ROASTED GUINEA PIG.
Food is one way that our ethnocentrisms are expressed. This Peruvian meal features cuy, or guinea pig, which is often raised in rural homes like we might raise chickens. However, most North Americans could never imagine eating guinea pig. What is the North American rule about the difference between "food" and "not food" that this ethnocentrism brings to light?
Credit: Barry D. Kass/Images of Anthropology

ETHNOCENTRISM AND CULTURAL RELATIVISM

Consider a group of people with a very different set of beliefs and behaviors from mainstream North American ones, such as the Efe people of the Ituri Forest in the Democratic Republic of the Congo. They regularly eat grubs as a source of protein. If you just thought "Eww, gross," you are probably not alone. It is a normal reaction to feel that the way *we* do things is normal and the way *others* do things is not. This idea – that our own customs are normal while other customs are strange, wrong, or even disgusting – is the notion of **ethnocentrism**. Ethnocentrism allows people to feel superior to others by denigrating differences in their behaviors, ideas, or values.

A degree of ethnocentrism is instilled in children at a young age. Members of a group are taught to love their country, identify with their city and state, and support their community. Pride in people and origins isn't a bad thing. It becomes a problem when ethnocentric ideas about the value of other people's beliefs and behaviors turn into hateful words or misguided actions. Thinking ethnocentrically does not allow people to fully understand other cultures because it blinds them to the intrinsic value in every way of life.

When undertaking research, anthropologists reject an ethnocentric mindset in order to understand people in the most objective way possible. Even if one does not agree with certain behaviors or values personally, it is the anthropologist's responsibility to observe, describe, and interpret those behaviors and values objectively. Anthropologists have a particularly important duty to keep ethnocentrism in check when studying other cultures, no matter how foreign those cultures are to our own.

In contrast to ethnocentrism, anthropologists use a model called **cultural relativism**. This is the idea that all cultures are equally valid, and that a culture can only be understood and interpreted in its own context. Because culture is integrated, as anthropologists we can understand any one aspect of culture only if we understand the whole. This holistic perspective allows anthropologists to study people's beliefs and behaviors without judgment.

CULTURAL ADAPTATION AND MALADAPTATION

One important quality of humans is that we are adaptable. In fact, our ability to adapt to changing circumstances is likely the reason that *Homo sapiens* are still here today. Although humans are still evolving biologically, it is the cultural adaptations that have not only allowed us to survive but also to thrive as the dominant species on earth. We have been able to expand across the globe into every environment possible: the desert, the tropical rainforest, or the snowy tundra. We've even been able to live in submarines under water and on the International Space Station. Our early *Homo* ancestors could never have dreamed as much.

Biological adaptations allow an organism to better survive in its present conditions, or to live successfully and reproduce in a variety of habitats. A good example is the hummingbird's long, thin beak and its wings that beat so rapidly it can hover. These physical adaptations allow the hummingbird to extract nectar from deep within a flower, finding nutrition where other birds cannot. Other easily recognizable examples of evolutionary fitness through adaptation include the buoyant, hollow fur of the otter, which allows it to float, and the long neck of the giraffe, which allows it to eat acacia leaves off the tops of trees. Humans, too, have some very useful biological adaptations (including bipedalism and a vocal tract that allows speech), but we are unique in developing advanced cultural adaptations.

Cultural adaptations include all the ways that humans use cultural knowledge to better adapt and succeed in their surroundings. Because humans have language, we can pass on knowledge orally and/or in writing. We use language to record, test, and develop knowledge. Rather than having to reinvent the wheel every generation, we compile and share our knowledge. The development of science and technology, including medicine, are products of culture. We can start a fire, use a blanket, buy a parka, or turn on the heat when we are cold, rather than having to evolve a fur coat.

There are cultural practices, however, that cause harm to certain members of a society. Any behavior that leads to a decrease in well-being or survival of the members of a culture is not adaptive, or **maladaptive**. Some cultural practices can favor certain members of society while harming others, as in the health disparities for communities of color discussed earlier in this chapter. The practice of female genital mutilation (FGM), which harms women's reproductive health, is maladaptive for women, even though the practice has existed in certain societies for as long as anyone can remember (see Box 8.3).

Today, some maladaptive practices are attractive to young people, leading to long-term health issues. For instance, the American Academy of Dermatology (n.d.) states that people under the age of 35 who use artificial tanning beds increase their risk of malignant melanoma by 59 percent. Melanoma kills over 10,000 people in North America each year. Not all culture changes are beneficial.

THE FUNCTIONS OF CULTURE

Culture has certain functions beyond providing the shared understandings that guide people's behavior. In any society, culture should provide for the basic needs of the group. Specifically, aspects of culture (beliefs and behaviors) should support the health and well-being of members and the survival of the culture itself.

BOX 8.3   Female Genital Mutilation (FGM)

Female genital mutilation (FGM) – also called female genital cutting or female circumcision – is widely practiced as part of a young girl's entry into the community and preparation for marriage. The practice involves the surgical removal of the clitoris, labia majora, and/or labia minora for nonmedical reasons. In more extreme cases, it involves "infibulation," the sewing together or cauterization of the labia minora, leaving only a small opening for urination.

Currently, FGM is common in 24 countries in Africa and the Middle East. More than 90 percent of women aged 15 to 49 have undergone the practice in three countries: Somalia, Guinea, and Djibouti (UNICEF, 2021). And migrants often continue the practice even after they have left their natal countries.

Although women willingly subject their daughters to this traditional practice, it is not an adaptive one. Some advocates for cultural freedom argue that undergoing the practice is a form of cultural identity and social belonging. If a girl is not "cut," then others in the social group may view her as "dirty," "rejecting tradition," or "unfit for marriage." Nonetheless, there are many reasons why FGM is seen as a maladaptive practice by the world's health and human rights organizations.

Primarily, the practice is maladaptive because it can lead to severe health problems. Immediate risks include hemorrhaging, bacterial infection, shock, and death. Long-term complications can lead to recurrent bladder or urinary tract infections, cysts, infertility, childbirth complications, and newborn deaths (WHO, 2020). Another argument against FGM is that it represents a severe form of gender discrimination against women and girls. It violates the fundamental rights of human beings to be free from cruel and degrading treatment. FGM has been denounced by UNICEF (2013) and international scholars as a dangerous and traumatic means of controlling women.

Since anthropologists take the perspective of cultural relativism, they avoid judging cultures based on their own set of values. Doing this would be ethnocentric and misguided. However, people who study culture can examine in an objective way whether aspects of a culture are adaptive or maladaptive. That is, if aspects of culture are adaptive, they should support the health and well-being of members. If maladaptive, they may lead to ill effects for the people or the longevity of the culture itself. To assess the adaptiveness of a culture, an anthropologist might examine the kinds of issues listed in Table 8.2.

If the responses to these questions are positive, then the culture is largely adaptive. If they are negative – that is, if people are not getting the things they need for their health and well-being – then cultural practices may be maladaptive. More likely, cultural practices may be adaptive for some – those who control wealth and power – and maladaptive for others.

TABLE 8.2

Assessing the Adaptiveness of Culture

Health	How is the physical and mental health of members? Do women get prenatal care to support infant health?
Demographics	What do birth and mortality rates say about the longevity of members?
Goods and Services	Can people get what they need when they need it? Is there access to clean, safe food and water?
Order	Do people feel safe? Are there systems in place for effectively dealing with violence?
Enculturation	How well does the culture get passed down to the next generation?

Culture changes that accompany modernization can lead to maladaptive results. For example, members of the Penobscot Nation of Maine have traditionally relied on fish from the Penobscot River as the main staple of their diet. In the twentieth century, non-Indigenous inhabitants of Maine built industrial paper mills along the river. As a result, the runoff from the paper mills polluted the river with dioxins, creating a toxic environment for fish and the humans who eat that fish (McKeon, n.d.). Dioxin is a carcinogen that poses severe hazards to the human reproductive and immune systems. Suffering from these effects, the tribe successfully lobbied the US Environmental Protection Agency to better monitor and filter the waste from the mills, which has substantially increased the health of the river. This case shows how an adaptive solution resulted from the fallout of a maladaptive practice for this community.

Raising Children

The values regarding how children are raised in a society are important to how culture and personality develop. Anthropologists who have studied child-rearing around the world have found two general patterns of enculturation: dependence training and independence training. Each type of child-rearing contributes to a different set of cultural values and different types of social structure. This is one way in which the integrated nature of culture can be seen clearly.

Dependence training is the set of child-rearing practices that supports the family unit over the individual. In societies with dependence training, children learn the importance of compliance to the family group. Typically, dependence training is taught in societies that value extended (or joint) families – that is, families in which multiple generations live together with the spouses and children of adult siblings.

Family members may work together in a family business, or they may pool resources. In horticultural and agricultural communities, this may mean that all members of the family are expected to work on the farm. Children may be indulged when they are young but learn quickly that they are part of a unit and must choose the family's needs over their own. Sometimes these cultural values are referred to as *collectivist* or *communal*.

Independence training refers to the set of child-rearing practices that foster a child's sense of individuality. It is typically found in industrial societies, like our own, and in societies in which earning an income requires moving to where the jobs are. The family unit in independence training societies is typically a nuclear family – that is, a family in which only two generations (parents and children) live together.

The individual is seen as an actor who has the right to shape their own destiny. Emphasis is placed on developing the skills and self-worth of each child so they can be competitive and successful in life. The sense of self is strongly linked to the individual over the group, which can create individual feelings of entitlement. These methods of child-rearing do not necessarily produce more independent children, as the name may suggest. Sometimes these cultural values are referred to as *individualistic*.

Anthropologist Dr. Susan Seymour (1999) studied changing family life in the state of Orissa, India, focusing on the roles of women in childcare. At the time of her fieldwork, some residents of the Old Town (a traditional village) of Bhubaneswar had resettled in the New Capital (a more Westernized part of town with secular schools and administrative careers). The division of the community into two separate socio-cultural environments had direct consequences on family life. While residents of the Old Town held fast to traditional dependence-training methods and values, residents of the New Capital adjusted to new opportunities, especially for women's advancement. Women's educational and employment opportunities resulted in a shift to more nuclear, rather than joint, families, and independence training. This trend is seen in many societies as modernization occurs.

A current example of the difference between these two modes of child-rearing can be seen in the acceptance of mask mandates as a response to the COVID-19 pandemic. Countries in East Asia (e.g., South Korea, Hong Kong, Vietnam, Thailand, Taiwan) tend toward collectivism and dependence training. Countries of the West (e.g., the United States, the United Kingdom, Italy, France) tend toward individualism and independence training. Maaravi et al. (2021) found several key correlations between COVID-19 deaths, adherence to mask mandates, and whether a culture has more collectivistic or individualistic values. Importantly, they found

Figure 8.7
GIRLS IN JUDO CLASS.
Young children in independence-oriented cultures are encouraged to pursue their interests, like the martial art judo, and develop their talents.
Credit: Stefan Schmitz/CC BY-ND 2.0

that "the more individualistic a society is, the more it suffers from COVID-19 related cases and deaths" (p. 3), and that "collectivistic orientation is associated with willingness to sacrifice for the common good by promoting the protection of one's environment from being infected via communal COVID related attitudes and intent of adherence to COVID health guidelines" (p. 4). Cultural factors guide people's decisions and should be considered for a full understanding of any global event like the COVID-19 pandemic.

FIELDWORK METHODS AND ETHICS

Cultural anthropologists study culture "in the field." That is, they live with another group of people for an extended period to learn firsthand how the group views the world and behaves within it. They immerse themselves in the culture and daily patterns of life such that they begin to understand how members think, feel, and act.

Ethnography isn't just description, however. What makes it social science is the ability of the ethnographer to situate people's ideas and actions within a larger context of practices and power relations. What are the larger systems that shape and limit these behaviors?

The process begins with a research question: Why do people do that? The anthropologist seeks funding to support the months or years of necessary field study and spends time preparing for their entry by reading all of the available material on the topic and area. They may learn the language or work with a translator. Then they begin their fieldwork by immersing themselves in the culture of the people they wish to understand.

Being in the field allows an anthropologist to produce an ethnography. Ethnography is both the process and the product, which is most often a document, book, or film. Ethnography is an artistic endeavor, because it must be written in a way that evokes the reality of the culture. It is also a scientific endeavor, because it must produce an authentic, rigorously researched representation of people and their behavior in a wider cultural context.

Participant Observation

In the field, an ethnographer uses a variety of methods to understand another group's way of life. The main method is called **participant observation**, a process in which a researcher lives with a people and observes their regular activities, often for a year or more. The ethnographer participates in daily life while at the same time reflecting and analyzing their observations.

Anthropologists believe there is no substitute for witnessing firsthand how people think and what they do. This is why they look forward to submerging themselves in a new environment, with all the messiness of life, and trying to make sense of it. Sometimes it feels as if learning a new set of cultural norms is like deciphering a code. The ethnographer's role is to observe, describe, interpret, and analyze behavior in order for this code to make sense. This is why anthropology relies more heavily on fieldwork than on surveys, which can provide a bit of the story, but maybe only the bit that people want to share.

While doing participant observation, an ethnographer seeks to understand a full picture of the culture. One can approach this goal by asking different kinds of questions. Questions may be one of three types:

- How do people think they should behave? (What are the norms and values in the society?)
- How do people say they behave? (Do they say they conform to these standards or not?)
- How do people actually behave? (This can only be discovered by long-term fieldwork and establishing trusting relationships with the people involved in the study.)

We can think of the difference between what people say and what they do as the contrast between **ideal behavior** and **real behavior**. Ideal behavior is the way people think they should behave, while real behavior is the way they actually behave. The differences between the two are interesting to anthropologists, as they show the contrasts between the values of society and the actual behavior of members. These kinds of questions elicit both cognitive and behavioral data.

Choosing Informants

For participant observation to result in desired research goals, the ethnographer needs to spend a lot of time talking to members of the community. These important individuals in the field study may be called **informants**, associates, or interlocutors, depending on the anthropologist's choice of terms.

Depending on the circumstances and the study goals, the ethnographer may choose one or more methods of approaching informants. In a **random sample**, the ethnographer's goal is to allow everyone an equal chance to be interviewed, which is done by selecting people randomly. This might best be employed in a small, homogeneous community or when an average is desired. A **judgment sample**, on the other hand, selects informants based on their skills, knowledge, insight, and/or sensitivity to cultural issues. Finally, a **snowball sample**, in which one informant introduces the ethnographer to other informants, can be very helpful.

The fieldworker will usually develop close ties to one or more informants who are chosen for their special insights and will spend a lot of time with them. These crucial contributors to the research are referred to as **key informants**, or **key associates**. They are often people with deep knowledge about the ethnographer's research topic. They can very often become close friends, with whom the fieldworker continues to correspond and collaborate beyond the field study.

Under the umbrella of participant observation, many different methods may be used. The fieldworker must be flexible and reflective enough to assess which techniques might work best, whether a technique is working to help answer the research question, and if the approach must be modified. These methods include, but are not limited to

- formal interviews (in which the same set of questions are posed to multiple informants)
- informal interviews (in which the fieldworker seizes an opportunity to ask questions)
- life histories or other oral histories
- case studies, in which a particular event is examined from multiple perspectives
- kinship data (a family tree or genealogy)
- photography

These methods each lend themselves to a certain type of data gathering. Depending on the circumstances and the study goals, one or more techniques may be used at the ethnographer's discretion.

Code of Ethics

Some people imagine that doing fieldwork among people is like being "a fly on the wall." In other words, the ethnographer would hang around unobtrusively, watching people go about their daily business while writing notes on a pad. In fact, the situation is generally the opposite, with the ethnographer getting into the mix of daily life and building relationships with people.

Ethnographers learn by doing (i.e., becoming participants in) what they want to learn about. For instance, cultural anthropologist Dr. Heather Paxson from the Massachusetts Institute of Technology worked alongside farmstead cheese producers in the United States in order to experience various cheese-making processes. Her ethnography, *The Life of Cheese: Crafting Food and Value in America* (2012), explores how crafting artisanal cheese adds economic and social value to the lives of the producers.

The ethnographer's presence during fieldwork is keenly felt, especially in the beginning, and may be distracting. The informants among whom one is working may not trust that this is an academic study – the anthropologist could just as easily be a government agent come to spy on them. In fact, anthropologist Dr. Napoleon Chagnon recalls that due to their suspicion of his motives, his Yanömami informants in the rainforests of Brazil and Venezuela told him lies about their relationships with family members. In the film *A Man Called Bee*, he admits to throwing out nearly all of the data gathered on kinship in the first year of his fieldwork (Asch & Chagnon, 1974). This suspicion is not without merit, since we know now that some early fieldworkers did share research with governments who may have used that information to target communities of people. Today, the discipline regrets being associated with politically motivated studies in which collaboration and respect for autonomy were not primary goals.

Once the hurdle of trust is overcome, other interpersonal problems may arise. For instance, intercultural communication is not always perfect, even if the anthropologist knows the language. The anthropologist's intentions may not always be clear, and they may read others' intentions wrong as well. There are plenty of possibilities for errors in judgment and poor decision-making in the process of fieldwork. Therefore, it is crucially important to have a set of guidelines that lay the foundation for interactions with others while in the field.

The largest North American organization of anthropologists, the American Anthropological Association (AAA), created such a set of guidelines called the

Figure 8.8
HUMAN TERRAIN TEAM MEETING WITH LOCAL LEADERS, AFGHANISTAN.
A US Army Human Terrain Team of soldiers and civilians meets with the local tribal elder and villagers in Koshab Village, near Kandahar Air Base, Afghanistan, in 2011.
Credit: US Army Europe, photo by Staff Sgt. Stephen Schester

Code of Ethics (2012). The Code of Ethics states that anthropologists must weigh the possible impacts of their actions on the dignity, health, and material well-being of those among whom they work. They must strive to "do no harm" as a result of their research. This seems clear enough on the page. Nevertheless, situations can arise when it isn't clear what the repercussions of a decision might be. It isn't always easy to navigate potential challenges in the field. This can lead to some controversies over whether anthropologists should be involved in certain endeavors at all.

This concern led to the AAA publicly opposing a US military program called the **Human Terrain System (HTS),** which embedded teams of academically trained social scientists with US Army brigades starting in 2007. Proponents of the project argued that providing a team of people with regional, linguistic, geographic, and anthropological knowledge would greatly reduce misunderstandings and misguided actions in the military theater (and have done so), thereby decreasing casualties. However, the AAA argued that the use of anthropologists in a war zone directly violated the AAA Code of Ethics. Embedded anthropologists could not ensure that their information would "do no harm" to the subjects of study, since those subjects were also often being targeted by the military. Due to these and other issues, the HTS program was discontinued in 2014.

APPLIED ANTHROPOLOGY

Finally, it is important to note that anthropologists are not only researchers and teachers. Many working anthropologists – cultural, biological, and archaeological – apply their knowledge of anthropological methods, theory, and perspectives to solve human problems. This field is called **applied anthropology**. Applied anthropologists work to find solutions for problems in the real world, rather than focusing solely on contributing to the body of research in the discipline. Some applied anthropologists also teach in universities, and some work outside academia entirely.

Applied, or practicing, anthropologists may work in corporate settings, for governments, or for non-governmental organizations (NGOs). They may work in any anthropological field, as consultants who are trained in participant–observation techniques, to seek solutions to problems. Often these anthropologists consult with organizations that are developing sustainable practices in countries in the process of modernization. They work toward solutions for drought or famine, such as helping provide clean water, or in medical clinics to bridge the cultural gap between Western and traditional medicine. Applied anthropologists help with surmounting the obstacles that come with modernization while remaining sensitive to a people's traditional values and identity.

Practicing anthropologists may work for a corporation interested in streamlining its organization. For instance, Dr. Elizabeth Briody, an applied anthropologist, worked for the automotive manufacturer General Motors for more than 20 years in the research and development sector. Briody explains what her job entails: "I conduct studies of GM culture. My role is to come up with ways to improve GM's effectiveness. In my research, I try to understand the issues that people face in doing the work they have been asked to do, and then offer suggestions to make their work lives better" (Fiske, 2007, p. 44). One project required her to examine the different corporate cultural norms and values among salaried and hourly employees, GM management, and United Auto Workers union leadership. Briody's work allowed collaborations to go more smoothly, with each group better understanding the others' assumptions about how to create an ideal work environment.

Applied anthropologists use a model of field research and implementation that is referred to as **participatory action research**. Because the goals of applied anthropology are to effect change in a community, the research prioritizes the needs and concerns of the people who desire it. As outsiders, the anthropologists can often bring to the table their understandings about relevant global systems. This provides a larger framework within which to understand a problem and seek potential solutions. Most importantly, applied anthropologists partner with community members throughout the process to ensure that the solutions are collaborative.

SUMMARY

This chapter has explored cultural anthropology, which is the study of culture, or the shared understandings that people use to guide their behavior. Mirroring the Learning Objectives stated in the chapter opening, the key points can be summed up as follows:

- Culture is shared, learned, integrated, and based on symbols; therefore, it is not instinctive or biologically based. In fact, even though people are different all over the world, biologically there is not enough difference for the human species to be separated into races. Racism, however, is very real and describes people's lived experiences.
- Rather than "race," anthropologists prefer to use the term "ethnicity," which encompasses all aspects of a person's culture.
- When anthropologists study culture, they take a culturally relative perspective, avoiding the biases of ethnocentrism and attempting to learn about people in an objective way.
- Cultural anthropologists may evaluate cultural practices to determine whether they are adaptive or maladaptive for the long-term health and well-being of the members of that society.
- Anthropologists may look at the functions of culture: how culture provides for its members such that people who share that culture get their basic needs met.
- Different ideas and practices of how to raise children result in different types of family values and are linked to larger social and cultural practices.
- When ethnographers go into the field to study any aspect of culture, there are a variety of methods that can be applied. All must be done with a commitment to ethical practices.
- Applied anthropologists use the skills and methods of anthropological research to collaborate with communities to solve real-world problems.

Review Questions

1. What are the components of culture?

2. What are the differences between an ethnocentric and a culturally relative approach to culture?

3. What are the criteria for adaptive aspects of culture?

4. What is a maladaptive cultural practice?

5. How do different child-rearing practices affect the development of personality and culture?

6. How should anthropologists in the field (whether face-to-face or virtually) ensure they are acting ethically?

7. How can anthropological understandings and perspectives help solve real-world problems?

Discussion Questions

1. How does the opening anecdote about the author's insect-eating dilemma illustrate the lens of anthropology?

2. What aspects of your own culture are maladaptive? Use the criteria in Table 8.2 to make your assessment.

3. What do you think about the argument that collectivist (i.e., dependence training) cultures are more likely to follow a mask mandate resulting from the COVID-19 pandemic?

4. If you received funding for a year of fieldwork, what community of people would you choose to study and what would be the focus of your research?

5. How might an applied anthropologist help solve a particular problem in your community?

Visit **www.lensofanthropology.com** for the following additional resources:

SELF-STUDY QUESTIONS **WEBLINKS** **FURTHER READING**

9

LANGUAGE AND CULTURE

LEARNING OBJECTIVES

In this chapter, students will learn:

- *that language is based on symbols.*
- *the differences between human language and primate communication.*
- *different hypotheses for the origin of human language.*
- *the steps a linguistic anthropologist would take to understand the components of a language.*
- *the components that make meaning beyond just words.*
- *the types of language that an ethnolinguist would study.*
- *how language is changing in the digital age.*
- *how languages have been suppressed and lost.*

> **Hello there! Yo, what's up? Have you eaten rice yet?
> Language and culture are deeply connected.**
> #LanguageAndCulture

INTRODUCTION

This chapter explores language, one of the most essential aspects of human culture. While human language has much in common with the communication of nonhuman primates, there are also many unique qualities that allow humans to think and interact in complex ways. There is an intimate connection between language and culture, seen in the ways we modify language use in different social, cultural, or political contexts. In addition, language allows humans to pass down oral and written knowledge, something no other species on the planet can do. Passing down cumulative knowledge allows the development of advanced science and technology. In essence, language has allowed humans to become who we are today.

Definition of Language

Human culture is heavily reliant on systems of communication that allow people to interact with one another in socially meaningful ways through voice, gestures, and written words. **Language** is a symbolic system because, through words, we refer to things that are not physically in front of us. We ponder ideas and concepts. We talk about things that have happened in the past or may happen in the future. Language is an infinitely creative world.

There is a huge amount of information that any person must process and produce to function fully in human society. This includes all the components of language that accompany our conversations, such as our tone of voice or hand movements, which also express meaning. Whether by sounds, gestures, or writing, language allows us to live with others in a cooperative and communicative environment.

A person's **speech** is influenced by multiple factors: biological, cultural, social, and political. Language is biological in that we use our mouths and throats to produce sound. Socio-cultural factors such as gender, socio-economic status, level of education, and geographic region influence the way we speak and provide additional information to listeners, who use all available information to pick up meaning. Language is also political because it is bound up in relationships in which power is constantly negotiated.

Because so much of what we say and how we say it is based on our cultural environment, it is clear that culture deeply influences language. In other words, who we are and where we come from shapes our speech and the way we interact with others. Even the way we greet one another carries meaning about history and values, as explored in Box 9.1.

The inverse is also true in that language shapes culture by reflecting the changes in society. Consider the rapid changes made by digital social media in the realm of

Figure 9.1
UPSET DRIVER USING HAND GESTURES.
This driver uses a hand gesture to express his annoyance with traffic.
Credit: State Farm/CC BY-SA 2.0

language alone. Texting shorthand, the use of hashtags, and words such as "selfie" or "unfriend" quickly entered an entire generation's daily conversations.

LANGUAGE AND COMMUNICATION: SIGNS AND SYMBOLS

Communication is based on **signs** – that is, something that stands for something else. In spoken language, there are two basic types of signs used in communication: index signs and symbols. Animals in the wild mainly communicate using **index signs**, that is, emotional expressions that carry meaning directly related to the response. For example, when a pygmy marmoset feels fear, it emits a high-pitched scream. A chimpanzee that has hunted and caught a monkey to eat will hoot loudly with excitement.

A **symbol** also stands for something else, like an index sign, but has no apparent or natural connection to the meaning. For instance, when we see a peace sign, we understand it to symbolize the concept of peace. The shape of the image does not reflect "peace" in a natural way, since peace is an abstract concept.

Language itself is symbolic. When we use sounds to speak, the sounds stand for the meaning of the words. When I say the word *peace*, the sound itself (/pees/) does not inherently capture the idea of peace, but rather refers to it. It is also true with gestures in signed languages and lines written on a page.

BOX 9.1 **Food Matters: Greetings! Have You Eaten?**

When a Thai friend stops by, they may greet you by asking *Gin khao reu yung?* In other words, "Have you eaten rice yet?" The speaker isn't really asking whether you've had lunch, but the phrase is used to mean "Hello" or "How are you?" In English, we similarly use phrases to greet one another that have other literal meanings, such as "What's up?" A non-native speaker unfamiliar with the informal greeting might wonder if something is actually "up." Even our mainstay greeting "How are you?" isn't supposed to elicit a list of maladies. It is just another way to say "Hi."

Several languages use a similar inquiry when greeting others. For instance, in South India, Malayalam speakers ask *Chorrunto?* ("Have you eaten rice?"). In Mandarin, *Chi le ma?* translates literally to "Have you eaten?" It's the same in Nepali with *Khana khannu bhaiyo?*

Why would so many languages ask whether someone has eaten as a greeting? The history of how greetings develop is different for every language and cultural context. For some, it's because asking whether someone has eaten and asking after someone's well-being is essentially the same. For other languages, the practice may have developed during times of food scarcity, when neighbors would inquire about how others were getting along by asking if they had a supply of a staple food. In every culture, there are norms of hospitality that make the exchange of food and drink customary when visiting.

The rituals of hospitality may be the biggest influence in the development of this greeting. Offering food and drink and a place to sit are very common, as are the culturally appropriate ways to accept them. In Western countries, accepting a glass of water when one is offered is seen as appropriate. In other countries, the norms of behavior require a respectful interchange of polite denial and insistence. For instance, in Iran, the custom called *taarof* requires that a guest accept the first serving of food or drink but refuse a second helping several times. The number of times and the level of insistence depend on the relationship between the host and guest.

Food is central to welcoming visitors into one's home. A combination of cultural context and history, along with norms of hospitality, has brought the question "Have you eaten?" into many languages as a way to greet others.

Humans have expanded the system of symbolic communication in ways that nonhuman primates cannot. Nonetheless, interesting research over the last several decades appears to show that the human and nonhuman uses of symbolic language are of degree rather than completely distinct. Indeed, there are examples of individual nonhuman primates, especially apes, that show each of these basic skills. However, we can say that humans have been able to fully develop these particular language features to an extent that surpasses any other primate species.

1) Humans use symbols freely. The first way in which human language distinguishes itself from animal communication is that we have the ability to talk about something in a symbolic way, or about a time besides the present. Planning is one of the brain's executive functions, along with abstract thought.

2) Humans use words to deceive. Humans possess the ability to say something that isn't true on purpose. We can even build a complex lie with many aspects to it.

3) Human language is infinitely creative. People can talk about things today they have never talked about before. Human language can create new phrases and sentences in nearly infinite forms. One modern example is the use of the prefix "e-" or "i-" (such as eBay or iPhone) to identify something digital or computer-related.

LANGUAGE ORIGINS

How did language begin? Evidence suggests that human language likely began as a system of gestures. Many primatologists see the origins of language early in our primate lineage, in primates who did not have the ability for verbal speech, but who used gestures and calls to express meaning to one another.

Walking upright would have allowed hominins to communicate silently through gestures while walking, even over long distances. Thus, gestures would have conferred a selective advantage on our bipedal ancestors. The advantage would be even greater if specific sounds could be made, so that the two individuals wouldn't need to be in visual contact to pass information between them.

One thing that most researchers can agree upon is the level of trust necessary for the shift from gestures to words. Why trust? Nonhuman primate communication is largely based on signals that are hard to fake, such as facial expressions of anxiety or a cry of fear. In contrast, words are symbolic; they represent something in a nonphysical and arbitrary way. Trust must be present for communication based on words rather than signals, since words may relate to something that is not immediately present. If gestures and sounds were made simultaneously, it is possible that the sound alone eventually became a trustworthy marker of meaning.

Nonhuman Primate Language

Many people want to know why, if apes are related to humans, apes can't speak. One part of the answer has to do with the brain. Nonhuman primates share similar – though not as developed – structures in the brain to humans, although a primate brain lacks the strong neural connections that the human brain has that link the areas to one another.

The second part of the answer has to do with mouth and throat anatomy. Nonhuman primates' mouths and throats lack the intricate musculature that humans have. Sometime before 50,000 years ago, the human tongue descended, the mouth got smaller, the larynx dropped, and the neck elongated. These changes allowed humans to develop an incredible amount of control over their breath and their ability to produce sound.

The changes leading to human vocal physiology had an evolutionary advantage as well. Controlling sounds meant that an individual could be better understood in social situations, leading to a higher level of cooperation and, therefore, survival. The advantages of speech came at some risk for humans, however. Because our larynx sits so low in the throat, we are at risk of choking on food as it reaches the esophagus, whereas nonhuman primates are not (see Figure 9.2). In evolutionary terms, the advantages of speech were more important for human survival than the risk of choking.

Humans are the only primates who are born with fully developed brain structures for acquiring and processing language. Three structures in particular help the human brain process language. Wernicke's area is thought to be where the brain primarily processes spoken language. Broca's area produces language. A third zone, called Geschwind's territory, allows the brain to understand different qualities of language simultaneously (spoken and written), which may help to classify sounds and words. Neurologists are still studying the ways in which these three areas are linked to one other. These structures are the same in both hearing and hearing-impaired people, demonstrating that the same brain processes are at work when people speak and sign.

Figure 9.2

THROAT ANATOMY WITH EMPHASIS ON DIFFERENCES IN LARYNX IN CHIMPS AND HUMANS.

The larynx of chimpanzees sits higher in the throat than it does in that of humans. A lower larynx allowed our human ancestors to develop the kind of control over sound that led to speech.

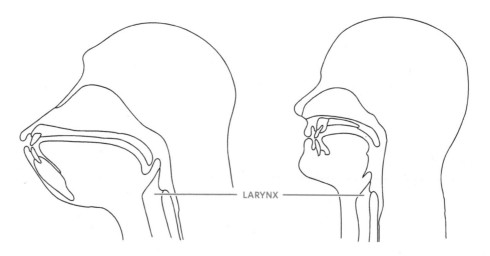

LARYNX

STUDYING LANGUAGE THROUGH THE LENS OF ANTHROPOLOGY

What Does a Linguistic Anthropologist Do?

While a linguist (a scholar who studies language) may focus on the units of construction of a particular language, a linguistic anthropologist is most interested in the cultural context in which a language is used. Linguistic anthropology studies the ways in which language, social life, and culture are intertwined. Linguistic anthropologists are interested in the different ways that people talk in different situations, how language helps define a group's worldview, and how gender affects language.

Since language is also one of the main ways that people assert power over others, language is political. Speech becomes coded with meanings that are negotiated by participants in a conversation. Linguistic anthropologists attempt to tease out and understand these meanings in order to understand their many layers.

Recording a Language

Imagine you are the first anthropologist to study a small, traditional society. In order to understand its culture, you must first learn its language. But with no textbooks or dictionaries, how do you begin? The first step would be to break down the language into components.

The first thing you would do is to listen to the sounds of the language. This is called **phonetics**, or the study of the sounds in human speech. Learning the sounds of the language allows you to understand which sounds are possible. For instance, the sound *-tl* is a sound used in the ancient Aztec language, Nahuatl, as in *tlatoani* (political leader) or *tomatl* (tomato). This sound combination is not found in English. Distinguishing which basic sounds are used and which are not is a good starting point.

Once you know the sounds of a language, you would want to know how those sounds convey meaning by understanding the **phonemics**. A phoneme is the smallest unit of sound that confers meaning. For instance, in English, the word *ox* refers to one specific thing (a particular type of animal). If a *b* is added, as in the word *box*, it changes the meaning of the word. Therefore, *b* is also a phoneme.

Then, you would want to learn about how words are structured to make meaning. A **morpheme** is the smallest part of a word that conveys meaning. For instance, the word *textbooks* contains three morphemes: *text* (a book used for instructional purposes), *book* (that which is read), and *s* (a marker to show it is plural). Morphemes differ from phonemes in that a single morpheme may contain several sounds.

Next you would decipher the **syntax,** or how units of speech are put together to create sentences. Grammatical rules govern speech in all languages, both spoken and signed. Statements are often organized differently from questions, for example.

In French, a statement places the subject (in this case, *vous*) before the verb: *Vous allez au marché*. A question reverses the order of the subject and verb, placing the verb (in this case, *allez*) first: *Allez-vous au marché?* Knowing the grammar rules of a language allows a person to be understood by others.

Semantics is important to understand how words and phrases are put together in meaningful ways. It won't do much good in the field if you can say single words but don't know how to put them together to make meaning. Semantics studies signs and symbols (described above) and the meaning derived from body language, facial expressions, and other nonverbal means of communication. Without understanding these other clues, words become empty of most meaning and often impossible to interpret accurately.

As an anthropologist, you would ultimately want to understand the **pragmatics**, or context, of a language. Every **utterance** depends on the context within which it is spoken. If someone says, "I love you," context is crucial for interpreting that statement. Who is the speaker? What is the listener's relationship to the speaker? What is the time and place of the utterance? All this extra information is included when we listen to and analyze another's speech.

NONVERBAL COMMUNICATION

Paralanguage

Human language goes far beyond just the words we speak. We use the term **paralanguage** to refer to all of the ways we express meaning through sounds beyond our words. Paralanguage is a subset of semantics, since it gives us information about meaning. The way someone speaks can give clues about the identity of the speaker within the first few utterances. This includes information about regional background or socio-economic class. Of course, the sounds of a person's speech don't always reflect their histories so simply. For example, a person with bilingual parents may display a mixture of speech habits. A person wanting to work in a job that requires a more "standard" form of mainstream speech, such as a radio DJ or newscaster, may have deliberately worked to erase a certain accent.

There are two main types of paralanguage. First, speech contains **voice qualities**. These are the background characteristics of a person's voice, including its pitch (how high or low a person speaks), the rhythm of one's speech, the articulation of their words, and types of lip movements. An angry person who says "I'm happy you're here" with pinched lips and little change in inflection sounds very different from a person with a wide smile and variable pitch who is genuinely happy. The same sentence said two different ways carries completely different meanings, which we interpret using our understanding of paralanguage.

Figure 9.3
RADIO DJ.
Radio disc jockeys (DJs) will sometimes train themselves to speak in a standard or dominant form of their native language to be marketable to a wider audience.
Credit: Ignatiev/iStock

The second type of paralanguage is called **vocalization**. These are intentional sounds humans make to express themselves but are not actually words. For instance, when an American English speaker says "Uh-oh," it signifies a problem, while "Ahh" lets someone know we understand, especially when accompanied by a nod of the head.

Voice qualities and vocalizations are culturally variable. Each language has its own set of meanings attached to its paralanguage. Even dialects of the same language can be extremely variable. Consider the differences in voice qualities and vocalizations between an English speaker from the Yukon Territory in contrast with one from Quebec, or a Southerner from Georgia in contrast to a New Yorker.

One variant of American English, African American Vernacular English (AAVE), may have its roots in the exposure of early American enslaved people to a variety of British dialects. Although some may mischaracterize it as "substandard" English, it has a set of consistent grammatical, lexical, and pronunciation rules that are widely shared among speakers. AAVE is as complex and rule-based as any other dialect.

Silent Language

Making meaning in a language involves more than words, and even more than sounds. The nonverbal cues that accompany speech, known as **silent language**, also contribute to meaning. Silent language refers to the very specific set of nonverbal cues, such as gestures, body movements, and facial expressions, that is acquired by speakers of a language. Sign language also uses facial expressions to provide meaning.

While paralanguage tends to develop based on a person's experiences (such as gender, education, or occupation), silent language is shared among members of a culture. For instance, North Americans nod their heads to mean "yes," and shake their heads from left to right to mean "no." In contrast, Indians shake their heads on a horizontal axis to agree, disagree, or simply for emphasis. One mustn't assume that even the most basic movements have the same meaning across cultures.

Because silent language can be entirely different in different cultures, it can easily be misunderstood if used incorrectly. A gesture with a positive connotation in one region of the world may mean something offensive in another. World leaders are not immune to making these kinds of mistakes. For example, in the early 1990s, US President George Bush Sr. was touring Canberra, Australia. He meant to show solidarity with protesting farmers by flashing a "peace" or "V for Victory" sign, with two fingers in the air. Unfortunately, he made the mistake of turning his palm backward, which communicated to the farmers that he wanted them to go "screw themselves." He later apologized for the error.

Space

Several features of silent language are especially important for understanding meaning in a cultural context. One of these is **proxemics**, or the cultural use of space. This field looks at how close members of a culture stand to one another based on their relationship. How far away do friends normally stand from one another? When strangers approach one another to speak, are they close enough to touch? It also examines how space is organized in homes and cities. Does household space assume members want privacy or that most activities will be done together? Does the community landscape offer places for people to gather, or is it structured for efficiency in getting to work and coming home?

One of the pioneers of the study of proxemics, Dr. Edward T. Hall, first classified the informal zones of personal space, as shown in Table 9.1. *Public space* is the largest zone that extends the farthest away from a person. In this large space, activities are felt to be relatively anonymous. Closer in, *social space* is where a person conducts regular business with strangers and acquaintances, and then closer still, *personal space* is the area into which friends may pass. Closest of all is *intimate space*, in which only the closest friends, lovers, and family may enter comfortably.

These comfort zones will vary widely between cultures. The North American comfort level for distance between people is quite large compared to that of some other cultures. The need for personal space may cause some foreign visitors to North America to feel that people are "cold" or "unfriendly," because we tend to stand at a distance.

TABLE 9.1

Proxemic Zones for People in the United States

Intimate	6 inches
Personal	1.5–4 feet
Social	4–7 feet
Public Distance	12–25 feet

Source: Hall, 1990

On a sparsely occupied train in North America, a person tends to put several rows between themselves and others. In contrast, a North American who boards a train in India or Japan at rush hour may find themselves in very close proximity to a group of strangers. This can cause anxiety for the person who is out of their comfort zone.

Movements

Another aspect of silent language is **kinesics,** or cultural use of body movements. In Mexico, touching one's elbow is a way to call someone stingy, since the word for "elbow" (*codo*) and "stingy" (*codo*) are the same. In Puerto Rico, people in conversation will crinkle their noses at one another. This gesture is shorthand for "What do you mean?"

Although the meanings of some gestures, such as smiling, are nearly universal, people of different cultures use different gestures to signify different things. Gestures like the "thumbs up" or "OK" sign, which are positive affirmations to North Americans, are severe insults in other regions of the world. The meanings of kinesic gestures are not always the same cross-culturally.

Touch

Related to kinesics is the cultural use of touch. Social life requires greeting others in culturally appropriate ways. North Americans often shake hands or hug (depending on age, gender, and status). Mexicans often kiss on the cheek once, and the Swiss kiss three times. In Eastern Europe (e.g., Serbia) and the Middle East (e.g., Turkey or Lebanon), men may kiss each other's cheeks in greeting, although this same-sex greeting may carry sexual connotations in other parts of the world. Table 9.2 summarizes the categories of touch.

TABLE 9.2
Categories of Touch

Functional/ Professional	Touching another in the course of one's work, such as that of a doctor or manicurist.
Social/Polite	Touch that is part of a greeting or hospitality, such as shaking hands.
Friendship/Warmth	Touch between friends to express mutual appreciation or support.
Love/Intimacy	Touching another to express nonsexual love and affection.
Sexual/Arousal	Touching in an intimate context.

Source: Heslin, 1974

Figure 9.4
"100 WORDS FOR LAWN" BY SPEED BUMP.
Many anthropology students learn that Inuit know over 100 words for snow, an often-cited example of linguistic determinism. This comic reverses the ethnographic gaze in a humorous way to refer to stereotypes of North American culture. In fact, according to Dr. John Steckley in *White Lies About the Inuit* (2008), Inuktitut (the Inuit language) has about as many words for snow as English does. The language is simply constructed differently.
Credit: Dave Coverly

These categories vary widely cross-culturally, especially in ethnic groups where a high value is placed on women's modesty, such as in Arab cultures. Certain religious groups, such as Orthodox Jews, prohibit all touch between men and women who are not married or blood relatives. In personal interactions, differences in the type and frequency of touch may be examined for cultural and gender distinctions. For instance, in the United States, women tend to touch one another more frequently than do men.

ETHNOLINGUISTICS

Ethnolinguistics is the study of the relationship between language and culture. It is generally considered a subset of linguistic anthropology. An ethnolinguist would be interested in how people's cultural environments shape their language use, or how language shapes the way people organize and classify the world.

It is understood that culture directly influences language. At the most basic level, a person is born into a given culture and acquires the language(s) necessary to interact with others. Because humans are born with the capacity for language, but not instinctively knowing any particular one, language is essentially a by-product of culture.

Looking at this in the opposite way, to what degree does the language we speak shape our perception of the world? One of the first linguists to research this question was Benjamin Lee Whorf, under the guidance of his academic mentor, Dr. Edward Sapir, in the 1930s. Although many people today refer to this idea as the Sapir–Whorf hypothesis, Whorf himself referred to it as the **linguistic relativity principle**.

The principle considers language to be intimately connected to culture, such that people who speak different languages may in fact experience the world in distinct ways. An extreme version of this argument asserts that one's language directly determines one's worldview. That is, the structures of a language lock people into seeing the world in certain ways. Few linguists today would argue for this type of total linguistic determinism. Nonetheless, the close correlation of language and culture for which Whorf argued clearly exists.

Consider the Nuer (who call themselves *Naath*), a pastoral people living in the Nile Valley of Sudan and Ethiopia. Their main mode of life centers on their herds of cattle. Cows and oxen are essential to the Naath economy, with people's wealth and status measured in the size of their herds. Naath people drink cow's milk and eat whey and cheese. Young men and women take on "cattle names" that identify them with their favorite animals, and gifts of cattle are given upon marriage. Because of the importance of these animals to the Naath people, the Naath language has many specific ways to talk about cows and oxen. For instance, anthropologist Sir E.E. Evans-Pritchard (1940) recorded over a hundred descriptive color terms for cattle.

An influx of culturally resonant terms can also appear suddenly due to a significant shift in cultural practices. This is what we saw happen during the COVID-19 pandemic. Within just a few months, the risks of becoming ill from the virus caused a near-global shift to the practice of *social distancing*, wherein people who didn't live together were urged to remain six feet (two meters) apart. *Quarantine* became a word that we began to use daily, along with creative derivatives such as a *quarantini* (any alcoholic beverage made to drink during *virtual happy hour*) and *quarantine pod* (a group of people with whom we could be physically close without wearing a mask, often used to refer to a group of children who attended home school together). The name of a business communication company, Zoom, was suddenly a noun and a verb ("I have to *zoom* at 11:00 a.m. so don't use the Wi-Fi"), and people working from home suddenly had to worry about family members or roommates *zoom bombing* a meeting. So many new words came into our daily lives, in fact, that the Oxford English Dictionary declined to choose just one "Word of the Year" for 2020.

Both Naath cattle vocabulary and the sudden influx of pandemic-related words in 2020 illustrate how language is deeply correlated with the cultural environment.

Color Categories

Most English speakers take the rainbow for granted. When one appears in the sky, seven colors are clearly distinguishable. Schoolchildren often learn a mnemonic to recall the seven colors, such as ROY G. BIV, with each letter standing for a color of the spectrum.

When anthropologist Dr. Victor Turner (1967) did fieldwork among speakers of the Ndembu language of Zambia, he found that they used only three primary color terms: white, black, and red. Other colors are either derivative (i.e., gray = "darker white") or descriptive (green = "water of sweet potato leaves" or yellow = "like beeswax"). Many other languages, like Vietnamese, specify more color terms than three, but have only one term to refer to both blue and green (*xanh*). Speakers define the color they want to identify by association. Is it *xanh* like the sky, or *xanh* like a leaf?

Figure 9.5

Sanga is the term used to describe the breeds of cattle native to sub-Saharan Africa. These Ankole-Watusi cattle are close relatives to the Abigar breed favored by the Naath.

Credit: Dennis Jarvis/CC BY-SA 2.0

What does this mean? Does language shape reality so much that, because the Ndembu have only three terms to talk about color, they visually see only three colors? No. Physiologically, their vision is the same as ours. They use one term for related colors because their cultural environment does not demand more precision. The use of metaphor and description fills in any gaps.

Cultural Models

Language contains a set of **cultural models** that reflect our thought patterns and guide our behavior. These models are widely shared understandings about the world that help us organize our experiences. The models also determine the metaphors we use to talk about our understanding of the world. In this case, metaphor is not merely poetic, but is a fundamental part of the way we organize our experience of the world.

For example, a cultural model that guides English speakers is the notion that illness is like war. We *build defenses* against illness, and we get sick because our *resistance* is low. We *fight* a cold, *combat* disease, *wage war* on cancer, and have heart *attacks* (Atkins & Rundell, 2008).

The importance of cultural models for anthropologists lies in understanding the worldviews of others. Another ethnic group, speaking a different language, may have developed an entirely different set of cultural models. In contrast to the idea of illness as war, the Diné (whom we commonly refer to as Navajo) understand

BOX 9.2   **Talking About: Outer Space, "The Final Frontier"**

"Space: the final frontier." This famous line opens the Star Trek television series, in which a starship and its crew "boldly go where no one has gone before" (changed from "no man" in 1987 to keep up with changing gender norms). The popular show debuted in 1966 at the same time the United States was racing to put the first man on the moon and lay claim to a brand-new world.

The idea of space as the last frontier of exploration and discovery became ingrained in our language, recalling the early days of White American **imperialism,** in which the Wild West was the frontier. The language of exploration and settlement is the most common cultural model used by people to talk about human space exploration. These are words with powerful and complicated histories. *Explorers* seek out new worlds and exploit them for their own purposes. *Settlement* implies it is ours for the taking.

Consider "Occupy Mars," the slogan that SpaceX CEO Elon Musk uses to promote his goal to put one million people on Mars by 2050. The term *occupy* means taking something by force that belongs to someone else, such as through conquest or annexation. Even if the slogan is meant to be tongue-in-cheek, the choice of language leads us to visualize colonies of humans who will use the resources of Mars for their own purposes and gain. It does not inspire, say, a commitment to scientific inquiry or awe for the beauty and fascination of another world.

It also begs the question of who exactly will be doing the occupying. Will a colony on Mars reproduce the inequalities of society that we witness on Earth? Who will the colonizers be, and how will they be chosen? Will Elon Musk be the king of Mars? These questions tend to be less popular than a narrow focus on the technology needed to get to the frontier.

Even though the space frontier is a place where most of us will never go, the word "frontier" compels us to explore. A frontier is a place that the adventurous spirit *should* take us, lest we lose the human drive toward "progress." The implication is that we must go, essentially, to save humanity.

Figure 9.6 OCCUPY MARS.
A woman promotes the SpaceX slogan "Occupy Mars" on her cap and t-shirt.
Credit: Oleksii Leonov/CC BY-SA 2.0

illness as a disruption in the harmony of the universe. With the support of family and community, a healer creates sandpaintings for a healing ceremony. Diné people believe that their gods, the Holy People, are attracted to the painting. When the sick person sits on the completed sandpainting, the Holy People absorb the illness and provide healing. The person's health becomes reconnected to the Holy People and thus realigned with life forces. In this case, we might argue for a musical cultural model of health among the Diné with ideas of being *out of tune* with the *harmony* of the universe. The two different models reflect an entirely different cultural approach to medicine.

Gendered Speech

Sex and gender also shape language use. This area of study is known as **gendered speech**. Men and women learn to use different speech patterns based on the cultural expectations of the sexes. Sometimes this leads to misunderstandings between men and women that are more about gendered speech patterns than about the individuals in conversation.

Of course, when we talk about categories such as "men's speech" or "women's speech," we are making generalizations that do not hold true for everyone. The data included are based on evidence but are variable among members of any society. Furthermore, these trends may or may not apply to the speech patterns of gender nonconforming people.

Georgetown University linguist and popular author Dr. Deborah Tannen (2007) argues that men's and women's conversational styles develop differently, leading to problems of "cross-cultural" communication. Women tend to emphasize emotion and empathy ("rapport talk"), while men tend to emphasize information or status ("report talk"). Because of these different expectations in a conversation, women may be left feeling like men "never listen," and men may feel like their girlfriends or wives are "demanding and needy." Gendered speech can cause frustrations in relationships without either partner being aware of the root of the problem.

During the second wave of the women's rights movement that began in the United States in the 1960s, Dr. Robin Lakoff (1973) wrote about the apparent sexism inherent in the English language. She argued that discrimination – mostly unconscious – against women was embedded in vocabulary choice, sentence construction, and speech practices. Consider the following two sentences Lakoff uses to illustrate this point: "Oh dear, you've put the peanut butter in the refrigerator again" and "Shit, you've put the peanut butter in the refrigerator again" (p. 50). Lakoff argues that we would attribute the first sentence to a woman and the second one to a man due to the degree of expletive. She argued that women are generally expected to speak in a "ladylike" (i.e., nonconfrontational and nonaggressive) manner.

While this is still true to a certain degree more than 40 years later, Lakoff saw that, even in the early 1970s, it was becoming more acceptable for women to use stronger language in public. These changes correlate with women moving into public positions of employment that have been traditionally held by men. The reverse is not true, however. It is acceptable for the less powerful group (women) to take on both language and behaviors of the more powerful group (men), but not vice versa. Diminished power or status is not desirable.

Another aspect of gendered speech that Lakoff addresses is women's frequent use of tag questions added to the end of a statement. The tag turns the statement into a question, decreasing its forcefulness. For instance, "The weather is terrible, isn't it?" Or, "Your sister seems happy, doesn't she?" By including the tag question, an utterance comes across as more accommodating. The tag asks for confirmation that what is being said is a valid opinion. Women's speech often uses tag questions, while men's speech tends more toward declarative statements indicating authority.

Speech Communities

A **speech community** is a group that shares language patterns. People who live in the same area may share language patterns and vocabularies, but speech communities also form in subcultures of people who share the same interests without necessarily living in the same area. Speech communities occur naturally as part of membership in a given social group but also may be manipulated consciously to signal one's membership to the group or to others.

For instance, *lavender linguistics* is a term used to refer to the speech patterns of members of the queer community. Gender and sexual orientation are essential to social identity and therefore can become important in the formation of language patterns. Linguists conclude that most gay speech patterns are not natural in a biological sense but rather are socially constructed.

Because gay subcultures are very diverse, certain features of language may be used to signal membership to others in the larger gay community. For gay men, clues signifying membership may include intonation, certain vowel and consonant modifications (e.g., "YASSS" instead of "yes"), or slang vocabulary. Some terms that signify insider status for gay men are *bear* (a large, bearded gay man), *slay* (to impress), and *queen* (a flamboyant gay man). For lesbians, insider status may also be signaled by the use of slang – such as the understanding of the terms *butch* (a masculine lesbian) and *femme* (a feminine lesbian) – but also by forms of nonverbal communication such as a more androgynous or masculine style of dress and hair.

Code Switching

In January 2009, then US president-elect Barack Obama went out for a chili dog at the world-famous Ben's Chili Bowl in Washington, DC. Ben's Chili Bowl has been a staple of the Black community in DC since the 1950s, run by the Alis, a well-known African American family. Of course, the president doesn't eat out alone – videos of the visit show him interacting with his entourage while waiting for his meal. When Obama gets his chili dog, he gives the (Black) cashier a large bill. The cashier offers to get his change, and Obama replies "Nah, we straight." In this seamless move, Obama has shifted from one type of language to another – from standard American English into African American Vernacular English – signifying his insider status in the Black community. This is an example of **code switching**.

Participants in two or more speech communities can move easily between them when the context calls for it. It's called code switching because it shifts between speech styles known to each group. For instance, when students address their professors, they often use more formalized speech patterns and vocabulary, but after class, when at the cafeteria with friends, their style of speaking and use of vocabulary fall into more relaxed patterns. Similarly, Obama's "Nah, we straight" created a connection between himself and the cashier that would not have been made had Obama addressed him formally.

Different levels of speech formality are called **language registers**. Many people use multiple registers in social interactions daily. Generally, languages have a formal and an informal register. In Spanish, for example, a speaker uses the formal *usted* when addressing a teacher, doctor, or other professional, as well as strangers who are older than the speaker. The informal *tú* is used to speak to friends, family members, and children. The use of registers may also be used deliberately in social situations in ways that are typically incorrect to invoke sarcasm or change the meaning of an utterance.

Some languages, such as Japanese and Korean, have multiple levels of formality in order to show respect to those with higher social status. In Japanese, language registers are called **honorifics** (linguistic ways to show honor or respect) or *keigo*. Using an honorific inaccurately signifies outsider status, and that a speaker does not fully understand the social implications.

Code switching also occurs between different languages when multilingual speakers converse. Words or phrases may be switched from one language to another in a single sentence. An interesting feature of code switching is that it is grammatically correct according to the rules of the dominant language of the sentence. For example, in the question "This dress, *es muy largo*, isn't it?" the speaker switches effortlessly from English to Spanish, then back again. The word *largo* is an adjective modifying the masculine Spanish word for dress, *vestido*. Even though the English word "dress" was used in the sentence instead of *vestido*, the ending of the adjective *largo* (with an "o") correlates to the masculine noun *vestido*.

This is done unconsciously; that is, the speaker seamlessly switches when it makes sense in the sentence to switch. Dr. Don Kulick, who studies language in Papua New Guinea, asked a Papuan informant why speakers switch from one language to another. His informant answered simply, "If *Tok Pisin* comes to your mouth,

Figure 9.7
JAPANESE MEN BOWING, OKINAWA.
Bowing is an integral part of Japanese expressions of respect. It often accompanies honorific language use when speaking to an elder or person of higher social status.
Credit: Akuppa John Wigham/ CC BY-SA 2.0

you use *Tok Pisin*. If *Taiap* comes to your mouth, you use *Taiap*" (Verhaar, 1990, p. 206). In other words, a speaker doesn't think about code switching, it just happens.

LANGUAGE IN THE DIGITAL AGE

Digital Language

The widespread use of personal digital devices, such as smartphones and tablets, has created a variety of new ways to communicate. In fact, it has become more and more common to write to one another (through email, instant messaging, or texting) than to talk on the phone or face to face. Due to these changes, writing has undergone a radical transformation in the past 20 years. Users of electronically mediated communication (EMC) in languages all over the world have developed creative new ways to write and talk.

A topic of interest to linguistic anthropologists is the widespread use of text messaging. Users text as if they were having a conversation, constantly inventing innovative ways to communicate using "fingered speech" (McWhorter, 2013). For instance, shorthand and abbreviations allow users to text rapidly. In English, *SMH*, *IDK*, *AFK*, and *LMAO* are just a few examples. Since users perceive texting as an extension of speech, these same abbreviations also find their way into spoken language. The shorthand becomes a word in its own right, and thus enriches the language. *OMG* (or *oh-em-gee*) as a single word now carries a different meaning than its original referent, "Oh my God."

Because texting lacks the nonverbal context that is essential to speaking face to face, challenges arise in expressing one's intent clearly. The use of emoticons and emojis, capital letters, and varied punctuation allows users to express complex levels of meaning. Consider the difference in meaning between "going to dinner with cousins 😊" and "going to dinner with cousins 😖." In this case, the emoji functions in place of silent language, such as a facial expression.

Those who use texting as one of their main modes of communication incorporate nuances and new meanings into EMC so rapidly that the language changes constantly. The abbreviation *LOL* (spoken as *el-oh-el*) for "laughing out loud" began as an authentic response to something funny. Now, after years of use, it has evolved into other meanings, such as *lol* (*lahl*) or *lolz* (*lahls*) to express irony, sarcasm, or just as a written placeholder to let the other person know that you're (sort of) paying attention. Table 9.3 lists some of the ways that EMC users express laughter (whether authentically or sarcastically) in different languages.

Research finds that students have clear rules for EMC, including levels of formality and appropriateness. Just as in spoken language, students use a different register when writing emails to friends than they would when emailing a professor.

TABLE 9.3

Laughing Online around the World

Thai	555	the Thai word for 5 sounds like "ha," so 555 sounds like "hahaha"
Japanese	www	warai = "laughing"
French	mdr	mort de rire = "dying of laughter"
Spanish	jajaja	the letter j sounds like an h, so jajaja sounds like "hahaha"
Korean	kkk	keukeukeu = "laughing"
Swedish	asg	asgarv = "intense laughter"
Nigeria	LWKMD	"laugh wan kill me dead"
Brazilian Portuguese	rsrsrs	risos = "laughter"

Sources: Garber, 2013; McCann & Brandom, 2012.

While "c u in class lolz" might be appropriate for friends, students code switch and use the more formalized register with teachers: "See you in class, Professor." There appears to be little support for complaints that texting is "ruining" language or preventing students from learning to spell. On the contrary, it is an exciting avenue for studying language change.

LANGUAGE CHANGE AND LOSS

As culture changes, so does language – as we see in the texting example above. Moreover, social changes can have large-scale impacts on language use. The results of contact, colonization, and assimilation force new modes of communication on speakers, especially in the language used by the less powerful group. Often languages will merge to some degree.

Pidgin languages develop when culture contact is sustained by people who don't share a common language. A pidgin language uses two or more languages for communication by mixing certain features together. The dominant language often supplies most of the vocabulary, while the subordinate language maintains features of its grammar. This particular mix is likely because pidgins develop quickly, out of necessity for communication. Words are more important for meaning and generally easier to incorporate than syntax changes.

When a pidgin language remains relevant and, in the next generation, becomes the dominant language of a group, we refer to it as a **creole** language. *Tok Pisin* is a language that first developed as a pidgin due to sustained contact with North Americans in Papua New Guinea. Today, it is one of the official languages of Papua New Guinea, with millions of speakers.

According to *Ethnologue* (Lewis, 2013), fewer than 7,000 languages are currently spoken in the world. At the time of this writing, about 2,000 of them are listed as "in danger." Because 90 percent of those endangered languages have fewer than 100,000 speakers, the twenty-first century may see a severe drop in languages spoken.

Many forces contribute to language loss, just as there are complex reasons for the loss of an animal or plant species in nature. A tragic possibility is that a language is left with no living speakers due to genocide. This occurred when the English colonized the island of Tasmania, off the coast of Australia, at the turn of the nineteenth century. Disease and attacks on the Indigenous population left few survivors. The last remaining Tasmanian Aboriginal woman, named Truganini, died in 1876. None of the Indigenous languages of Tasmania have any Native speakers left.

Another, less violent, cause of language loss is that some languages evolve completely into other languages. The sacred language used to write ancient

BOX 9.3

Talking About: Gaming

Wherever there is human culture, there will be anthropologists. This means that anthropologists are also online, studying virtual communities. The internet has become a virtual ethnographic site to learn about communities such as **fandoms** and underground subcultures, and the languages they use to communicate. The language that the online gaming community uses is like any other subgenre of language, similar to a dialect or specialized "lingo" used by insiders of that community.

Gamers acquire the language as they become more and more involved as members of a game's community. *Noobs*, or players new to the game, will find themselves on the outside of this interaction until they have enough experience to learn the language. While there is a set of terms that all gamers know and use (see Table 9.4), there is also highly specialized vocabulary in specific games in which users interact.

World of Warcraft is an MMORPG (massively multiplayer online role-playing game) originally released in 2004, with 26 million users as of 2021. The complexity of the game requires an extensive vocabulary list – primarily ways to talk about the game using just a few letters. This allows players to type quickly while commenting to and about each other. For example, to refer to the different player classes, users write abbreviated terms: dk (death knight), lock (warlock), dru (druid), pally (paladin), wiz (mage), and dh (demon hunter). Players even insult each other with specialized slurs such as "huntard" (bad hunter) and "noobid" (bad druid). Note: All lowercase letters are used for in-game conversations because of the practical need for speed while typing.

TABLE 9.4

General Online Gaming Lingo

Term	Definition	Usage within Regular English-Language Context
lag	Glitching, or unusually slow speed of game response	"I couldn't get there because I have *lag*."
xp	Experience points	"You don't have enough *xp* to enchant that sword."
kr	Kill ratio	"Her *kr* is sick."
npc	Non-player character	"I married an *npc*; they're so boring."
spawn	Beginning point; hub	"Meet you at *spawn*."
afk	Away from the keyboard	"Going to the fridge, will be *afk*."
noob	Newbie, someone new to the game	"What a *noob* mistake!"

Zoroastrian religious texts, Avestan, had already disappeared in its oral form before the development of its written language in 3 **CE**. Original spoken Avestan had become several languages, including Old Persian and probably also Pashto, spoken in Afghanistan. The language was re-created in a written form in order to preserve the ancient prayers.

Another way a language may disappear is due to deliberate suppression by a dominant culture after contact. The Ainu of Japan are an ethnic group who live on the Japanese island of Hokkaido. They have experienced severe discrimination by non-Ainu Japanese, beginning in the fifteenth century with invasion and enslavement. Brutal treatment of the Ainu decreased their numbers considerably. In the mid-twentieth century, the Japanese claimed Ainu land, prohibiting hunting and fishing. The use of the Ainu language in schools was prohibited, forcing children to learn Japanese instead. Estimates of remaining Ainu speakers today range from just 20 to 30 individuals.

Violent language suppression tactics occurred close to home as well. Starting in the 1800s and continuing until the late 1900s in some regions, the governments of the United States and Canada made attendance at special schools (called "boarding schools" in the United States and "residential schools" in Canada) for Native American/Indigenous children mandatory. The children were removed from their homes and forbidden to speak their traditional languages under threat of punishment. They suffered terrible abuses and even death at the hands of boarding school staff, who were infamously encouraged to "kill the Indian in the child." This violent and oppressive treatment caused the decline of not only Indigenous languages but also Indigenous culture.

A language is of critical importance because it encodes all of a culture's information. For instance, there are many words and phrases that cannot accurately be translated into another language without a lengthy description. Even then, native speakers will say that the translated description does not capture the essence of the original term.

One way that Indigenous languages adapt as culture changes is by rejecting English loan words. For instance, the Languages Commissioner of Nunavut (Canada's northernmost territory) chose the word *ikiaqqivik* to represent the word "internet" in the Inuktitut language (Soukup, 2006). It translates to "traveling through layers," which is the way Inuktitut shamans describe their experience of traveling through space and time on a quest. In a similar way, an internet user travels through multiple locations (sites) with information written in and about the past, present, and future. This is an example of how traditional concepts can integrate into modern ones, preserving original cultural elements.

Some languages that were critically endangered are now being revitalized. One example that shows a reversal is from the Māori people of Aotearoa (New Zealand). English settlers outlawed the Māori language (*reo Māori*) in schools by 1867. A hundred years later, it was clear that the language was dying out. Only about 18 percent of the Māori population could speak it in the 1970s, and most of those individuals were over 50 years old (Tsunoda, 2006). With the awareness that reo Māori was a dying language, school programs known as "language nests" were established beginning in the 1980s. These programs provide an early childhood foundation in Māori language, values, and culture for children ages zero to six years old. The success of the language nests led to a demand for Māori-language primary and secondary schools. Today, there is revitalized interest in traditional Māori language, and the number of reo Māori speakers is increasing.

In an effort to appeal to young language learners, some advocates for Indigenous languages are pushing for more online use and visibility. For example, many Indigenous leaders are working to translate internet interfaces into their native languages. The First Peoples' Cultural Council of British Columbia offers more than 100 Indigenous language keyboard interfaces via apps on their website at www.firstvoices.com. Users wishing to text, send email, or search the web can now do so in any Indigenous language of Canada, Australia, or New Zealand, along with many Native American languages of the United States.

SUMMARY

This chapter discussed language as one of the main characteristics of human culture. Mirroring the Learning Objectives stated in the chapter opening, the key points can be summed up as follows:

- Although nonhuman primates (and some other species) can communicate with rudimentary forms of symbolic language, humans are the only species to have developed advanced forms of symbolic language.
- How human language began is a question that has been addressed in many ways. Some ideas emphasize the role of trust and the need for people to trust that a sound means the same as a recognized gesture.
- In order to record an oral language they are studying, linguistic anthropologists break the language down into components, including the units that comprise sound, grammar, and meaning.
- Linguistic anthropologists are interested in the associated body language, facial expressions, and other silent language that lend meaning to an utterance.

- When looking at speech patterns from an anthropological perspective, speech communities and cultural contexts (such as gender) provide rich cultural data.
- Although some may fret that languages used today on the internet and for social media are being irreparably damaged, anthropologists actually find EMC an exciting area of creative language change.
- Unfortunately, some languages have been lost forever due to severe cultural oppression. Nonetheless, today many languages are in a process of revitalization.

Review Questions

1. Do other primates or animals use the same kind of symbolic language that humans do?

2. Why do anthropologists argue that language is much more than verbal speech?

3. To what degree do anthropologists today believe in the validity of the linguistic relativity principle?

4. What are some of the ways that language changes in social situations?

5. What are some of the interesting aspects of EMC?

6. Why do languages lose speakers?

7. How are languages becoming revitalized today?

Discussion Questions

1. What special vocabulary (or "lingo") do you know by virtue of your membership in a subculture or specialized social group?

2. What words entered into your personal vocabulary for the first time as a result of the COVID-19 pandemic?

3. Do you think that texting is ruining written language? In your experience, how has texting changed the way you talk or write?

4. Have you encountered people while traveling who had different zones of kinesics, proxemics, or touch?

Visit **www.lensofanthropology.com** for the following additional resources:

SELF-STUDY QUESTIONS	WEBLINKS	FURTHER READING

10

FOOD-GETTING AND ECONOMICS

LEARNING OBJECTIVES

In this chapter, students will learn:

* *about the connections between how people get their food, how they organize themselves socially, and how their food-getting methods impact the environment.*

* *the differences between food foragers and food producers.*

* *about distinct types of foraging based on the resources of a given area.*

* *which forms of economic production, distribution, and consumption are found in different types of societies.*

* *the characteristics of food-producing societies, including horticulturalists, pastoralists, intensive agriculturalists, and industrialists.*

* *that many diverse diets based on nutrient-rich foods can be healthy for the human body.*

> **Going to the grocery store? Harvesting a meal from the farm? Foraging for water roots? How you get your food is fundamental to how you live your life.**
> #FoodGettingAndEconomics

INTRODUCTION

How do people get the resources they need to survive? Do they grow or raise their own food, forage for it in the local area, or purchase it at the grocery store? Each of these different food-getting strategies lays the foundation for a very different type of society. The ways people acquire their food dictate their daily schedules, inter-actions with the environment, modes of cooperation and competition, and the expectation of the division of labor among genders.

Food-getting strategies also provide a foundation for the **economics** of a soci-ety, or how goods and services are produced, distributed, and consumed. Economics examines how food is found, grown, or harvested; how people get that food; and how that food is eaten. Anthropologists also look at how material goods are made and by whom, how they get into the hands of people other than the producers, and how those items are used. In addition, economics looks at resources such as land and water. How are those natural resources used?

Anthropologists divide the many different types of food- and resource-getting strategies into general categories. The largest division is between food foragers (those who find food) and food producers (those who grow food). Within this broad division, procurement strategies are separated into categories: foragers are in one category, while the category of food producers includes horticulturalists, pastoral-ists, intensive agriculturalists, and industrialists. Often a society will practice one or more of these strategies. Each will be explained in detail in this chapter.

ADAPTIVE STRATEGIES: FOOD FORAGERS AND FOOD PRODUCERS

When anthropologists examine different food-getting strategies, they find it useful to distinguish between those who use what the land produces and those who

Figure 10.1
FORAGERS.
Foraging peoples represent a way of life that humans have practiced in varying forms since the beginning of our species. In other words, we became human while living in small foraging groups.
Credit: Ariadne Van Zandenbergen/Alamy

deliberately manipulate the environment to produce food. Those who seek their food supply among available resources are called **food foragers** or **hunter-gatherers**. Groups that farm, keep food animals for their own use, or otherwise transform the environment with the goal of **food production** are referred to as **food producers**. Depending on the means by which the food is produced, they may practice horticultural or agricultural techniques, engage in animal herding, and/or rely on others to produce, distribute, and make food products available. A culture's **foodways** are fundamental to the structure and functioning of its society.

Food-getting strategies are flexible and nonexclusive. No society is locked into one settlement or economic pattern, and all societies have a dynamic relationship with their environment and with other societies with whom they come into contact. Several strategies may be used at one time, with one generally being dominant. For instance, a pastoral herding society may also plant crops part of the year and trade or purchase certain food items at a local store.

Food-procurement methods are subject to change from internal pressures and external sources. These range from environmental change, the invention or adoption of new **technology**, peaceful trade, or violent conquest. Furthermore, as new resources make themselves available, groups can and will become change agents. They will act on their own behalf for better economic opportunities, revitalizing traditional foodways or working toward increased sustainability for local food systems. Even small-scale Indigenous communities are involved in global processes of change. Their products and services are linked to not only local but also regional, national, and international economies.

FOOD FORAGERS

It is estimated that humans have spent 99 percent of their existence hunting and gathering for survival. It's important to look at the traits of foragers to understand not only techniques of food procurement but also the types of social networks upon which human society is built. Even though most people today no longer forage for a living, our basic humanness is defined by the cooperation and social connections between people that foraging fostered.

While the lifestyles of various **foraging** peoples share many traits, there are also major differences. The environments, gender roles, supernatural belief systems, and other features of foraging groups may be distinct. Food-getting strategies are always embedded within a set of unique cultural values, beliefs, and practices. To get a sense of some of these differences, this book examines several different foraging lifestyles among the Hadza of Northern Tanzania; Ju/'hoansi (pronounced *zhut-wasi*) of the South African Kalahari Desert; and **Inuit** of the Canadian Arctic.

Forager Foodways
Depending on the ecosystem, foragers' daily food may consist of wild plants, animals, or fish. The types of wild plants are highly variable, and certainly more than just "nuts and berries." Land-based plant foods include a wide variety of wild fruit and vege-tables, roots, seeds, tree sap, and nuts. For those groups with access to lakes, rivers, or the ocean, aquatic plants, including algae and seaweed, provide excellent nutri-tion. Hunting brings in local game, including small and large mammals, reptiles, amphibians, and birds. Insects can be gathered for an easy source of protein, and animal products such as honey and eggs may also be gathered. In marine environ-ments, foragers' daily meals may consist primarily of fish, marine mammals, and crustaceans. For instance, Inuit of the Canadian Arctic hunt caribou, seal, and sea birds in the winter, and they supplement their diet with a variety of fish and whale in summer months when the ice thaws.

While hunting is often portrayed in popular films as the primary source of food among foraging groups, in fact ethnographic and archaeological studies show it was the opposite in certain areas. Living among the Ju/'hoansi, anthropologist Dr. Richard Lee discovered that the group could identify over 90 different plant foods in the desert environment of the Kalahari, which provided them with a wide range of vitamins and minerals, including fat and protein from plant sources.

Hunting brought in only about 20 percent of the group's calories, while the gathering of plant materials and insects supplied the bulk of their diet. Although we tend to think of hunting and gathering as eking out a meager living at **subsistence** level, the Ju/'hoansi diet was undoubtedly more nutrient dense than our limited diets today.

The Hadza of Northern Tanzania is one of the last remaining groups on earth in which up to 40 percent of members still hunt and gather exclusively as their main food-getting strategy. Approximately 1,000 Hadza remain in their ancestral homeland in the area of Lake Eyasi, bordering Serengeti National Park. The Hadza primarily hunt game that comes to their water holes to drink, and forage for tubers, berries, and baobab fruit. They also trade for foodstuffs such as maize (corn), millet, and beer. Like many other foraging peoples across the world, the Hadza especially prize honey as a source of energy.

Why do the Hadza still primarily forage? Anthropologist Dr. Frank Marlowe (2010) argues that it is mainly due to poor ecological conditions for farming and pastoralism. The soil is largely unsuitable for **agriculture**, and infestations of the tsetse fly prevent the successful herding of animals. Therefore, Hadza people stay isolated as foragers as a choice. They know that they cannot survive by farming the depleted soil, and they choose not to work for others. Because they don't want to give up their autonomy, they continue hunting and gathering in a close-knit community.

Social Organization

Foragers live and travel together in small groups that anthropologists call **bands**. These groups can vary in size based on seasons. For instance, among the Hadza, band size has varied little over the past 100 years, with the average around 30 people. During berry season, however, band membership can temporarily grow to 100 people. The Ju/'hoansi lived in bands of 30 to 40 people moving across the landscape before being settled into camps in the 1970s. Foraging Inuit live in extended families, from a dozen to over 50 people, depending on the geography of the area.

What are the advantages of staying in small bands? In a harsh environment where survival depends on cooperation, it is important to minimize problems and stay together. Fewer interactions cause fewer opportunities for conflict and division. The measure of these interpersonal conflicts is referred to as the **social density**, or the frequency and intensity of interactions among group members. Maintaining small numbers minimizes the density, making social life easier than if there were several hundred individuals living together.

Nonetheless, where there are people, there is conflict. An often-used solution to interpersonal conflict is for individual members to join another group, either temporarily or permanently. This causes the numbers in a band to fluctuate occasionally. It also keeps the bands from breaking apart.

In general, tasks are divided by gender. Although a **sexual division of labor** predominates, it does not mean that men are necessarily restricted from gathering or women from hunting. However, due to women's role in pregnancy and childcare, along with a multitude of tasks they perform at the campsite, it is often more efficient for men to hunt. Some tasks are open to all group members, and among a few groups, hunting tasks are done by women, especially accompanied by dogs. Among the Hadza, men, women, and children gather honey, their most prized food item.

Bands have no social classes. Life in an **egalitarian** society means every member gets immediate rewards from foraging. Sharing the same access to resources limits status differences. All adults have some say in making decisions that affect the group, as there is no formal leadership beyond the respect afforded to the wise. In addition, being nomadic requires that everyone carry all their possessions on their backs when moving from camp to camp. This limits the number of belongings a person can have to what they can carry. Both the lack of **specialization** and lack of ownership help maintain the egalitarian nature.

How does everyone get approximately the same amount of resources if some families have more able-bodied members, or certain hunters are more skilled than others? Bands are **cooperative societies**, in which sharing is a key strategy for survival. When groups of hunters return with game, or women return from gathering, the food is divided among members of the group. Among the Ju/'hoansi, an individual member may have 10 or more sharing partners in a network that may be called upon when needed. This **reciprocity** network creates a safety net in times of hardship. According to Marjorie Shostak (1981), when Ju/'hoansi people are angry with one another, they may call someone stingy – a terrible insult in a cooperative society.

Foragers and the Environment

Foragers are **nomadic**, meaning they move frequently. The Hadza and Ju/'hoansi are examples of foragers: they move frequently and process food on site. In contrast, Inuit are foragers with base camps: they bring their fishing catch or other marine foodstuffs back to their base camp for processing. Because Inuit generally live in tundra environments, they use domesticated husky dogs (or, today, snowmobiles) to pull sleds for transportation to these sites.

Each group knows its home territory well, over which it moves annually to locate seasonal foods. Each band has some historical connection to its route and some rights over it, although bands do not have a conception of owning land or water as in a capitalist society. Once food resources in a given area are sufficiently used, the group moves to the next site. Inuit may have winter and summer base camps, moving between them twice a year as the seasons change.

Until the twentieth century, foraging territory was very large. Therefore, by the time a group returned to any previous site, the food resources would be plentiful again. Today, land is in short supply. Foraging as a primary means of sustenance is possible for only a tiny percentage of people in the world.

Most foraging peoples today have mixed diets, with foods coming from many different sources, including local commodity stores. Inevitably, local stores introduce processed foods, which are lower in nutrients. For instance, anthropologist Dr. Polly Wiessner (2002) found that store-bought items in the Ju/'hoansi village at Xamsa from 1996 to 1998 included sugar, flour, bread, soup, candy, chips, and beer. Nevertheless, most of the store-bought items are shared among camp members, in an extension of the traditional exchange economy.

ECONOMIC RESOURCES: WHO GETS WHAT AND HOW?

A major part of understanding food-getting strategies has to do with how that food gets distributed to others. Examining processes of exchange shows us how food and other resources, including items of cultural, religious, or symbolic worth, are distributed among group members. Based on a model created by economist Dr. Karl Polanyi (1944), anthropologists classify the types of exchange in the world's societies in one of these three ways:

1) Reciprocity
2) Redistribution
3) Market Exchange

Reciprocity is practiced in all types of societies. Redistribution is found specifically in societies with central governing authorities, such as farming, pastoral, or industrial societies with official leaders. Market exchange is found in agricultural and industrial societies in which surpluses are produced. All three are processes of distribution, or getting things into the hands of people.

Reciprocity

Reciprocity is a set of social rules that govern the specialized sharing of food and other items. Sociologist Dr. Marcel Mauss (1925/1966) originally referred to these items as gifts, including the gift of one's time or effort in addition to actual physical items. However, gifts are not given in a vacuum. Strict social rules dictate the requirements of sharing among members of a group, especially when the group relies on reciprocity to survive.

BOX 10.1   **Talking About: Hunting**

In the ethnography *"We Are Still Didene": Stories of Hunting and History from Northern British Columbia*, anthropologist Dr. Thomas (Tad) McIlwraith (2012) examines talk about hunting among the Iskut people. His study is primarily interested in how language is embedded in social contexts. By looking at everyday speech, he concludes that "hunting is the central metaphor that informs Iskut culture" (p. 12).

Iskut villagers are *didene* ("native") and have ancestry in the area dating back thousands of years. Because the village is of recent, twentieth-century construction, not all villagers have the same backgrounds, experiences, or dialects. However, the one thing that all Iskut people have in common is their reliance on and identification with hunting.

In focusing primarily on communication among hunters, McIlwraith found that hunting was rarely spoken about openly and directly. Nonetheless, the use of hunting metaphors and allusions in everyday speech allows Iskut people to establish personal connections between hunters through shared experience, uniting the community. Hunters from different areas even use common structural elements while retelling stories about their hunting experiences. Talking about hunting affirms the relevance of traditional Iskut life in a modern context. Further, it allows them to assert their difference from non-Indigenous outsiders, such as government workers or anthropologists, who may claim ownership over their histories or **Traditional Ecological Knowledge** (TEK; see Chapter 14).

Cultural models at work in Iskut life establish the idea that animals, nature, and people are connected in a close-knit web. Animals should not be spoken of poorly or treated with disrespect. Similarly, the land should also be treated well; otherwise, punishment may follow. For this reason, Iskut villagers have stood their ground in an ongoing political battle to protect their territory from government-sanctioned mining exploration and development. As McIlwraith discovered, talking about hunting privileges personal relationships over economics and stewardship of the land over exploitation. In this way, talking about hunting is essential to Iskut identity.

Figure 10.2 DIDINI KIME, "YOUNG CARIBOU CAMP." This typical Iskut hunting camp has been used by multiple generations of people from the same family. Occupied as a base for caribou- and moose-hunting expeditions in the late summer and fall, this camp is visited throughout the year.
Credit: Thomas McIlwraith, 2012/University of Toronto Press

Parties involved in a reciprocal exchange enter into a social and economic bond. Once a gift is given, the two parties are connected in an ongoing relationship. If one side of this relationship does not reciprocate with a gift of some type that is roughly equal in value, then the bond between them is damaged. Failing to reciprocate can destroy social, political, and economic relationships between individuals, families, or entire communities.

Resource Distribution: Generalized Reciprocity

Having built upon Mauss's idea of "the gift," Dr. Marshall Sahlins (1972) applied the concept to three kinds of reciprocity that he saw during his fieldwork among the Ju/'hoansi: generalized, balanced, and negative. Friends and family often practice a loose form of reciprocity we call generalized. The value of a gift is not specified at the time of exchange, nor is the time of repayment. However, the parties involved have the responsibility to reciprocate at some time and in an equal way. Because every society has a circle of people they trust, **generalized reciprocity** can be found in every type of society. Familiar examples in modern urban society might include throwing a party for close friends with the expectation that they will invite you in return when they host a party, or caring for a sick parent who has done the same for you many times.

In a foraging band, such as the Ju/'hoansi, hunting is governed by the rules of generalized reciprocity. Hunters begin preparations by equipping their quivers with others' arrows. Killing a large animal such as a giraffe takes multiple arrows, likely from each member of the hunting party, who must track and follow the animal for days as the poison debilitates its system. Therefore, responsibility for the kill is already shared from the moment the hunters set out. If the hunt is successful, the hunters will divide the animal in such a way that all members of the band receive some meat. Contrary to what we might imagine, only a small portion goes to the hunters and their families. However, by entering into a relationship of sharing with each member of the group to whom they have given meat, the hunters have solidified an ongoing bond. The debt of food will be repaid to those hunters at a later time, when others have brought home meat. Foraging groups include nature in their reciprocal networks of giving and sharing, knowing that as long as they treat nature well, nature will continue to provide for them.

FOOD PRODUCERS

Horticulturalists

The last 15,000 years have seen a human population explosion. As numbers of people grew, land became scarcer and resources decreased. Some foraging groups with lands suitable for planting began supplementing their foraging lifestyle with small-scale farms or gardens. As was explained in Chapter 6, there may have also been social and political reasons for the changes. Anthropologists refer to these groups as **horticulturalists**.

Because small-scale farming requires daily maintenance, groups that plant must settle in one area. Their villages are often small and occupied year-round. Hunting and gathering trips fan out from this central location. Small-scale farming is done with the use of simple hand tools, such as digging sticks and other garden tools fashioned from objects in the environment. These groups rely on rainfall for water.

Horticulturalist Foodways

Horticulturalists are food producers. While they may practice some hunting and gathering, they get a substantial percentage of their calories from crops they have planted, tended, and harvested. Crops vary widely, depending on the demands of the environment. Often there is some reliance on roots and tubers, possibly grains, and a selection of fruits and vegetables appropriate to the region.

How does a major change in food-getting strategies occur, such as the change from foraging to planting? Economic anthropologists see the answer to this question in the relation between group size and the food items available at any given location. The number of people who can be sustained with the existing resources of a given area is called the **carrying capacity** of the land. Among foragers, a group will remain in one place until the resources needed to feed and shelter all members of the group are used. Then, they move on to the next campsite. If the human population in the area is so large that available food items are never enough, a group will be forced to seek a new strategy to feed its members. This process appears to be the origin of most **horticulture**.

The Kaluli people are horticulturalists who live in the tropical rainforest of Papua New Guinea. They occupy communal homes called longhouses. They refer to their longhouses and their social group with the name of their land, signifying a deep connection to their physical environment. The Kaluli mainly gather wild sago, a starch from the sago palm, and supplement this with a variety of produce from small, family-maintained gardens. Small game and fish add animal protein to the diet. Kaluli food procurement strategies are largely cooperative, even though men and women pursue separate activities. Men clear the land for swidden farming and plant crops. Women tend gardens, gather small game for extra protein, process food, and look after the village's pigs.

Social Organization

Among horticulturalists, food-getting tasks are most often divided between men and women. For instance, among the Yanömami of the Amazon rainforest, men clear and prepare fields and plant and harvest crops, including plantain, sweet potatoes, cotton, and tobacco. They also hunt and fish, controlling the group's food resources. Women's work among the Yanömami is entirely domestic. In contrast, among the Jivaro of Peru, women are responsible for planting, tending, and harvesting crops, including the sweet potato, manioc, and squash that provide the bulk of the Jivaro diet. Men hunt game and fish as supplements.

Moving from a nomadic life of foraging to a sedentary village life of tending gardens rearranges the most basic patterns of social life. No longer are people keeping only what they can carry on their backs; now they can accumulate goods and store them in their homes. This fundamental shift in behavior brings with it a challenge to the traditional egalitarian values of the group. Where sharing and cooperating were the most essential practices, now inevitably some individuals and families will have more than others, based on the production of their gardens. The tensions created by these new practices need to be reconciled, as sharing is an intrinsic part of their value system.

In order to maintain an equal level of status among all members of the group, a society will practice some sort of **leveling mechanism**. This is a social and economic obligation to distribute wealth so no one accumulates more than anyone else. Settled societies develop rules for how and when goods get distributed, with the wealthiest members of the group experiencing the most pressure to share with others. Between individuals, leveling may take the form of "demand sharing," in which members of the group may request items on demand. Among the Ju/'hoansi, it is perfectly appropriate to demand or take meat or other food items when hungry.

There are also social institutions that more formally distribute wealth, such as the **cargo system** found in Maya villages and towns in southern Mexico and Central America. A man undertakes obligatory volunteer service, using his personal wealth to support local events. The more wealth he has, the longer he is pressured to volunteer, and the more prestige he earns. Ideally, leveling mechanisms such as the cargo system help to keep the socio-economic system of a horticultural society aligned with the traditional values of a foraging one.

Horticulturalists and the Environment

A sustainable method of farming when there is plenty of available land is known as **swidden** (also slash-and-burn) **cultivation**. This is the primary technique used in many different locations around the world to grow crops ranging from bananas to rice. Using swidden cultivation, farmers prepare a plot of land by clearing fast-growing trees and other plant material from an area and burning the debris directly in the plot. Ash from the fire acts as a soil conditioner and fertilizer, containing high levels of potassium, calcium, and magnesium. Gardens are planted in the nutrient-rich ash. After harvesting crops from that plot for a time, farmers move to another area and begin again.

The movement from place to place on large areas of land allows the used plot to lie fallow and "rest." Wild plant material eventually regrows. Depending on the amount of land available, a group can farm many plots in this way before returning to the first, allowing land to lie fallow for up to 10 years or more.

Done correctly, swidden farming works with an area's natural ecosystem. The swidden technique mimics what happens after fires burn a landscape: after several years, plant life flourishes again. Done poorly, however, it can erode the soil. This is the result when plots are not left to lie fallow but are used continually without the micronutrients in the soil being replenished.

Resource Distribution: Balanced Reciprocity

Because horticulturalists live in small-scale societies, as foragers do, they also practice reciprocity. Their main methods of distributing food within the village are generalized reciprocity (in which they share with family and close friends) and **balanced reciprocity** (in which they trade with others outside their trusted circle). Balanced reciprocity is an exchange in which both the value of goods and the time frame of repayment are specified. Trading partners who need to ensure that items or payment will be delivered on time use this type of exchange. Because the value of the items is known, as is the delivery time, failing to come through is a major social transgression.

Off the coast of Papua New Guinea, a system of balanced reciprocity exists called the **Kula Ring**. This involves the circulation of gifts among trading partners in the archipelago of the Trobriand Islands. On an agreed-upon date, and with all of the necessary magical and ceremonial preparations complete, a man sails out to a designated location between islands to meet his long-time trading partner. At that time, he passes on a gift of a red shell necklace (*mwali*) or white shell armband (*soulava*). As part of the same ceremony, he receives the opposite item from his trading partner. The necklaces move in one direction around the islands and the armbands in another. It may take up to a decade before the items return to this same person. Having the items in one's possession gives a man status, but more importantly, the history of each object remains with it. Trading exchanges like these maintain alliances between groups on different islands, sometimes preventing them from going to war with one another. The man who has at one time owned, and then given away, many Kula items enjoys a great amount of prestige.

Another example of a gift-giving ceremony with great social and cultural significance is the **potlatch**, an event common to Indigenous groups of the Pacific Northwest Coast extending from Alaska southward through British Columbia, Washington, and Oregon. The basic elements of the potlatch include a host group (a kinship group) inviting guests to witness an event of significance.

The potlatch typically includes the reciting of oral history, feasting, dancing, and gift-giving. The acceptance of gifts following the event signifies the acceptance of the event.

Prior to the arrival of Europeans in the region, it was likely that potlatches were quite rare for any particular group to host, being reserved for such events as a person's formal assumption as chief. Neighboring groups would be invited, and the potlatch would last weeks or even months.

In 1885, the Canadian government imposed a legal ban on potlatches, with imprisonment as punishment. They saw the potlatch as wasteful, harmful to economic growth, and an impediment to social progress. As a result, thousands of items used in the ceremonies, many sacred, were confiscated and ended up in private and museum collections. Even more detrimental was the disruption of an essential aspect of coastal First Nations life, with an entire generation unable to participate unless the ceremonies were underground. The ban on potlatches in Canada was lifted in 1954.

Potlatches continue in contemporary times and today are much more common than in precolonial times. In addition to being organized to validate a person's assumption of the position of chief, for example, potlatches may be held today for a variety of reasons, including a person obtaining an Indigenous name (and all the rights and responsibilities that go with that), marriages, and mourning the loss of a community member. Acceptance of gifts signifies that guests agree that the host has the right to their position, rights, and responsibilities.

Although we include the potlatch in a chapter on subsistence and economics, recently many anthropologists have taken issue with the tendency to reduce it to those functions alone. While the explicit function of the potlatch is to validate an event of significance, it has other social and ideological functions as well. Potlatches provide opportunities to recite and validate oral history, validate myths and other stories through performance, affirm identity and status, and maintain alliances. In addition, some anthropologists are critical of the lack of focus on how colonialism affected the practice, including, but not limited to, legislation.

Pastoralists

Not all foragers find it most efficient to settle in villages and plant gardens. Herding pasture animals is most successful in areas where the ecological conditions are poor for farming, such as in desert environments. The way of life that revolves around herding animals is called **pastoralism**. Depending upon the region, animals suited for herding may include goats, sheep, camels, yaks, llamas, reindeer, or cattle. Social and political motivations may also contribute to adopting livelihoods that shift from foraging to pastoralism.

Pastoralist Foodways

In pastoralist societies, **animal husbandry** is the main mode of sustenance. Animal herds provide food staples such as milk, blood, butter, yogurt, or cheese. Occasionally, an animal may be slaughtered for symbolic or ritual purposes, but the utility of live animals far outweighs the benefits of slaughtering animals for meat. Although pastoralists generally don't farm, some groups may practice a more diversified economy that includes some cultivation. They also trade with neighboring groups for food and other items.

The Basseri are pastoralists who live in Southern Iran. Today, there are approximately 16,000 Basseri occupying 3,000 tents in a region that extends from mountains to desert. The group is divided into networks of families that migrate together (occupying a handful of tents in the winter and up to 40 in the summer). They are nomadic, moving their herds of sheep and goats along a route called the *il-rah* (tribal road). The road is the property of each tribe at a specific time of year, allowing full access for all groups.

Men generally ride horses while migrating along the route, while donkeys and camels carry women, children, and possessions. To maintain an adequate standard of living, each household strives to keep at least 100 sheep and goats; some may have up to 400. Milk and milk products (buttermilk, butter, and cheese) make up the bulk of the Basseri diet, supplemented with meat. The Basseri occasionally forage, hunt, and cultivate for additional dietary items, although the majority of external items comes from trade or purchase at the marketplace.

Figure 10.3
BLUE-VEILED TUAREG NOMADS, NORTH AFRICA.

These Tuareg men, notable for their blue head scarves, live a pastoral lifestyle in the Sahara Desert of North Africa. They ride and herd camels, moving across national borders to access pasture lands.
Credit: Barry D. Kass/Images of Anthropology

Pastoralists are nomadic, since herding animals requires going to where the grazing is good. Therefore, male herders may leave their families at a home base and be away for months at a time. Most pastoralists, such as the Basseri, use horses to cover a lot of territory and help with herding the animals. During the warm months, the group may move anywhere from once every three days to as often as once a day. During the cold season, the base camp may remain stationary for longer, with herders making forays out to pasture.

Each tent houses an individual family that is relatively autonomous, although the larger social group consists of all the families that migrate together. The division of labor requires that, generally, men and boys herd animals, haul wood and water, and roast meat back at the tent. Women generally take care of the majority of food production and other domestic duties such as washing and sewing.

Pastoralists and the Environment

Nomadic pastoralism is sustainable in environments that are unsuitable for farming. Land that may be unproductive for cultivation can serve as excellent grazing lands for herd animals. Pastoralists may seasonally move back and forth over long distances to productive pastures, a migration movement known as **transhumance**. In addition, grazing may actually help the environment in that it encourages the **biodiversity** of native plants.

Pastoralists attempt to utilize every part of the animal and minimize waste. Beyond food products, animals also provide material goods. For example, animals' hair, wool, and hides can be woven into clothes, shoes, and tents. Dried organs, such as stomachs, can be used to carry water. The manure of grazers is highly fibrous, allowing animal dung to be used as fuel for fires.

Resource Distribution: Reciprocity

Like foragers and horticulturalists, pastoralists practice reciprocity. When people know one another well, the **social distance** is minimal. Thus, they are likely to practice generalized reciprocity. When items are traded between lesser-known or unknown members of different communities, greater social distance requires the use of balanced reciprocity. In this case, items are generally exchanged on the spot for an agreed-upon value.

Intensive Agriculturalists

Large populations that can produce more than just the amount of food required for a subsistence economy practice what we call **intensive agriculture**. This type of planting is intensive because the land has a short (or no) fallow period, meaning fields are planted year-round with different crops. The intensity of this planting

Figure 10.4
WET RICE CULTIVATION, IRAN.
Women plant and harvest rice in the province of Mazandaran, Iran, on the southern coast of the Caspian Sea. Plentiful water is essential to wet rice cultivation.
Credit: Mostafa Saeednejad/ CC BY 2.0

may deplete the soil more rapidly than horticultural methods, which typically allow fields to recover for a time. Therefore, agriculture requires more preparation and maintenance of the soil through fertilizers, crop rotation, and water management. This type of intensive cultivation generally also requires more highly developed tools, such as plows, irrigation, and draft animals. All these inputs cost more in human labor, but also make agriculture more productive per acre than horticulture.

Intensive Agricultural Foodways

The earliest evidence of agriculture is from approximately 9,000 years ago in the Middle East. Populations living between the Tigris and Euphrates Rivers in Mesopotamia settled on the rivers' floodplains to make use of fertile land and water resources. They dug irrigation canals to bring water to their crops, relying heavily on grains such as wheat and barley. Over the next several thousand years, agriculture appeared independently in other locations across the globe: the Indus Valley (Pakistan), the Yellow River Valley (China), the Nile Valley (Egypt), the Andes (Peru), and Mexico. Box 10.2 describes some of the unique features of Aztec agriculture.

Where agriculture arose, populations grew with a steady supply of food from crops. One might assume that since agriculture led to population growth, farming supported better nutrition. In fact, the opposite is true. Studies of the bones and teeth of people in farming societies, compared to those in foraging societies, show that health suffered under an agricultural lifestyle. This is because most agricultural societies depend heavily on just a handful of crops, especially grains, reducing the variety of vitamins and minerals in their diets.

There is also evidence for **domestication** of animals such as cattle, goats, and sheep in the same time period. Animal domestication refers to the process of shaping the evolution of a species for human use. This is done through choosing the traits most suited to human needs and breeding animals for those traits. Domestication develops companion animals to accompany hunters and working animals for farms, as well as provides alternate sources of nutrition from animal products.

BOX 10.2   **Food Matters: Ancient Aztec Foodways**

The Aztec Empire was built on intensive agriculture. However, it began with the migration of nomadic peoples called the *Mexica* (me-SHEE-ka) into the Valley of Mexico. Settling on an island in today's Mexico City, the Aztecs founded Tenochtitlán in 1325 **CE**. By the time of the Spanish conquest of Mexico City in 1521, approximately 200,000 Aztecs inhabited a series of islands linked by waterways and canals (drained by the Spanish as part of their conquest strategy). Agriculture laid the foundation for the growth of the Aztec population, although they still supplemented their diet with foraging, hunting, fishing, and swidden cultivation.

Maize (*Zea mays*) was a staple crop and played a revered symbolic role in Aztec political and religious life. Sacred deities personified the phases of the lifecycle of corn, including Xilonen ("fresh, tender corn," a virgin goddess); Cinteotl ("deified corn," or "Maize Cob Lord"), and Chicomecoatl ("7 Serpent," the personification of mature, dried seed corn) (Gassaway, n.d.).

Other important crops included beans (*Phaseolus vulgaris*) and squash (*Cucurbita* varieties). Together with maize, the Aztecs planted these crops, called the "Three Sisters," in close proximity: the maize stalks provide support for bean vines, and squash plants suppress weeds, growing low to the ground. This type of companion planting can produce high quantities of calories per acre. In addition, beans are "nitrogen-fixing" plants, which replace the nitrogen in the soil used by the maize. These three staple crops support soil sustainability and still provide nutrition for millions in the Americas today.

In addition to practicing traditional agriculture, the Aztecs employed an ingenious method of increasing farming acreage using the waterways surrounding their islands. They developed floating gardens, called *chinampas*. The chinampas were built by piling mud into a shallow area of water and planting willow trees in the corners. Roots of the trees would anchor the garden to the bottom of the lake, creating an artificial farming platform.

Social Organization

Intensive agricultural cultivation requires a fully settled population that can work the land throughout the year. Because a shift to grains as the staple crop can feed a large number of people, agriculture allowed populations to grow and settlements to expand over wide areas of land. Large populations result in more complex social, economic, and political systems. This complexity is reflected in the way settlements expand into a tiered structure, with high-status people living in the central area and lower-status people living in villages on the periphery. Because the central settlement is heavily populated, it is referred to as a **city**.

No longer is farming a way of life for everyone, as in smaller-scale societies. Therefore, occupational specialization begins. Agricultural laborers do not own

BOX 10.3

Food Matters: Sharing "Spread" in Prison – Reciprocity and Social Capital

A couple of hours before lights out in county jails and prisons, inmates start to feel familiar pangs of hunger, unsatisfied physically and emotionally by the tray meal served as dinner. At this time of the evening, many inmates begin making a "spread" to share. On the surface, it's a simple food dish made up of whatever the inmates can cobble together. Individually, it's a way to connect with one's identity and foods of comfort while getting needed calories. In an economic sense, however, sharing spread is an important way to build one's social status and share membership in small networks of people behind bars.

Spread is a makeshift prison meal with an instant ramen noodle base. Anything can be added – beef sticks, Cheetos, corn chips, pickles, sweet jam packets – to simulate the flavors of foods enjoyed on the outside. All the ingredients are pounded together and "cooked" with hot water so the starchy noodles absorb all the flavors. The cook's imagination is limited only by the ingredients that can be found on the regular meal trays or purchased in the commissary.

In their research, anthropologist Dr. Sandra Cate (2008) and photographer Robert Gumpert found that inmates craving Mexican flavors might add tortilla chips, jalapeño-flavored cheese product, hot sauce, and chili beans to their noodles. Others may prefer an "Asian stir fry," with ramen soaked in peanut oil saved from a lunchtime peanut butter sandwich and mixed with leftover vegetables and meat. By simulating tastes from their lives outside of prison, inmates gain a degree of comfort from these meals.

Sharing spread is essentially an act of balanced reciprocity, since those inmates who add ingredients or materials (such as a garbage bag to "cook" larger quantities) most often are the ones to partake in the meal. This provides a sense of fairness over who gets a seat at the table and who doesn't. However, generalized reciprocity allows some charity to

their farms but work for others. Owners of the land reap the benefits of their labor, as well as the wealth produced from selling the crop surplus at the marketplace. Others pursue a multitude of occupations, such as artisan, trader, merchant, soldier, or scribe. Some occupations are more highly valued than others, as reflected in a social and economic hierarchy.

This type of complex society requires the control of a centralized governing body, with the power of an officially recognized politico-religious leadership. A class of **nobles** develops, which is able to harness the labor of workers to farm, build, or fight. The **peasant** class supports the growth of the settlement by providing labor, often under oppressive conditions. Agricultural societies force the development of a social hierarchy in which those who control resources have power over those who do not.

inmates who might not have anything to share that night, or who don't get commissary money sent to them from the outside. Like any exchange in a tight-knit community, the exchange must be reciprocated at some point, or the relationship will end.

Figure 10.5 **INMATES SHARING "SPREAD."** Inmates at San Francisco County Jail #5 give thanks before sharing a meal of spread.
Credit: Robert Gumpert

French sociologist and anthropologist Dr. Pierre Bourdieu (1986) discusses these kinds of real and symbolic exchanges in the context of **social capital**. Like economic capital (money), social capital is the set of resources accessible to a person by virtue of their membership in a social group. High-status groups provide insiders with social capital, while low-status groups have less to offer. In the case of these makeshift prison meals, those with the knowledge, ingredients, or materials to make spread become part of a community. Membership in that community continues as long as the exchanges continue.

In prison, these "communities" are generally ethnically based. Thus, African Americans, Latinx, White, or other groups of inmates form social and political networks around ethnic identities. The sharing of spread helps to maintain the bonds between members and delimit the boundaries of who is in and who is out.

The world of agriculture changed dramatically before the invention of modern agricultural practices with the plantation system, in which British and other Western European companies captured African people and brought them across the Atlantic Ocean to work as enslaved laborers. Roughly 12 million African enslaved people toiled in fields of sugar cane, tobacco, coffee, cotton, and other profitable products to bring wealth to the colonies' landowners in the Southern United States, Caribbean, and South America. Up to two million African people are thought to have died from the horrific conditions crossing the Middle Passage from West Africa to the Americas from the late 1500s to early 1800s. Human slavery is the most extreme version of any agricultural hierarchy, set in a much larger context of colonization and capitalism.

Intensive Agriculture and the Environment

Agricultural production leads to an entirely different relationship between people and land. While small-scale cultivation generally conserves future resources, the goal of large-scale agriculture is to maximize production. The intensity of year-round cultivation requires the use of more advanced tools. Draft animals suited to the area (such as oxen, zebu, or yaks) are used to pull plows to till the soil and create trenches for planting.

Agriculture takes many forms based on the needs of different crops. The most common crops are maize, wheat, rice, millet, sorghum, and barley. Rice, first domesticated in China approximately 9,000 years ago, is one of the world's most commonly cultivated staple grains. Different varieties of rice are suited to different growing methods – such as dry rice cultivation, wet rice cultivation, and deep-water rice cultivation – depending on the ecology of the area. Highland areas may be **terraced** to accommodate the irrigation needs of rice or other crops on mountainsides.

Resource Distribution: Redistribution and Market Economy

Societies that have developed central authorities, such as religious or political leaders, have more control of resources. They can demand taxes or tribute or hold festivals for religious deities that require donations of food or money. Two ways that societies with centralized governing bodies can get food and other resources to their members are through redistribution or the market economy.

Redistribution

Redistribution is the process by which goods and money flow into a central entity, such as a governmental authority or a religious institution. These goods are counted, sorted, and allocated back to the citizens. Taxes and tribute are forms of redistributive processes. For instance, modern industrial societies require that citizens pay taxes annually. The monies collected are then redistributed through public works such as road repaving, brush clearing on public land, or other infrastructure upgrades. Tribute items are those that are required by a central governing body at regular intervals, in addition to or in lieu of taxes. Tribute items include food items or material goods.

The Aztecs of ancient Mexico demanded vast quantities of tribute from their subordinate regions. Regularly, tribute chiefs called *calpixque* would collect foodstuffs, jaguar pelts, paper, copal incense, and ceremonial items to bring to the Aztec capital of Teotihuacán in exchange for protection. Some scholars argue that it was precisely the vast quantities of tribute that helped rally the people dominated by the Aztecs to the Spanish cause. Many of those groups turned against the Aztecs when given the opportunity.

Redistribution is also used in religious practices when offerings for gods or ancestors are brought to a place of worship. After the items are made sacred in a ceremony in which the gods are thought to partake of them in a nonearthly way, they may be divided among the worshippers. The Hindu *puja* is a form of worship in which members of the religious community bring offerings to the temple. After the ritual, food may be divided and shared with members.

Market Economy

Large and complex populations develop a **market economy,** which is a more formal and bureaucratic system. The laws of supply and demand set market rates for food and other goods, which must be traded or purchased according to a set price. The price remains the same for all consumers, some bargaining notwithstanding, since most buyers and sellers no longer know one another personally. (Informal economic exchanges also persist in market economies. People make reciprocal exchanges between family and friends.) A market economy is the foundation of a **capitalist system**. In order to participate as members of society, individuals pay taxes to the government, which then redistributes the money in public works, such as infrastructure improvement.

Intensive agricultural and industrial economies are built in the marketplace, in other words, in the buying and selling of goods and services. Because farmers are producing a surplus, a central location for exchange draws people to negotiate the cost of items. In general, the laws of supply and demand set prices. In other words, when there is a lot of something, it will fetch a low price, but when there is little of something that many people want, it will fetch a high price. Staples such as grain will be accessible to all, even the lower social classes. Only the upper classes will be able to afford exotic goods.

The market economy is based on the use of **money** for buying and selling goods and labor. Today, we think of money in terms of dollars and cents. However, throughout history, money has taken many forms. It can be anything that is used to measure and pay for the value of goods and services. Money must be portable, so it can be brought to the marketplace for transactions. It must also be divisible, such that it can be measured to the appropriate amount, and change can be given. Trading land for a cow is fine if the value of the plot of land equals that of the entire cow. Change can't easily be given, though, unless the cow is butchered (which can be messy). It is much easier to weigh out bags of salt or yams to the exact amount.

Other examples of items that have been used as money throughout history, and that are more easily divisible, are shells, teeth, jaguar pelts, bones, beads, tobacco, and metals. Foodstuffs such as salt, rice, cacao beans, peppercorns, and alcohol have also been commonly used as money. Teeth, bones, and shells are referred to as

special-purpose money, in that these items were used only to measure the value of things in the marketplace and lacked another use beyond a symbolic one.

In contrast, salt and cacao beans are **multipurpose money** (also called commodity money) in that the commodity can be used for other purposes besides simply as money. In other words, the item has value in itself. For instance, salt is an essential mineral for human bodies and is used to preserve and flavor foods. Societies that have used special-purpose and multipurpose money for generations suddenly find themselves "poor" in a market economy. One result of colonization is that it changes the social and economic value of items and makes cash the only valued mode of payment. This often forces people to find jobs, usually menial or low-wage, within the cash economy.

Ancient Aztecs and Maya greatly valued cacao beans and used them to make a sacred drink used in religious rituals by the elite. Cacao beans were so valued that they were included in the list of tribute that Aztec-controlled regions paid annually to the empire. Mesoamericanist Dr. Michael Coe (2013) found that the Aztecs demanded a total of 980 loads of cacao beans annually, each load weighing 50 pounds.

Early Spanish settlers to the area were even tricked by counterfeit cacao beans. The beans were actually removed from the pod, which was then stuffed with dirt to give the right weight and feel before its use as money! This would be an example of **negative reciprocity,** in which the seller is deceiving the buyer as to the real value of the object. Sometimes negative reciprocity can be expected, such as when trading partners with complicated histories meet to exchange items. However, in this case, the Spanish may not have expected the deception. With dirt inside the cacao pod, buyers are certainly not getting what they paid for.

Industrialism

Industrialism is a way of life in which highly mechanized industry produces food. This was the second major shift in food-getting technology. The first shift was working the land, rather than simply relying on its bounty. The second shift took agriculture out of the hands of many workers and placed it in the hands of fewer, who rely on advanced technology. The productiveness of a farming operation on a massive industrial scale relies on organization and management, the power of machinery, the effectiveness of **chemical inputs** into the soil, and, today, information provided by the internet. The main goal of using technology to produce food is to create a viable product at the lowest cost possible.

Industrial Foodways

Around 1800, a slow but steady industrial revolution began changing the way people in Western countries did their work. New machinery took over small-scale or home-based production, completing products much faster and more efficiently. Steam-powered engines fed by coal were put into wide use for transportation and power generation. Larger-scale wind- and water-powered technology, such as windmills and water wheels, allowed farms to grow in size and to produce more food for more people at a lower cost. Since the middle of the twentieth century, mechanized production has moved toward tractors and combines that are powered by gasoline. Agriculture is now also heavily reliant on biochemicals such as pesticides, herbicides, and fungicides to help manage the success of crops on such a large scale.

Fields with thousands of acres may today be planted with a single crop, such as corn or soy, to maximize profit. **Monocultured** crops are more susceptible to loss from a single type of soil-borne illness or insect pest than are naturally resilient mixed ecosystems. Monocropping also depletes certain nutrients from the soil, especially when done year after year. Companies that produce and control seeds, fertilizers, and chemicals are constantly seeking new technologies to make their products more attractive than those of others. Unfortunately, agricultural products are consolidated into just a handful of multinational companies, which tend to make a profit at the farmers' expense.

An example of one country's challenges with industrial agriculture is India's "Green Revolution." In the 1960s and 1970s, farmers in the Indian state of Punjab were the first to adopt North American technologies to increase crop yields using high-yield seed varieties, chemical fertilizers, mechanized irrigation, and, later,

Figure 10.6
MECHANIZED
COTTON
HARVESTING.
A cotton picker mechanically harvests and rolls cotton into modules in this monocropped field in Mississippi.
Credit: Faungg's Photos/CC BY-ND 2.0

genetically modified (GM) seed. At first, these technological advances in agricultural methods increased yields and supported the incomes and health of farming families. But these gains were not sustainable. Farmers who adopted the intensive input technologies found their soils stripped of nutrients. Purchasing seed and fertilizers annually became a major burden for rural farmers. In addition, heavy water requirements ended up tapping the groundwater wells dry.

In tragic, worst-case scenarios, farmers have been taking their own lives in order that their families can receive insurance money on which to survive. This has been occurring with alarming regularity, with official reports of 12,000 Indian farmers committing suicide in 2015 alone. A recent study by Carleton (2017) looking at trends over 47 years directly links rising global temperatures and resulting crop failures to these suicides.

Social Organization

For examples of an industrial society, all we need to do is to look around us. Industrial food production operates in our cities and towns, and it links food producers to consumers on a global scale. Ironically, the ability to feed more people than ever through mechanized technology has created a situation in which fewer people than at any time in our history are involved in the production of their food.

Most food production takes place in rural areas on private lands owned by corporations. These farming operations can stretch over thousands or hundreds of thousands of acres. The general public is not allowed on these private lands, so consumers do not see crop or meat production. Food animals are raised in **confined animal feeding operations (CAFOs)**, which operate differently than farms of the past did. In CAFOs, thousands, or even millions, of animals are fenced or crated to create maximum profit in a minimum of space. Because these conditions are often unhealthy, **conventional** production demands that antibiotics be given to food animals at every meal to prevent illness.

In addition, food production is a hierarchical activity, with landowners and land managers in control of a large workforce. Farm workers are referred to as "unskilled" laborers due to their rote tasks and low wages. This ranked structure has its roots in the plantation slave economy discussed above.

Agricultural workers have seen some improvement in working conditions in more developed nations. However, conditions faced by the vast majority of farm workers who labor for others are largely unregulated. Often without legal status, these immigrants work for low wages, live in difficult housing conditions, and lack medical care and job security. Immigrant workers are a crucial part of farm labor and industrial food production, which would effectively collapse without them. Economists and academic researchers agree that international immigration has little to no economic impact on native-born populations and, on average, benefits them (Card et al., 2012).

The process of food distribution in industrial societies is complex. Conventionally produced fresh food goes on a long journey from "farm to fork," traveling an average of 1,500 miles (Pirog et al., 2001). Produce must be picked before it is ripe and transported to several distribution centers before arriving at the store to be sold. Shipping produce long distances requires that varieties be selected for color and durability rather than solely for taste.

Most industrial food products are highly processed and require preservatives to ensure a long shelf life. Processed foods with added chemicals are less healthful for the human body than fresh meats or produce, and they are sold at much lower cost. Therefore, the very cost structure of food for purchase challenges people with limited income to regularly eat healthful foods.

Industrialism and the Environment

Industrial food production creates a number of environmental concerns. Most of these relate to pollution of the area surrounding farming operations and beyond. Pollution may be caused by animal waste or biochemical inputs such as pesticides and herbicides.

Since factory farms concentrate an enormous number of animals in a very small area, the farms generate far too much manure to be absorbed by the land. Excess

Figure 10.7
INDUSTRIAL FACTORY FARM.
Factory farms and confined animal feeding operations have drastically changed the way people produce food. Farmers are now managers of a largely unskilled work force where profit maximization is the goal. Even with government subsidies, it is hard to make a good living as a farmer in the United States today.
Credit: MENATU/Shutterstock

manure is stored in huge holding tanks or manure lagoons, and it is often overapplied to fields, releasing hazardous gases into the air. It often contaminates local groundwater and waterways with pathogens and excess nutrients. According to the United States Environmental Protection Agency, agricultural practices are responsible for 70 percent of all pollution in US rivers and streams.

Pesticides are also responsible for illnesses in people who are exposed to them through farm labor, through spraying around their homes, or in the foods they eat. Even when pesticides are used correctly on farms, they still end up in the air and in the bodies of farm workers. Pesticide exposure is associated with dizziness, headaches, nausea, vomiting, and skin and eye problems. Long-term exposure is associated with more severe health problems such as respiratory problems, memory disorders, miscarriages, birth defects, and several types of cancer.

THE HUMAN DIET

With all these different methods of food procurement and production, it follows that human diets are widely diverse. The Maasai and Samburu (*Lokop*) people of Kenya and Tanzania can live mainly on blood, milk, and occasionally meat from their cattle; Hindus eat a vegetarian diet of mostly grains, pulses, and vegetables; and Inuit can mostly subsist on fish, seal, whale, and other marine life. How can all these populations be healthy?

The human body has the incredible ability to get the nutrients it needs from many different sources of carbohydrates, fats, proteins, and a range of vitamins and minerals. The environment can also help nutrient synthesis. An equatorial climate helps the skin synthesize vitamin D quickly from the sun. Vitamin D is essential for growth and development, and Vitamin D from sun exposure may compensate for a lack of vitamin D–rich foods.

One thing that seems to occur with regularity is a decrease in the level of nutrition when people shift from a diet based on locally sourced and home-prepared foods to one that is heavily based on store-bought and processed foods. Local foods offer a diverse array of nutrients, while processed foods rely heavily on white flours, processed soy, and chemicals. Sources of sugar in the local environment, such as fruit or honey, often provide better nutrition and a lower **glycemic index** than processed sugars.

This shift in diet and the associated health problems, including obesity, is referred to as **nutrition transition** (Popkin, 2001). It has occurred all over the world as a result of globalization. For some Indigenous groups, such as the Akimel O'odham and Tohono O'odham of Arizona, the transition has taken a debilitating toll on their members' health. Obesity is common, and the incidence of diabetes is 15 times higher than in the non-Native population. A community-wide initiative to return to local foods (such as tepary beans, cholla buds, and cactus fruit) has been successful in some areas where people have committed to changing their lifestyle, and in fact it has restored health for Native individuals (Nabhan, 2002).

Industrial food production in the developed world today provides consumers with a mind-boggling variety of choices. There is so much food choice that consumers can adopt a specific diet based on personal goals. Cultural trends and a quest for good health may cause us to limit our choices and take on a label for our eating style, such as vegetarian, vegan, pescatarian, Paleo, or raw-foodist.

Studying human diets throughout the past several hundred thousand years has made it clear to anthropologists that humans evolved as omnivores. The bulk of our human diet has come from plant material and wild animal protein. Of course, the ratio of plant to animal food items was dependent on what the environment offered. It appears to be most adaptive for humans to eat whole foods that are as close to the natural forms in which they grow as possible, avoiding highly processed food products.

Conventional food production, with its focus on high yields, results in lower levels of nutrition within food items today than in the past. Conventional food production has other risks as well including unsanitary factory and farm conditions and environmental degradation. Although the human diet evolved as omnivorous, the negative aspects of industrial production cause some individuals to choose alternative diets that reduce the impact of their food choices on the environment.

SUMMARY

This chapter has examined the different ways in which people procure their food and access the resources they need to survive. Mirroring the Learning Objectives stated in the chapter opening, the key points can be summed up as follows:

- Depending on the limitations of the immediate environment, a society's technology (including knowledge and skills) will be different.
- The largest difference in technologies is between those people who seek their food (food foragers) and those who manipulate the environment to grow and raise it (food producers). Not only are the food procurement techniques different, but the social structures and environmental impacts are different as well.

- The population size and complexity of societies tend to grow as people move from foraging to horticulture or pastoralism. Intensive agriculture allows a society to support a very large population. The mechanization of industrial agriculture can feed people all over the globe.
- Different types of economic systems – including reciprocity, redistribution, and market economy – are used in societies with different population sizes and levels of complexity.
- Each food-getting technique creates a different relationship between people and their ecosystems, with some of the most aggressive practices occurring on a large scale today with the use of monocropping in agriculture.
- It seems clear that as long as humans eat a diet based on whole foods, they can be healthy and thrive on a wide variety of foods.

Review Questions

1. What characteristics distinguish food foragers?

2. What characteristics tend to correlate with the five basic food-getting strategies?

3. Which food-getting strategy has the least environmental impact, and which has the greatest environmental impact?

4. What are the three basic types of economic systems?

5. What are some of the major changes that have accompanied industrial farming?

6. Since humans everywhere eat different kinds of foods, what seems to be the main requirement for a healthy diet?

Discussion Questions

1. What type of "tool kit" does a member of a modern industrial society need to survive?

2. Since foraging is the least disruptive to the environment of all the food-getting strategies, why can't people just live by foraging today?

3. Do alternative food movements today have any similarities to any of the traditional foodways?

4. What are some of the varied responses of modern consumers to the industrial food system?

Visit **www.lensofanthropology.com** for the following additional resources:

SELF-STUDY QUESTIONS	WEBLINKS	FURTHER READING

11

MARRIAGE, FAMILY, GENDER, AND SEXUALITY

LEARNING OBJECTIVES

In this chapter, students will learn:

- *about the variety of stable marriage and family patterns across cultures.*
- *about different cultural rules for taking one or multiple spouses.*
- *the correlates of different kinds of family and residence patterns.*
- *about marriage as an economic exchange between families, requiring compensation.*
- *how different societies trace their family lineages.*
- *that ideas about gender are cultural constructions, while the idea of sex is a biological one.*
- *that a variety of gender identities and sexualities exist on a spectrum.*

> **Families, gender, sexuality, and sex are all on a spectrum, like nearly everything else about humans.**
> #MarriageFamilyAndGender

INTRODUCTION

Expectations about how men and women should behave in society are shaped by culture. Our roles are learned early within the family, where the fundamental structures of social life are organized. The roles and responsibilities of marriage partners, parents and children, siblings, and extended family members teach us the appropriate ways to behave as a male or female, and the limits placed on that behavior.

This chapter examines family life and marriage patterns across cultures – for instance, why some societies prefer arranged marriage or accept multiple husbands, and why others do not. Although North Americans may be most familiar with a particular type of family structure – that is, parents and children living together in a single-family home – this is not the most common arrangement. In fact, the small, two-generation family unit is a relatively new development in cooperative living. Anthropologists studying marriage and family patterns around the world recognize that there are diverse family arrangements that lay the foundation for a stable society.

Furthermore, the male/female gender divide that North Americans take for granted is not universal. In the two-gender system, males are "masculine" and females are "feminine" (according to Western concepts of gender). However, gender roles vary widely across cultures in many different and accepted ways. This chapter will also examine how culture, not biology, dictates the social norms and expectations of men and women across the **gender spectrum** in society.

MARRIAGE

The first thing that may come to mind when we think of **marriage** is love, which is something that modern Westerners accept as an important bond between people who chose one another as partners. In fact, even though people across cultures and throughout history have experienced marital love, it is not always the basis for marriage. Until recently in human history, social, political, and economic reasons have been the main reasons to bring families together through marriage. Indeed, love itself is a cultural construction; that is, culture shapes the way that people think about love and behave while "in love."

It follows that while marriage and family are found in some form across cultures, practices differ widely. How do we identify marriage with all of these variables? To recognize marriage across cultures, anthropologists look for three main characteristics: (1) sexual access between marriage partners, (2) regulation of the sexual division of labor, and (3) support and legitimacy of children in society.

1) Sex: Sexual relations within a marriage partnership are sanctioned by society. This doesn't mean that extramarital affairs are not expected. In fact, in some societies it is understood that husbands and wives will take lovers outside their marriage. Among the Ju/'hoansi, for instance, taking lovers is a common practice as long as it is discreet.

2) Division of labor: Marriage regulates the tasks that women and men are expected to perform in society. Some of these are biological, such as childbirth and nursing of infants. However, most of the expectations about what work men and women will do are based on cultural values of what is appropriate. This sexual division of labor contributes to the group's survival because, although men and women perform different tasks, resources will be shared.

3) Children: Children need care and support to grow up physically and emotionally stable. A socially sanctioned marriage and family structure – in whatever

Figure 11.1
CATHOLIC WEDDING CEREMONY IN MANILA.
This couple is getting married in a Catholic church in Manila, Philippines. Marriage is set within a complex set of traditions that include expectations regarding religious practices, family patterns, and social life.
Credit: Barry D. Kass/Images of Anthropology

form they may take – helps provide the kind of environment that supports child development. In addition, children born from a marriage union are considered to be legitimate heirs to family property. They will take on the benefits and responsibilities that come with inheritance.

SPOUSES: HOW MANY AND WHO IS ELIGIBLE?

All societies have clear rules for marriage, though they may differ greatly. We can think about marriage in human society as a partnership between families who join together in a relationship that is based on an exchange of partners. This way of thinking may not be familiar to modern Western people who choose partners based on love and trust. However, it makes sense when we consider economic patterns of exchange. Family bonds can contribute to survival and mutual support by sharing resources such as land, food, and money, or non-tangibles such as childcare, time, and labor.

Family roles will differ based on whether members are related by marriage or by blood. Two people joined in marriage create a web of economic and social relationships between their **families of orientation**, including their blood-related parents, siblings, grandparents, and more distant relatives. If the pair has their own children, we refer to this as the **family of procreation**. Each of these family members has a role based on their position in the social group.

Monogamy

Marriage between two people is referred to as **monogamy**. Marriages that are most common in the world are opposite-sex unions, between one woman and one man. Societies with high divorce rates practice **serial monogamy**. This is marriage to one partner at a time with a succession of partners.

Industrial Western nations practice monogamy due to historical and social factors. First, European colonizers imposed strict religious rules wherever they conquered and colonized native peoples. Catholicism requires monogamy, based on a series of laws established in the fourth century **CE**. The Roman Catholic Church created a series of prohibitions against multiple spouses, divorce, and adoption. Childless couples, having no other recourse, were then forced to bequeath their land to the Church. For this reason, the Roman Catholic Church became the largest landowner in Europe.

Second, monogamy works best in countries that rely on independence training to raise children, such as in Western countries. One of the results of independence training in formative childhood years is that children will grow to be individualistic adults with a drive to act in their own best interests. Therefore, choosing one's own spouse is a natural result of this kind of upbringing.

Figure 11.2
FAMILY AFTER WALK EVENT, CALGARY.
This North American family joins the small majority of nuclear families headed by one man and one woman. However, there are an increasing number of different family arrangements in North America, including inter-ethnic, LGBTQ+, single-parent, and non-married families as well.
Credit: davebloggs007

Although monogamy is most often practiced between one man and one woman, many societies in the world and throughout history have supported and accepted same-sex unions. While heterosexual partnerships support the biological reproduction of the species, the successful rearing of children can result from a multitude of different family types. Indeed, "anthropological research supports the conclusion that a vast array of family types, including families built upon same-sex partnerships, can contribute to stable and humane societies" (American Anthropological Association Executive Board, 2004).

Polygamy

The marriage practice of having two or more spouses is called **polygamy**. This is a gender-neutral term that can refer to either multiple wives or husbands. Polygamous marriages have social, economic, and political functions, including benefits for producing and raising children, keeping land holdings together, and distributing labor. Having multiple partners may also be referred to as polyamory, with or without marriage.

Polygyny

The most common type of polygamous marriage arrangement is **polygyny**, or having two or more wives at the same time. Many societies in the world accept polygynous marriages; however, not all men in those societies have the wealth to take on more than one wife. (To clarify: the fact that a great many societies, including small societies, accept polygyny does not mean that most of the world's nations or most people practice polygyny.) Expensive gifts, such as animals or food items (and, today, modern appliances, cars, or gold) are given to the bride's family upon marriage. Not all men can afford the expense or the maintenance of a larger household.

From a biocultural perspective, it makes sense that polygyny is popular for the survival of the species. Multiple wives allow families to grow rapidly, whereas having multiple husbands doesn't afford the same benefit. There may be a surplus of women in any given society since more men die of violence than women in raids and wars. In addition, women's life expectancy tends to be longer in places where women and men enjoy equal access to health care.

In the Turkana tribe of East Africa, wealthy men marry multiple wives to keep up with growing herds of goats and cattle. It's important for the success of a man's family to have multiple adult women to watch the different herds as well as take care of domestic responsibilities: building the huts, preparing food, and caring for children. While we might imagine that women do not want to share their household with other wives, Turkana women accept the help. In fact, in many polygynous societies, a woman can divorce her husband if she can prove he has the resources to take on a second wife and refuses.

Polygyny also exists in small numbers in North America, as practiced by members of the Fundamentalist Church of Jesus Christ of Latter-Day Saints (FLDS) and offshoot groups. One such group lives in the settlement of Bountiful, British Columbia. Consisting of approximately 1,000 members, the Bountiful population practices polygyny according to Mormon fundamentalist values. While members assert their rights to practice polygyny due to religious freedom, court cases have also supported the rights of members, especially women and children, to be free from abuse in these and other settlements.

As mentioned above, there are downsides and dangers for women in polygynous marriages. Many polygynous societies do not afford women the same rights as men. Women may be treated as property and essentially sold to husbands who can be more than twice their age. Because they often move far from their families of orientation, wives who suffer mistreatment have little recourse. Tragic circumstances may ensue for women who are forced into marriages where they are abused. Under the best circumstances, co-wives can live together as sisters in a household where all adults respect one another. Under the worst, women may be abused or killed or commit suicide rather than be forced to exist under oppressive conditions.

Polyandry

Polyandry, the custom in which a woman takes two or more husbands, is less common, but it is still practiced in a smaller number of the world's societies.

Figure 11.3
ORTHODOX JEWISH WOMEN PRAYING AT THE WESTERN WALL.
These Jewish women are praying at the Western Wall in Jerusalem, in a separate location from the men. Jerusalem's Orthodox Jewish community has social rules, such as endogamy, that support the continuation of the culture.
Credit: Janet Kass/Images of Anthropology

From a biocultural perspective, it limits the number of offspring of each husband, passing on fewer genes. Furthermore, it is unlikely that a society will have a surplus of men due to their shorter life expectancy, emigration for work, or war casualties.

Polyandrous marriage may be beneficial in places where limits on population growth aid survival. In Tibet and Nepal, women may marry brothers to keep their land holdings intact. The scarcity of land available in the Himalayas makes this a better choice than splitting the land into smaller, unsustainable parcels at marriage. In Tibet, polyandry may also allow husbands to support the family by each joining a different sector of the economy: herding, agriculture, and trading (Levine, 1988). Brothers and their wife who remain in the same household continue to share access to family resources and build their shared wealth.

Exogamy

Further rules exist to narrow down the eligible pool of potential spouses. **Exogamy** is the practice in which marriage partners must come from different social groups. Depending on cultural norms, a person's choice of marriage partners may have to be from outside one's clan or lineage. Socially, this practice links families from different communities together, creating alliances. Bioculturally, it acts to broaden the gene pool of any intermarrying group. This limits the possibility for inbreeding and leads to more genetic diversity.

For instance, Pacific Northwest Tlingit society is divided into two large categories that are referred to in Tlingit social life as **clans**. Women who belong to the Raven (*Yeil*) Clan seek partners from the Eagle (*Ch'aak'*), also called Wolf (*Ghooch*), Clan. Members of the same clan address one another as "brother" and "sister." Therefore, it would not be appropriate to marry someone from one's own clan, since they are socially classified as siblings.

Endogamy

In contrast to exogamy, some societies require marriage partners from the same social group. This is called **endogamy**. Endogamous marriage can be seen in societies with strong ethnic, religious, or socio-economic class divisions and may be due to geographic or social isolation.

One of the best-known endogamous marriage patterns is the **caste system** of India. All Hindus are born into one of four major castes with membership in a subcaste, or into a category outside the system as a Dalit or "untouchable." Castes are unlike social classes because membership is based on descent; there is no movement from one caste to another. Until very recently in India, marriage partners were only chosen from within the same caste, often the same community, or even the same extended family (such as first cousins).

While endogamous practices serve to maintain a homogeneous group, close endogamy reduces genetic diversity. Reproducing within a limited group of people for multiple generations increases the risk of the expression of harmful recessive genes. With severe inbreeding, genetic diseases become hard to escape, such as the hemophilia passed from Queen Victoria through European royalty.

A number of societies have practiced sibling endogamy, especially among the ruling classes. Historically, sisters and brothers married to preserve the royal bloodlines of ruling families. For instance, the Egyptian pharaoh Cleopatra was married to her brother Ptolemy XIV. Sibling marriages keep the ruling power within the family.

Incest Taboo

The **incest taboo** is a result of the hazards of this kind of close endogamy. It is an example of a cultural universal, that is, a cultural practice that is found across cultures and throughout time. Cultural universals often point to something that underlies the universal needs of human beings.

Even in cases in which closely related royalty married one another, as mentioned above, the wider society did not share those tendencies. How cultures define kinship and relatedness affects how the taboo operates. Nonetheless, there are a number of reasons anthropologists believe that societies everywhere shun incest.

The first reason the incest taboo is universal is psychological, in that children raised together develop sexual aversion toward one another. This is referred to as the Westermarck effect, based on the work of Finnish researcher Dr. Edward Westermarck. He found that unrelated children raised together on Israeli *kibbutzim* very rarely develop sexual relationships or marry.

The second reason is social, due to the need for clear-cut roles in society. If a woman marries her son and they have a child, is the infant her son or her grandson? How do people involved in and related to these partners behave around one another if each person has multiple roles? The "role confusion" that results undermines successful social interaction.

The third reason is political, because marrying outside one's own family creates relationships with others. Inter- and intra-group alliances contribute to the stability of the larger society. Forcing people to seek partners outside their family leads to the benefits of exogamy, including social and economic stability and political alliances.

Finally, the fourth reason the incest taboo exists universally is biological. Reducing the gene pool for generation after generation causes a loss in genetic diversity. This loss leads to a higher potential for genetic diseases and a risk to the longevity of the species. These risks occur because more deleterious conditions emerge fully when both mother and father pass on an afflicted **allele** to offspring. It is likely that the biological threat to survival lies at the heart of the taboo, with social and psychological reasons developing to support it in human society.

FAMILY RESIDENCE PATTERNS

Where does a newly married couple live? Let's begin with the definition of a **household** as a domestic unit of residence. In a household, members contribute to child-rearing, inheritance, and the production and consumption of goods. Members of a household do not need to physically live under the same roof, yet they still contribute to the needs of the whole family.

A household is most commonly synonymous with a family unit, but not always. For instance, in the Amazonian rainforest of Brazil, members of Mundurucú communities send their sons to live in the village Men's House (*eksa*) at around 13 years old. From then on, they become contributing members to the residential unit of teen and adult males, rather than to the house of their mothers and sisters. Living among older males teaches the boys men's knowledge, such as hunting, mythology, and men's religious rituals.

Nuclear Family

In industrial societies such as our own, a couple is generally eager to start a new household after marriage, away from their households of orientation. They are more likely to reside in an independent household in a **nuclear family**, with two generations living together (i.e., parents and children). We refer to this type of residence as **neolocal**. There are many variations on the nuclear model, since sometimes grandparents, a single parent, or another caregiver provides for children.

Living in nuclear families makes sense in industrial societies, due to independence training. Child-rearing that encourages independence also stresses self-fulfillment and ambition. Nuclear families are generally smaller units than extended families, and adults can pursue work opportunities with relatively easy mobility. In other words, it's easier to move four people than it is to move fifteen. Moving to where the jobs are ensures food security, since city dwellers rarely produce their own food. Although neolocal residence is familiar to us in North America, it is actually the least common residence type in the world's societies.

Extended Family

The type of family structure that is most common across cultures is the **extended family**. In an extended family, blood-related members will bring their spouses to live with their family of orientation. This creates a household that is a mix of people related by marriage (**affinal** kin) and by blood (**consanguineal** kin). Depending upon the custom, wives may live with their husbands' families, or husbands may join their wives' families. Until recently in human history, all people lived in extended families for cooperation and protection. This family model may be most advantageous for the constant care of children, with many role models of both sexes.

There are several different residence patterns for extended families. Here we will concentrate on the two most common. When husbands join their wives' families of orientation after marriage, it is referred to as **matrilocal** residence. The extended family in this type of household includes sisters, their parents, their husbands, and their children. When men marry, they move to their wives' homes. The Hopi, an Indigenous nation that lives in the American Southwest, practices matrilocality. Therefore, a Hopi groom will leave his family of orientation upon marriage and take up residence with his wife's family of orientation.

Matrilocal residence is common under conditions in which land is held by the woman's family line. Married women remain in the home of their family of orientation so land does not get divided. In these societies, women's cooperation in subsistence is crucial, such as in horticultural societies where women do the bulk

Figure 11.4
HOPI WOMEN
PARTICIPATING IN
THE KAIBAB FOREST
RESTORATION
PROJECT.
These Hopi women are
collaborating with the
US Forest Service in a
2014 restoration project
in Castle Springs, Kaibab
Forest, Northern Arizona.
Credit: Kaibab National Forest/
CC BY-SA 2.0

of the labor. And of course, in polyandrous societies, residence at the wife's household would be required if she has multiple husbands.

When wives join their husbands' families of orientation after marriage, it is called **patrilocal** residence. This is the most common type of residence pattern in the world's societies, wherein extended families are made up of brothers and their wives, their children, and the brothers' parents. When women marry, they move to their husbands' homes. This is the traditional arrangement, for instance, for Han Chinese families. The Han are the majority ethnic group in China. Therefore, most Chinese brides will leave their home of orientation to live with their husband's family of orientation.

In societies where men play the predominant role in subsistence, such as pastoralist or intensive agricultural societies, patrilocal residence is common. This way property may be accumulated and passed down through the man's family line. It is also found in societies in which men's cooperation in central government and warfare is important. Polygyny also requires patrilocal residence so multiple wives can live together in their husband's household of origin.

A type of extended family that may not be strictly consanguineous or affinal is a **family of choice**. This term originated with the LGBTQ+ community, in which people on the gender or sexual spectrum may experience rejection from their family of orientation and seek a "chosen family." However, the term may be used to refer to anyone who seeks supportive partners, kin, or close friends to form a desired household in a location of their choice (neolocal).

MARRIAGE AS ECONOMIC EXCHANGE

Fundamentally, the union of individuals from two different families is an economic exchange. Not only is the bride or groom "given" to the other family, but a series of gifts is also given from family to family to cement their bond. This is most true for extended families in which rules for residence after marriage are clearly delineated. **Marriage compensation** depends upon the cultural context: Who is losing a family member and who is gaining one?

Bride-Price and Bride Service

If the bride leaves her family's household of origin and becomes a resident in her husband's household of origin, the husband's family compensates the bride's family. The compensation is called **bride-price** (also called bridewealth), which refers to the valuables that a groom and his family are required to present to the bride's family. The young bride is not only an additional resource for support and labor in her new household but is also expected to bear children that will extend the husband's family line (and provide more potential resources and labor).

Bride-price is often paid in a series of gifts, such as those required upon marriage by the Trobriand people of the Kiriwana (also called Trobriand) Islands in Papua New Guinea (see Table 11.1). The gift-giving begins before marriage and continues until sometime after the actual ceremony, sometimes annually for a long period of time. Complying with the exchange on time and correctly signals each family's responsibility and commitment to one another.

Another way in which compensation may be given to the bride's family for the loss of their daughter is to offer **bride service** to her family. Rather than a gift of goods or money, the groom spends a period of time working for her family. Among the Ju/'hoansi, early marriage of children ensures a long period of bride service in which the young husband hunts for his bride's family. Ideally, this is an adaptive practice supporting the bride's family. However, in certain circumstances, it may also be used for the personal gain of the bride's parents.

TABLE 11.1

Trobriand Islanders' Exchange of Marriage Gifts

These are the gifts given by family members prior to, upon, and after marriage as described by Bronislaw Malinowski in the unfortunately titled *The Sexual Life of Savages in North-Western Melanesia* (1929).

Marriage Gift (in the order that it must be given)	Consisting of ...
(1) *Katuvila*	Cooked yams, brought in baskets by the girl's parents to the boy's family.
(2) *Pepe'i*	Several baskets of uncooked yams, one given by each of the girl's relatives to the boy's parents.
(3) *Kaykaboma*	Cooked vegetables, each member of the girl's family bringing one platter to the boy's house.
(4) *Mapula Kaykaboma*	Repayment of gift 3, given in exactly the same form and material by the boy's relatives to the girl's family.
(5) *Takwalela pepe'i*	Valuables given by the boy's father in repayment of gift 2 to the girl's father.
(6) *Vilakuria*	A large quantity of yam-food offered at the first harvest after the marriage to the boy by the girl's family.
(7) *Saykwala*	Gift of fish brought by the boy to his wife's father in repayment.
(8) *Takwalela vilakuria*	A gift of valuables handed by the boy's father to the girl's father in payment of gift 6.

Sometimes very young girls are married to older men in exchange for financial support to her family.

Dowry

Societies in which the groom goes to live with the bride's family of orientation have the opposite form of compensation. **Dowry** is a gift of money or goods from the bride's family to the groom's family to compensate for the loss of their son. This dowry is essentially the portion of the bride's inheritance given early, to ensure the new couple will have some financial resources.

Dowry can also be seen in societies in which neolocal residence occurs. These goods help the new couple begin their life together once they separate from their parents. In the colonial United States, dowry consisted of goods the girl sewed or crafted throughout her young life, collected in a *trousseau*. While there is an expectation of neolocal marriage in modern North America, the pressure to provide a dowry has considerably lessened. The parents of the bride may still pay for wedding expenses as a legacy of this practice.

Perhaps surprisingly, we also see dowry in societies that practice patrilocality. This is due to cultural values that encourage families to marry their daughters into a higher class (or caste, in Hindu India) than their own. The focus on "marrying up" forces the bride's family to promise expensive gifts to ensure a good marriage for their daughter. When families have multiple daughters, dowry gifts can be a burden on the family's resources.

Promised but undelivered dowry gifts can cause major problems for a woman living in her husband's household. In some circumstances, an unscrupulous family-in-law can extort gifts or money from the bride's family. Under severe circumstances, it can be worse. Each year, thousands of women commit suicide or are killed in India and Pakistan over dowry. The news often reports the death and disfigurement of women from acid or kerosene burning. This practice is common enough to be known as **dowry death**. Both India and Pakistan have passed laws officially outlawing the request, payment, or receipt of a dowry, but the practice still exists in these countries, embedded in cultural patterns of marriage and family life.

Arranged Marriage

Throughout time and across cultures, the most common way to ensure a suitable union has been for parents to arrange the marriages of their children. In an **arranged marriage**, parents will generally seek a match for their son or daughter from the same (or higher) caste or community, socio-economic class, or religion. This ensures they will pass down their values and material wealth to the next generation while joining together in a multifamily arrangement that will be mutually beneficial. Arranged marriage was the norm for most of human history. Marriage based on romance and love is a relatively recent development.

Arranged marriage may sound strange or even terrifying to those raised in independence training societies. However, modern arrangements for this practice generally take into consideration the wishes of the young people involved, especially among more educated families (see Box 11.1). For instance, a son or daughter may have the power to refuse a particular match. The arrangement is not the same as **forced marriage**, in which a young person has no say, or **child marriage**, in which young girls are betrothed to older men.

BOX 11.1 Arranged Marriage in Mumbai

Arranged marriage is still an option among young, educated, middle-class women in India's biggest city. However, arrangements happen in very different ways today. I (author Laura González) was fortunate to spend time in Mumbai, India, learning about modern arranged marriage.

Today, girls meet prospective partners through social events called "marriage meets," community registries (called "marriage bureaus"), and online matrimonial sites. More liberal parents may even accept matches that are initiated by the girls and boys themselves – once condemned as "love matches." (Indians refer to unmarried people as "girls" and "boys," regardless of age.)

Until the mid-twentieth century – the generation of these girls' parents – nearly all Indian marriages were arranged by extended family connections. Traditionally, a matchmaker, who may have been a family friend or relative, would solicit possible matches on behalf of the family. After photos and background information were exchanged, a potential match would be identified. The boy's family would be invited to the girl's house for a "bride viewing." While the parents talked, the girl would serve tea, speaking only when addressed and not being allowed to make eye contact with the boy. After this initial meeting, the girl and boy might not see each other again before the wedding day, depending on what parents allowed.

From the perspective of a person raised in an independence culture, one might wonder how it's possible for modern, educated young men and women to accept a marriage arranged by their parents. Nonetheless, there are several reasons the system is still desirable, even with the possibility of dowry problems. First, the dependence-training model used in India creates a highly interdependent extended (or "joint") family unit, with the clear responsibility of parents to find a spouse for their children. Second, the bride-viewing model is no longer used formally in most cases. Third, contemporary arranged marriage in Mumbai allows some "dating," which gives the young couple a chance to get to know one another. In addition, young people know clearly what their families' expectations are for their future marriage partners and have internalized these guidelines. In her fieldwork, Dr. Serena Nanda (2000) found this is because Indian girls trust their parents to make good decisions for their futures.

Today, it is not as likely for a marriage to be strictly arranged with no sense of individual needs being met. It is also less likely for a marriage to be strictly self-initiated and self-managed by the couple until the wedding and beyond. A hybrid set of practices exists today in which urban, middle-class girls in Mumbai negotiate the traditional social expectations of Indian joint families and the modern tensions of urban life with its focus on self-fulfillment and female empowerment (González, 2013).

Figure 11.5
AUTHOR LAURA GONZÁLEZ WITH STUDY PARTICIPANTS, MUMBAI.

Author González with girls from Mumbai's St. Xavier's High School after an interview on marriage.

Credit: Laura González/Image used with permission of all participants

KINSHIP

All human groups face certain issues such as how to regulate sexual activity, how to raise children, and how divide the labor necessary for subsistence. **Kinship,** or family relations, provides a structure for doing so. Since all societies recognize kin, every society has rules linked to family and household organization. Dividing labor along gender lines is one way to ensure that labor is distributed. Another way is through family descent groups, or **lineages,** in which each side of the family has different responsibilities.

Although kinship implies the relatedness of people through blood or marriage, there are forms of kinship that extend beyond these boundaries. The practice of adoption brings individuals who are not biologically related into a kinship relation. Adoptive families (and their variants, such as step-, foster, or surrogate families) generally experience the same social norms and expectations as biologically related families, although sometimes the laws governing different types of families are different.

Fictive kinship is the term sometimes used to refer to a family of choice, referenced above, or a constructed "family" of unrelated individuals, such as members of an urban gang who rely on each other for social support, economic resources, and protection. Relationships built upon mutual caring and attachment may be called **nurture kinship**, such as between a mentor and mentee or godparent and child.

A **descent group** is a group of people who trace their descent from a particular ancestor. Descent groups form connections from parents to children, tracing their lineage through their father, mother, or both parents. Dividing the extended family in this way allows different rights and responsibilities to be assigned to different family members. Certain members of the descent group might act as godparents to newborn children or be responsible for harvesting crops when they ripen.

Some societies trace their genealogy through both the mother's and the father's line, called **bilateral descent**. The English language underscores this equality: we use the same term to refer to the same relatives on our mother's side and on our father's side (aunts, uncles, or grandparents). These kinship terms represent generally equal expectations of both our father's and mother's families.

In some societies, descent is reckoned along one family line. This form of descent is called **unilineal**. There are two types of unilineal descent: **patrilineal** descent is traced through the father's bloodline, and **matrilineal** descent is traced through the mother's. One lineage only is responsible for the continuation of the family's name and possessions, such as landholdings or other inherited items.

Societies with unilineal descent encode differences in social roles with different terms for the same relations on either side of the family, such as "father's brother" and "mother's brother." The different terms mark clearly defined, and different, roles and expectations in the family. The languages listed in Table 11.2 are just a few of the many that use different terms to refer to people on different sides of the family.

TABLE 11.2

Talking about Families

Language	Aunt (Maternal)	Aunt (Paternal)
Arabic	Khalto	Amto
Farsi	Khaleh	Ammeh
Hindi-Jain	Mosi	Bua
Mandarin	Yí	Gū
Cantonese	Yîhmā (if older than mother)	Gūmà (if older than father)
	Yî (if younger than mother)	Gūjè (if younger than father)
Urdu	Badi Khaala (if older than mother)	Badi P'hupoo (if older than mother)
	Choti Khaala (if younger than mother)	Choti P'hupoo (if younger than mother)

Every culture has norms regarding how men and women should act and interact with others based fundamentally on their gender. Males and females are subject to social and cultural expectations of the correct ways to behave in virtually all areas of life, from child's play to education, friendships, and relationships. Gender roles especially play out in the family, where members share resources and raise children.

Young children receive messages constantly about what it means to meet the expectations for a boy or a girl in their society. Combined with aspects of an individual's physiology, a person's **gender identity** is constructed within social and cultural norms. Natural human development allows for gender and sexual identity along a spectrum, with many variations and fluidity. Through their study of human biology and the cultures of the world, anthropologists understand that a binary approach to gender, sex, and sexuality is inaccurate.

Defining Sex, Gender, and Sexuality

Although in English we often use the two terms interchangeably, *sex* and *gender* refer to different aspects of a person. **Sex** refers to our biological and physiological differences, including sex chromosomes, hormones, reproductive structures, and external genitalia. **Sexuality** refers to attraction, either romantic or physical.

On the other hand, the set of social meanings assigned by culture is referred to as **gender**. Gender is a person's internal experience of identity as male, female, both, or neither, as well as the expression of that identity in social behavior. Sex, sexuality, and gender have origins in our physiological development, although the cultural expectations of sexuality and gender will differ based on the society in which they are found.

Approximately 1 person in every 2,000 is born **intersex**. They have a combination of physiological or morphological traits that place them on the sex spectrum (and possibly also the gender spectrum) in a way that does not allow simple definition of male or female. As one might expect, cultural responses to intersex infants are widely variable. In some cases, doctors decide to manipulate the external appearance of genitalia and assign a child's sex, which has the result of also assigning a gender.

These choices are not always in the best interest of the individual in the long term, as biological sex consists of more than simply genitalia. Teens who have had their sex chosen for them at birth often face sex and gender identity challenges as they go through the changes accompanying puberty. As more awareness is created regarding intersex issues, parents are instead guided toward medical counseling and encouraged to allow children to align with their own gender identity as they grow.

Gender Roles

Gender roles are the culturally appropriate roles of individuals in a society. They express the cultural norms expected of a person of different sexes.

For instance, a female infant born in North America who grows up to play with dolls exhibits "feminine" (girl-like) gender characteristics. This is generally thought of as the normative behavior for a young girl. However, if she prefers toy guns or refuses to wear a dress, society may deem her "masculine" (boy-like) or **androgynous** (gender neutral). In this case, her **gender expression** does not fit neatly with social expectations of female behavior.

Different societies assess masculinity and femininity differently. For example, many Indigenous societies, such as the Quechan from the US Southwest, value women with upper body strength. Historically, the woman of the Quechan household was expected to grind dried corn to a fine powder for use in cooking. On the other hand, Maasai pastoralists of Kenya prefer women with a slender but strong build, as they will be responsible for milking cows and hauling water and firewood.

The characteristics in Table 11.3 come from the Bem Sex-Role Inventory, a questionnaire measuring masculinity, femininity, and androgyny developed in 1974 by psychologist Dr. Sandra Bem in her work with undergraduates at Stanford University. Survey participants checked the boxes corresponding to the traits with which they identified. This gave participants a result that indicated they were more masculine, more feminine, or more androgenous. More than 40 years later, the

TABLE 11.3

"Masculine" and "Feminine" Traits of North American Men and Women

"Masculine" Traits	"Feminine" Traits
Self-reliant	Yielding
Willing to defend own beliefs	Eager to soothe hurt feelings
Independent	Soft-spoken
Leadership ability	Does not use harsh language
Willing to take risks	Childlike
Analytical	Gullible
Assertive	Loves children

original survey is more interesting for its stereotypes of masculine and feminine traits from the 1970s than it is for the results.

Gender Identity

Gender identity is largely determined by biological factors and then formed through interactions with others in society. As an essential part of culture, gender identity begins to form as soon as others interact with an infant. No matter the society, children are dressed in the appropriate clothing for their gender and encouraged to behave in accordance with gender norms. In Western societies, for instance, girls are encouraged to wear skirts or dresses, and boys to wear pants.

Those who experience their gender identity as matching their **assigned sex at birth** are referred to by the term **cisgender**. For example, a person designated male at birth (by the attending physician or midwife) who experiences his gender as male (i.e., "feels masculine") is cisgender. Some people internally experience and/or express their gender identity as different from their assigned sex at birth. They may identify as **transgender**; that is, they have a gender identity and/or expression that more closely matches another gender, such as a person assigned a male sex at birth who experiences their gender as female. People who experience their identity or expression along a spectrum, that is, in a nonbinary way, may self-identify as **gender queer** or **gender fluid**.

In many societies around the world, nonbinary gender identity is accepted as a natural expression of gender. In Western societies, however, the influence of religious colonization determined how many aspects of gender and sexuality have been shaped throughout recent history. In the United States, for example, young people who experience their gender identity as nonconforming may experience a feeling of **gender dysphoria**, in which they feel distress due to their identity. Often this has to do with the level of tolerance in the society.

Third Genders

Across cultures, there are many examples of societies that recognize multiple genders. In these societies, an individual could choose to take on a social identity that was outside of the two-gender model. That is, a person could self-identify as one of three genders: masculine man, feminine woman, or other not-man not-woman. Some societies recognize four or more genders: masculine man, feminine man, masculine woman, and feminine woman. In many cultures, gender-variant individuals were and are accepted, sought for spiritual guidance and blessings, and thought to occupy special roles in society.

One gender variant was recorded across more than 120 Native American and First Nations cultures, including the Zuni (*A:shiwi*), Crow (*Apsáalooke*), and St. Lawrence Island Yupik. Although the older anthropological literature sometimes

RESEARCH in PAIN
BY SALLY CAMPBELL GALMAN

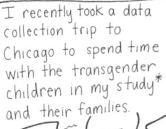
I recently took a data collection trip to Chicago to spend time with the transgender children in my study* and their families.

CHICAGO

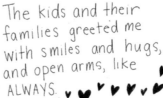
The kids and their families greeted me with smiles and hugs, and open arms, like ALWAYS.

We played...

"This kid at school says that the new president will kill all the transgender people."

... and we talked.

"She has been waking up at night, with nightmares, afraid the president is going to make her be a boy."

BUT THE THING IS: MY STUDY IS ABOUT RESILIENCE, not terror, not trans kids and families as victims of policy. NOT those SAD STORIES.

At least until last November — but I am still finding the same strength... EVERYWHERE the children are.

TRANS PRIDE

And it is still the story I am hearing, and telling.

I AM TRANS!

Here's me at the big protest!

BUT THE other STORY IS there and it HURTS

And I am not really used to talking about PAIN as a part of

RESEARCH, as part of the researcher's ROLE to absorb it, or as a part of subjectivity.

* generously supported by the Spencer Foundation

I left the kids' homes that night and turned on the radio and I cried the whole way home in the darkness.

AM I SAFE?

My child is also trans. So that is also part of it.

It is not pity or romantic adultism or unchecked, wild sentimentality. It is real PAIN. And I feel the guilt and the indulgence and the horror of it.

NO

I collected data from before the U.S. election, through the campaign...

The participant experiences were overwhelmingly positive— a rosy future ahead for transgender kids...

It felt like an instant— when it was interrupted...

NO   AM I SAFE

The positive trajectory wasn't destroyed—but the skies got a hell of a lot darker.

So, I got to learn about BEFORE and AFTER and DURING the

MOMENT OF IMPACT

and all that followed.

And now I go over and over it like some kind of Zapruder film of our HIGH SPEED COLLISION WITH HATE.

And even so, the shock waves keep coming endlessly.

And I worry about what will be consumed in their wake.

I AM TRANS!

Parents and I exchange tearful glances...

HOPE

Because we know. And we don't know.

FEAR

"If I don't do the RIGHT thing, my child could be in danger—but I don't know what the RIGHT thing is."

"It seemed like just yesterday everything was going to be okay—and now we are all GROPING in the DARK."

As a researcher, I went from HOPEFULLY documenting young lives of PROMISE, FEARLESS PRIDE, VISIBILITY AND HOPE...

HURT

TRANS ♥

to dolefully bearing witness to families' desperate and often fruitless but always BRAVE attempts to keep kids safe amidst post-election UNCERTAINTY and HATE.

PAIN

Bluebond-Langner's (1978) work on dying children is one of the most painful reads, but also the most powerful AND most helpful in navigating PAIN.

WORRY

She writes that, even more than to science, we have a responsibility to our child participants to get past our own pain and honor them as FULL and competent people who understand their worlds.

But to give pain still its space. Feel it.

"The anger soars. It wells up into an indictment of this country for its priorities on spending. It is thrown up to a God I am not quit sure exists, but who deserves to be blamed just the same. I have to blame someone, something... I must do something and what can I or anyone else do?"

Of course, the children in my study are not dying. They are resilient even in the face of these events. BUT they are not as safe as we thought, or as we imagined.

As researchers, many of us are more used to UNCOVERING and UNMASKING inequalities, shining LIGHT on OPPRESSION.

BUT now it is utterly UNMASKED and my training has not prepared me for what to do when the lion's mouth OPENS AND BARES ITS TEETH.

MAKE AMERICA FEAR AGAIN

As Douthat (2017) observes, this populism will certainly be brought down by its own chaos and incompetence — but watching children in its rising waters is painful.

I AM TRANS

Kenji Miazawa wrote, "We must embrace pain and BURN it as FUEL for our journey."

Under the best circumstances, Research takes stamina and FUEL, and we must also use pain to resist the

JUSTICE

tendency to NORMALIZE

And this is shaping up to be a long journey.

Bluebond-Langner, M. (1979). The private worlds of dying children. Princeton NJ : Princeton University Press.

Douthat, R. (2017, February 1). How populism stumbles. The New York Times. Retrieved from https://www.nytimes.com/2017/02/01/opinion/how-populism-stumbles.html

Dylan, B. (1991). Last thoughts on Woody Guthrie. [Recorded by B. Dylan]. On The Bootleg Series, Rare and Unreleased 1961-1991 [Audio Recording]. New York: Columbia Records.

www.sallycampbellgalman.com

refs to this group of individuals as *berdaches*, this is a derogatory term derived from the French word for "prostitute" and is no longer used. The Native American and First Nations communities instead identify with the term **two-spirit**. Indigenous languages also have terms for their two-spirit people, for instance, *aayahkwew* among the Cree, and *winkte* among the Lakota.

In Native American and First Nations cultures, two-spirit people may be assigned either a male or female sex at birth. They are socially accepted within the tribal community as a **third gender**, neither man nor woman, but with elements of both. The idea that both genders could inhabit the same body is accepted in regard to two-spirit people; either male or female clothes could be worn on a given day, and individuals could do both male and female work.

Another example of a third gender variant is the **hijra** of India and Pakistan. Hijras occupy a role that is between the sexes in Indian society – that is, a third

Figure 11.7
JEREMY DUTCHER.
Jeremy Dutcher is an award-winning Canadian composer and musician who identifies as two-spirit and queer. A member of the Tobique First Nation, he sings in the Wolastoq language, which has no gendered pronouns.
Credit: Artsandstuff1/CC BY 4.0

gender. Anthropologist Dr. Serena Nanda (1990), who has spent years researching hijra communities in India, explains that hijras consider themselves "not-man, not-woman."

Hijras are individuals assigned a male or intersex body at birth who adopt feminine behaviors and attributes such as dress, ornamentation, names, and mannerisms. Yet culturally, their occupations, behaviors, and severe social limitations distinguish them from Indian women. Hijras live in communal homes under the tutelage of a hijra *guru* (teacher). They are expected to survive on charity and payment for services performed, such as blessing babies or performing at weddings.

The hijra is essentially a spiritual role to which a person is called. Members of the hijra community become devotees of *Buhuchara Mata*, an incarnation of the Hindu Mother Goddess. Their devotion requires a vow of sexual abstinence. For this reason, hijras have surgery to remove their penis and testicles. Because the surgery is prohibited in hospitals, the risk of infection or death can be high. Nonetheless, this surgery is seen as a ritual transformation from which hijras take on their spiritual power.

Hindu belief regards pansexuality and gender fluidity as permitted expressions of human identity and desire. The pantheon of Hindu deities includes some who are sexually and gender ambiguous, combining aspects of maleness and femaleness, or who transform themselves from one into another. The deity Shiva, for instance, has both male and female characteristics, renounces sex, and yet is eroticized in Hindu mythology. Hinduism is comfortable with gender and sexual ambiguities in a way that Western religions are not.

Unfortunately, the existence of hijras in ancient Sanskrit scripture does not translate into acceptance for hijras in today's society. Although India legally accepted hijras as a third gender in 2016, they remain on the margins of society due to severe prejudice, social discrimination, and often violence. Hijras facing discrimination may be forced to make a living by begging for charity on the trains of Indian cities, extorting money from businesses, or performing as dancers or sex workers.

SEXUAL IDENTITY

Like gender identity, **sexual orientation** (the nature of one's romantic or sexual attraction to another person) may be formed socially or biologically. Although orientation is often referred to as "sexual preference," for most people, their genes determine sexual orientation; they do not feel it is a preference.

Nevertheless, sexual orientation is expressed within a set of cultural values and expectations. Human cultures differ widely in their acceptance of orientations other than heterosexuality. Some societies see a range of sexual orientations as a natural

BOX 11.2   **Talking About: Gendered Pronouns**

People who experience their identity as other than cisgender (that is, as other than a gender that matches their designated sex at birth) may choose to use a pronoun that reflects their identity and/or gender expression. In English, a person may use the pronouns he/him/his to reflect a masculine identity; she/her/hers to reflect a feminine identity; they/them/theirs, or other pronouns, to reflect a fluid or nonbinary identity. Although gendered pronouns are often referred to as "preferred pronouns," transgender and nonbinary individuals do not feel it is a preference, but an expression of their authentic self.

This can be challenging for speakers who are learning to use alternate pronouns for their gender nonconforming friends or family for the first time. It is sometimes hard to remember to use "they" as a singular pronoun, even though it has been used this way in English for hundreds of years, especially when the referent is unknown. For instance, if the doorbell rings, we might ask someone in the room, "Can you go see what they want?"

It can be even more challenging for nonbinary people whose native languages are gendered, such as Spanish, French, or Italian. It is essential in Romance languages to alter the forms of words to match the gender of the subject (person speaking) or object (person being spoken about). In an effort to make the Spanish language more inclusive, for example, some use a final "x" rather than o/a as in the words "Latinx" (rather than Latino/male, or Latina/female) or "Chicanx" (rather than Chicano or Chicana). In writing, those wishing to be inclusive may use the @ sign, as in ell@s (signifying ellos, ellas, and those in between).

However, applying an "x" to all adjectives while speaking can be confusing. Bilingual speakers find it useful to code switch into English (or another nongendered language) to describe themselves with adjectives. This is because the English words do not require a/o endings. See the following examples:

"*Estoy cansado.*" ("I am tired," spoken by a male.)

"*Estoy cansada.*" ("I am tired," spoken by a female.)

"*Estoy* tired." ("I am tired," spoken by anyone; removing the gender referent by switching into English.)

Code switching into English allows people to more easily remove the gendered descriptors from their language, although it is of limited use, suitable only for bilingual speakers in conversation with bilingual listeners. As acceptance of gender nonconforming identities continues in North America and around the world, more discussions about gendered language use will arise.

part of human life. Other societies have low or no tolerance, even to the degree that a person may be put to death for consensual same-sex acts under **sharia law**, found in the nations of Iran, Saudi Arabia, Sudan, Yemen, and northern states of Nigeria.

Heterosexuality is the term used to describe romantic or sexual attraction, or sexual behavior, between partners of the opposite sex, while **homosexuality** refers to that between partners of the same sex. Although heterosexuality is the most common sexual orientation, homosexuality and other non-heterosexual forms of attraction and behavior are common in cultures around the world and throughout time. (Note: Although used here as a category, the word "homosexual" is now an outdated term in speech and has been largely replaced by the word "gay.")

It is important to note that gender identity is independent of sexual orientation, as any person may be heterosexual, gay, or **bisexual**. Beyond the traditional binary division of sexuality, a person may self-identify as **pansexual** (attracted to anyone regardless of gender or sexuality), **polysexual** (attracted to people of multiple genders or sexes), or **asexual** (having few or no sexual or romantic feelings). In other words, a person of any gender may prefer men (**androphilia**), prefer women (**gynophilia**), both, some, any, or none.

Within certain cultures, homosexual practices are essential to the functioning of society. For instance, among the Etoro of the Southern Highlands province of Papua New Guinea, homosexual relations among males are linked to male development and power. In particular, the Etoro believe male oral intercourse ensures a man's physical growth and enhances his spiritual strength. Sexual relations with women tend to be focused on the goal of reproduction. Many social rules revolve around the importance of these practices, such as abstaining from sex with women during important times of year, like the harvest.

Several hypotheses have been proposed for determining whether a society will be permissive regarding homosexuality and gender variants. An ecological hypothesis argues that homosexuality is tolerated or accepted in societies that have historically experienced pressure from population growth or food shortages. Allowing homosexuality and nonbinary gender variants means that not every couple will produce offspring. Thus, population levels will remain stable, and food will remain sufficient. A socio-political hypothesis makes a correlation between societies in which abortion or infanticide is prohibited and homosexuality is punished or not tolerated. This may be a result of strict religious laws that came to be imposed based on colonial relationships.

Body Modification and Gender

Body modification is one of the most common ways people express their identity and make themselves attractive. Today's makeup, piercings, and tattoos that showcase modern individuality actually have deep roots in human history. Ancient Egyptian royalty used kohl eye pencils made from minerals to accentuate their eyes. Young Tlingit boys of the Pacific Northwest Coast wore multiple ear piercings to represent high status in the community. Young Mursi women of the Omo River Valley in Ethiopia traditionally wear a circular clay or wood plate in their lower lip

BOX 11.3   **Talking About: Gender in Politics**

The election season of 2020 brought more women into the US Congress than ever before, including the first female vice president in the nation's history, former senator Kamala Harris. Her confirmation came four years after a tumultuous 2016 national election, during which more women stood for candidacy for the presidency than in any year past. One such candidate, former Secretary of State Hillary Clinton, became the Democratic Party's nominee and went on to win the popular vote.

Although the United States has never had a female national leader, and Canada has only ever had one, Kim Campbell, for five short months in 1993, voters elected a record number of women in 2019 and 2020; women representatives accounted for 26.5 percent of the 117th US Congress and 28.9 percent of the 43rd Canadian Parliament. These results show that women are running for office and being elected more than ever in North American politics.

Nonetheless, research on perceptions of female political candidates has shown that people use different criteria to judge them compared to their male counterparts, with the criteria often having little to do with women's abilities or preparation. For example, clothing and general appearance continue to be hot topics relating to female office holders, with Kamala Harris's purple coat dominating the commentary over her swearing-in ceremony at the US presidential inauguration in January 2020. Similarly, former First Lady Michelle Obama's Harvard law degree was talked about far less than her fashion choices and defined biceps.

Criticism of female politicians' personalities is also gendered. During her 2016 campaign for president, the media criticized Hillary Clinton for being too aggressive, unlikeable, and unemotional, perhaps because these traits are considered "unladylike." Due to lingering stereotypes of supposed female strengths and weaknesses, popular opinion favors women in certain branches of office only, where they are deemed to be the more appropriate choice. Political science professor Dr. Monika McDermott argues that "women are seen as better on health care, education, and women's issues, and men are better on guns, national security, and foreign policy" (as cited in O'Donnell, 2020).

Women, including Black, Indigenous, and other ethnically diverse women, are now more likely to hold positions of power in North America than ever before. Nonetheless, women and gender nonconforming individuals have a long way to go to reach parity with men. As stated above, currently only 28.9 percent of the members of the House of Commons of Canada and 26.5 percent of the members of the US Congress are women. Put into context, the governments with the highest percentage of women in office are Rwanda (61 percent), Cuba (53 percent), Bolivia (53 percent) and the UAE (50 percent) (UN Women, 2020).

Figure 11.8
KAYAN LAHWI.
Kayan women's heavy brass coils compress the clavicle as a young girl grows. After many years, it gives the impression of a very long neck. This woman lives in Bagan, Burma.
Credit: Daniel Chit

in preparation for marriage. Many of these modifications signal that individuals have completed a **rite of passage**; that is, they have moved from one stage of life to another.

Māori warriors of Aotearoa, or modern New Zealand, had to earn the right to a full facial tattoo called a *moko*. The moko provided information about the wearer's tribal and family history and acted as a marker of male identity. Female adult Māori traditionally would also wear a partial moko, with their lips and chins inked blue. Although the images look similar, Māori designs are not like "tribal" tattoos that can be bought in any tattoo shop today. Moko cannot be bought – only earned through life experiences. In fact, Māori writers and public figures have protested the imitation and commodification of their traditional designs by Western wearers, who have no knowledge of the significance of these designs.

Scarification is another method of inscribing artistic symbols on the body as a marker of identity. To create scar designs, the skin is cut or burned, with irritants such as dirt added to create a keloid (raised scar). In societies in which scarification is performed, females will receive certain scar designs and males receive others. Among the Karo of Southern Ethiopia, females preparing for marriage will receive scars on their chests, which are rubbed with ash to create a keloid effect, considered beautiful. Karo males receive differently designed scars on their chests for having killed enemies, enhancing a man's status.

Kayan women (called *Kayan Lahwi* or *Padaung*) from Burma (Myanmar) wear coiled brass neck rings that give the illusion of an elongated neck, considered a sign of beauty. In fact, the neck vertebrae do not elongate, but the weight of the rings presses down and deforms the clavicle (collar bone) as a young girl grows. Neither Kayan boys nor men wear the rings.

After civil unrest in Burma in the 1980s and 1990s, a permanent refugee village was set up for fleeing tribal people on the border of northern Thailand. Because the Kayan Lahwi are unique, they draw foreign visitors to the area. Today, tourists visit the Thai village to pay a few dollars to see the "longneck giraffe" women.

Most of the income brought in by tourism goes to non-Indigenous managers or local governments, while members of the ethnic community receive a very small percentage.

Since the mid-2000s, some Kayan Lahwi have chosen to remove their rings in protest of the treatment of the Kayan people. Removing the rings causes discomfort for several days with residual bruising and discoloration. Eventually the muscles strengthen and discomfort subsides. Some women report feeling freer once the rings are off, and others report feeling sad at the loss of their cultural tradition.

SUMMARY

This chapter has examined how marriage, family, and gender roles structure a great deal of social life in human societies. Mirroring the Learning Objectives stated in the chapter opening, the key points can be summed up as follows:

- Many different marriage patterns and family types can contribute to stable societies, including those built on monogamy, polygamy, and same-sex partnerships.
- Throughout most of human existence, people have lived in extended family groups for the many benefits this structure provides. With the increase in industrial societies, nuclear families became the norm due to the need to move to jobs and changing family expectations.
- Because marriage is considered to be a joining of two families in most societies, rules exist to regulate marriage practices, compensation, and the responsibilities of descent.
- Although marriage among people in modern Western societies tends to be self-initiated, many societies in the world still practice arranged marriage.
- Families may trace the rights and responsibilities of their lineage through one or both parents' lines of descent.
- Gender roles are the culturally appropriate roles of a man or a woman in society, dictated by social norms. Sex refers to the biological assignment of a person's anatomy and physiology. Some people experience their gender identity as different from their assigned sex at birth.
- Gender, sex, and sexuality all exist on a spectrum.
- Gender identity is separate from sexual orientation, which refers to romantic or sexual attraction.
- In some societies, people who experience a different gender identity are accepted in a third (or other) gender role.

Review Questions

1. What are the different marriage and family types that exist across cultures?

2. What are the biocultural benefits of exogamy?

3. What types of compensation are given in different marriage exchanges?

4. What is the difference between sex and gender?

5. How are the identities of a gay person, a transgender person, and a nonbinary or gender fluid person different?

Discussion Questions

1. What makes a "good family"?

2. Are marriage tendencies in North America exogamous or endogamous? Why?

3. What do you think about the division of traits in the Bem Sex-Role Inventory? If you were to create a modern version of the Inventory, how might it be different?

4. From your perspective, how is reading the graphic panel called "Research in Pain" different than reading a textual description of the same thing?

5. What are ways in which colleges and universities have made campuses safer for LGBTQ+ students, faculty, and staff?

Visit **www.lensofanthropology.com** for the following additional resources:

| SELF-STUDY QUESTIONS | WEBLINKS | FURTHER READING |

12

POLITICAL ORGANIZATION

LEARNING OBJECTIVES

In this chapter, students will learn:

- *how societies maintain order and stability within their own borders and with other societies.*
- *why and how societies use power and controls differently.*
- *the differences between societies with uncentralized governments and those with centralized governments.*
- *the characteristics of bands, tribes, chiefdoms, and states.*
- *how power is used to create inequality based on gender and access to resources.*
- *about the different types of violent conflict within and between societies.*

> **Humans used cooperative strategies to survive for more than a hundred thousand years as foragers and hunters. When we settled down, suddenly there were things worth fighting for.**
>
> #PoliticsAndPower

INTRODUCTION

All societies, whether small or large, use a set of rules to guide their members' behavior toward one another. These rules may be official, such as a code of written laws, or unofficial, such as a set of social expectations. They may be embedded within a community's cultural or religious values ("Do not steal") or imposed on them from the outside ("No border entry"). The types of expectations, moral codes, policies, and laws will differ based on the size and complexity of the society.

This chapter examines different types of **political organization**, or the way a society maintains order internally and manages affairs externally. Through the lens of anthropology, politics refers to a wide range of actions and interactions that have to do with power. Power relations are negotiated among individuals; for instance, between a parent and child, teacher and student, chief and subject, master and slave, or two people in a bar watching a baseball game. On a broader scale, these interactions occur between larger groups such as communities, organizations, governments, and nations.

Anthropology focuses on the following questions in the study of political organization: How is power distributed and used within a society? How do societies regulate the power relations between their own and other groups? Furthermore, the study of political organization examines how safety and order are maintained

Figure 12.1
**HULI BIG MAN,
PAPUA NEW GUINEA.**
A Huli Big Man of the
Southern Highlands of
Papua New Guinea wears
an elaborately decorated
wig to demonstrate his
political role.
Credit: Lee Hunter/Images of
Anthropology

within a group. Is there a central authority, like a government, that imposes rules and punishes those who break them? Or does the group share the responsibility for making decisions?

POWER, AUTHORITY, AND PRESTIGE

Political relationships are managed by the use of power, authority, and prestige. A person, community of people, organization, or nation may use one, two, or all three of these strategies to control others. They may be used positively as "carrots" or negatively as "sticks."

Power is the ability to compel another person to do something that they would not do otherwise. It may be by threat of punishment ("stick") or promise of reward ("carrot"). Some degree of power exists in all social relationships, from the everyday interactions of neighbors to global relations between countries. The use (and misuse) of power is one of the means by which people become unequal in terms of resources and social status. Social, economic, and political inequality stem from uneven access to resources. Therefore, power is an important aspect of culture.

Power is used in essentially two ways: coercively or persuasively. *Coercive* power uses physical force or the threat of it. Examples of coercive violence are schoolyard bullying, rape, and war. There are many examples in history of the coercion of enslaved peoples to perform manual labor under threat of physical punishment. For instance, in the second century **BCE**, rulers of the Qin dynasty in China harnessed the labor of imprisoned slaves to construct the massive fortification of the Great Wall of China. During the construction of the different areas of the Wall, it is thought that up to a million workers died. For leaders who use coercive power, safety is a low priority since laborers must participate or suffer the consequences.

Persuasive power relies not on force, but on changing someone's behavior through argumentation using religious or cultural beliefs. Persuasive power offers a reward for compliance, rather than a threat. This reward may be measurable, such as wealth, or it may be personal, such as increased status, power, or emotional fulfillment. In 1997, a self-proclaimed prophet named Marshall Herff Applewhite created a religious revitalization movement called Heaven's Gate, based in Southern California. Applewhite convinced a cult of followers that they could escape the impending destruction of earth, which only he, as the earthly successor to Jesus Christ, could foresee. He taught that the promise of a better life was only attainable through suicide, at the precise time the Hale–Bopp comet passed overhead. In March 1997, 39 people dressed in identical black tracksuits and new Nike shoes willingly took their lives after being persuaded psychologically and emotionally.

Some individuals may have power over others, but few also have authority. **Authority** is the use of legitimate power. In large, complex societies with a centralized government, citizens grant the power of rulemaking or punishment to an individual or set of individuals, such as a ruler, congress, or police force. These entities have the authority to exercise power with the consent of their members.

Governments with democratically elected leaders bestow authority upon a prime minister or a president to lead the country. Nonetheless, the individual heading the government must also use persuasive power to convince the cabinet or congress to agree with policy decisions. This arrangement prevents all the power and authority from remaining in the hands of one person. When a leader rules by non-democratic means, such as in a dictatorship, then power is exercised legitimately in the eyes of the law, but it may not reflect the will of the people.

Prestige is a type of social reward that can only be given to a person by others. It refers to the positive reputation or high regard of a person or other entity merited by actions, wealth, authority, or status. It may be by virtue of birth into a particular family, personal achievement, or membership in a highly regarded social group. A celebrity may have prestige by virtue of being in the public eye.

The "**Big Man**," found throughout Melanesia, uses his prestige as an informal leader in his community. A clever and charismatic person, he represents his tribe to outsiders and mediates conflicts when necessary. The Big Man is generally wealthy and affirms his status with great shows of generosity. The role of the Big Man confers prestige and persuasive power; however, he has no officially recognized authority to make decisions for the group. Should he fail to represent the people well, a new Big Man will be sought.

SOCIAL CONTROLS AND CONFLICT RESOLUTION

Internalized Controls

Societies maintain order within their groups and in their relations with other groups using a series of controls. Some controls come from within, as part of the society's cultural values of what is right and wrong. These **internalized controls** guide a person toward the right behavior based on a moral system. They may be based on cultural standards or religious tenets, and they are often taught within the family. An example of an internalized control would be refraining from lying or stealing because a person believes that it is wrong to do so. Internalized controls may also come from a mixture of sources, as seen in the incest taboo, which is likely a combination of genetic, social, and psychological avoidances (see Chapter 11).

Maintaining order through a belief system is very different from a state system that uses a set of codified laws and punishments. Internalized controls embedded in belief seem "natural." They are entrenched in the way people think about the world. For instance, people who inhabit the world's forests have deeply held beliefs that their own health and survival is intimately connected to the health of the forest. It is "normal" and "right" to them to protect the ecosystem on which they depend from deforestation.

Cultural patterns develop around a social focus on either shame or guilt, both of which function as internalized controls. **Shame cultures** are those in which conformity to social norms stems from wanting to live up to others' expectations. The criticism of others, especially those of higher social status, is to be avoided at all costs. Even suicide is considered as an alternative to "losing face" when a person is dishonored. Suicide rates among adult males can be high in shame cultures if a man feels he has brought shame to his family by losing his job or by being caught publicly in a compromising situation. Often shame is more prominent in communal, or dependence-training cultures.

In contrast, **guilt cultures** focus on one's own sense of right and wrong and the punishment that can result from breaking the rules (especially when that punishment is meted out by supernatural entities). In general, Western cultures tend to be guilt focused, wherein members suffer emotionally due to failing to meet their own expectations rather than those of others. This is linked to Western cultures' reliance on individualism and independence training. Both guilt and shame operate to some extent in most societies as internalized controls to keep people from transgressing social norms.

Externalized Controls

Externalized controls are imposed from the outside. Rules regulate behavior by encouraging conformity to social norms. Authority figures enforce these rules within which a person, organization, community, or nation operates. External controls vary in degree from community gossip to the death sentence. Sometimes just knowing the controls exist can be enough to deter someone from breaking the rules.

Sanctions are the punishments that result from breaking rules. They may be informally meted out by community members or formally enforced by authority figures. Informal sanctions can be preventative (grounding teenagers to keep them from getting in trouble) or retributive (spanking a child). Gossip is an effective negative informal sanction, especially in smaller communities, that is both preventative and retributive. More formalized sanctions may be legally imposed and include punishments such as fines, prison, exile, or death. Countries linked by trade agreements also impose sanctions by limiting or preventing the movement of goods or funds.

Figure 12.2
"SACRIFICE" BY SPEED BUMP.

Sacrifice of animals was done with the intention of appeasing a group's deities, spirits, or ancestors. Sacrifice of humans was more likely an external sanction, meant to punish prisoners and terrorize enemies.

Credit: Dave Coverly

Inuit of Arctic Canada traditionally used song duels to solve problems in the community. This is an example of an externalized control, because others determine right and wrong. Song duels resulted in informal sanctions, because they were part of what is known as "customary law" (in contrast to "government law"). Song duels were held at festive community gatherings, during which the aggrieved parties would sing humorous and deprecating songs about each other. Community members present identified the best song and presentation, declaring the winner of the song duel. In this way, differences would be aired openly in a community forum, and solved in a publicly accepted way.

In the state of Burma (Myanmar), pro-democracy leader Aung San Suu Kyi suffered under formal legal sanctions imposed upon her by the Burmese ruling junta in 1989. She was imprisoned in her home and kept under house arrest for 15 years. Although these sanctions were imposed on Suu Kyi to quiet the call for democracy in Burma, they did not have the result the military government desired. Impressively, during this time she attained international recognition and won several awards, including the Nobel Peace Prize in 1991. She was finally released in 2010 and became a political leader in the country again. Since 2012, however, her status as a global leader of democracy has suffered due to her failure to denounce Burmese military actions targeting the Rohingya people, a Muslim minority. She is now experiencing the effects of informal sanctions, with world leaders accusing her of allowing **ethnic cleansing**.

In contrast to negative sanctions, rewards shape behavior positively by applauding good behavior and encouraging it in the future. Informal rewards might include recognition from community members for killing an animal for the first time on a hunt or for achievements in school. Formal rewards might include a gift of cattle upon completing a male puberty ritual (such as among the Maasai) or a military honor for bravery in battle (such as the US military's Purple Heart medal).

TYPES OF POLITICAL ORGANIZATION

There are two major types of political systems among the world's cultures: those that make decisions collectively, and those that concentrate power in the hands of a few. In general, smaller societies rely on the group to make decisions, and they use informal controls to maintain order. Larger, more complex societies require centralized governments and thus use formal controls.

Uncentralized Systems

Uncentralized systems have no central governing body. Therefore, community members impose sanctions on those who break the rules. This type of system is found primarily among smaller, more homogeneous societies, such as foragers or horticulturalists. In uncentralized systems, social rights and responsibilities are organized along family lines. That is, a person's place within the descent group and lineage dictates their role in society. Kinship relationships serve to govern people's relations with one another.

When problems arise within these groups, they will seek informal leaders to help mediate or negotiate. Informal leaders may be respected and wise, have prestige, and are often elders with experience. However, they have no official title or real authority to enforce judgments. When people do not comply with the judgments of elders, social mechanisms serve to humiliate or coerce them back in line through gossip, loss of reputation, or social ostracism. Informal sanctions operate widely in uncentralized societies.

Centralized Systems

In **centralized** political systems, a ruling body of one or more people is given the authority to govern. This occurs in larger, more complex, and heterogeneous societies. In such a society, not all members of the population know or are related to one another; the kinds of relationships governed by kinship ties or bonds of reciprocity are thus lacking. Therefore, the governing body creates a formal code of oral or written laws by which the population must abide, no matter whether the offender is a family member or a stranger. The ruling individual or group has both the power to control others and the legal authority to do so.

Cultural anthropologist Dr. Elman Service (1962) developed a typology to classify the different types of political organization in the world's societies. His ideas are based on the anthropological perspective called **cultural materialism**, in which a society's organization is directly related to whatever adaptations are necessary to survive in its environment. This is not the only perspective that anthropological theorists use to explain how power emerges in societies; however, it is one that is widely known and utilized. According to Service's classification, there are four types of human societies: the band, tribe, chiefdom, and state.

Each type has a different structure of social, economic, and political organization. The more homogeneous types – band and tribe – have uncentralized political organization, while the more heterogeneous types – chiefdom and state – have centralized political organization. Even though we present these four types as if they were unique and clearly distinguishable from the others, in truth, societies operate on multiple levels at the same time along a continuum. That is, in a tribal society, one may find social controls that also operate in band societies. At the same time, the tribe may act politically as part of a state society.

Bands

Bands are groups of approximately 50 to 100 individuals who rely on hunting and gathering as their main means of subsistence. A band will camp together while foraging, creating temporary structures for shelter and protection, but will move frequently to seek out the next desirable location. Since bands are small, a majority makes the decisions. This includes when and where to move the group next or what the outcome of interpersonal conflicts will be. There is no centralized government or other coercive authority.

The band's uncentralized power is reinforced by the egalitarian status of the group: no one member has more access to resources or authority than any other. Any leaders are temporary, based on good decision-making, charisma, or their ability to communicate well. They can attempt to persuade others but have no authority to enforce decisions. Positions of respect and high esteem are not hereditary.

As such, only informal sanctions may be used in band societies. Mediation and negotiation among antagonistic community members help to resolve differences. Gossip and ridicule can keep people in line with social expectations. Fear of reprisal from supernatural forces may also serve to guide people's behavior.

The Ju/'hoansi of the Kalahari Desert have a band organization. In this foraging society, an informal leader may be sought to settle a domestic dispute between husband and wife. If a solution cannot be found, this same individual can also grant a divorce with the support of the community. In situations where a solution cannot be found, an individual or couple may leave their band temporarily or permanently and join another band where they have relatives. The ability for individuals to move fluidly between bands allows the group as a whole to remain cohesive.

Tribes

Tribes are groups with higher population density than bands. They are horticulturalists or pastoralists, living in separate villages spread out over a wide area. The villages are linked by clan membership to a common ancestor, which may be real (a historical person) or mythic (an animal or deity). Although villages of a single tribe are separate, they are tied to one another by clan membership, their real or fictive kinship, and a common language. Often these strong links can be useful when the tribe needs to come together to solve larger issues (see Box 12.1).

BOX 12.1   **Kayapo and the Belo Monte Dam Project**

Tribal peoples often unite their villages for a common cause. Sodalities, or pan-tribal associations, help bring people with similar interests and goals together across a region. Since 1988, the Kayapo (*Mebengokre*) people of the Amazonian rainforest have been uniting to protest the building of a hydroelectric dam by the Brazilian government in the Pará territory. The dam, finally completed in November 2019, will create an enormous lake that will flood the Kayapo and other peoples' native lands around the Xingú River. Flooding will force 20,000 people, including thousands of Indigenous peoples, to relocate and leave everything they know behind.

When Kayapo leaders first protested the dam construction, high-status chiefs encouraged their fellow chiefs of Kayapo villages and other Amazonian peoples to show support. Thousands of Indigenous people made the days-long trip out of the forest by boat to show strength at a meeting with Brazil's electric company. During the meeting, Indigenous men and women whose land was under threat gave hours of speeches underscoring their claims to the land as the original inhabitants. The international media covered the highly publicized event. Even Western celebrities who advocate for environmental causes, such as the musician Sting, lent their support to the Kayapo. The show of force against the dam effectively postponed its construction.

In 2006, protests began again with the new Belo Monte Dam project proposal. Kayapo chiefs extended their political reach to bring in new allies, even meeting with representatives from the World Bank. Brazilian construction methods have come under fire from the UN Human Rights Council, and international human rights groups around the world have criticized the way in which some of the Indigenous communities have been overlooked or coerced. Unfortunately, these political and humanitarian efforts to stop the dam had little effect.

As of 2019, the Belo Monte Dam project is complete, making it the fourth largest dam in the world. The tribal peoples of the area will be relocated, effectively losing their land, the resting places of their ancestors, and potentially their way of life. Repercussions for the thousands of Kayapo and other Indigenous people of Pará, whose identities and survival are intimately connected to the forest, have yet to be seen.

Figure 12.3 **KAYAPO CHIEF RAONI AT PROTEST.** Protests against the construction of the Belo Monte Hydroelectric Dam in the Amazon rainforest of Brazil have gone on for decades. Kayapo Chief Raoni Metuktire, seen here at a protest, was nominated for the 2020 Nobel Peace Prize as a leader and spokesperson for Indigenous peoples of the area that will be affected.
Credit: © Gert-Peter Bruch / Wikimedia Commons, CC BY 3.0

Tribal power is also uncentralized, in that there is no central government to impose rules or punishments. However, leaders arise based on their skills and experience, or due to their birth into a noble or high-status clan. The Melanesian "Big Man" discussed above is an example of this type of leader. Disputes can be solved through mediation or through unofficial court-like resolution methods in which the village comes together to hear and discuss issues. Tribal societies may act along egalitarian lines, like bands, or they may tend toward a **ranked** system, in which hereditary positions of status and prestige are passed down within families.

Another way that tribes remain united over a wide area is through links between individuals that cut across village lines. A **sodality** is a group that brings people together through common concerns, age, or interests. For instance, children of a particular age-set across multiple villages may go through a rite of passage at the same time. Boys and girls of the Dinka (*Jieng*) tribe of South Sudan may undergo an initiation ceremony together, a painful facial scarification of forehead lines marks their passage from childhood to adulthood.

Chiefdoms

A **chiefdom** is found in more populous societies in which intensive agriculture is practiced. These preindustrial societies have a more complex structure, with villages linked together by districts. Due to the complexity and large population, a centralized government is required with formalized leadership. There is an officially recognized chief at the top of the chain of command and a bureaucracy of greater and lesser chiefs in place to manage the different levels of governance.

The chief is generally a hereditary office, not an elected one. He (a chief is most often male) comes from the wealthiest family in the chiefdom. Every society will have rules that govern inheritance of the chief's position. The seat is most commonly passed to the son of the chief (patrilineal) or the son of the chief's sister (matrilineal). A chiefdom is an example of a ranked society, in which one's family line dictates whether a person will have prestige and status.

In Samoa, an island nation in the South Pacific Ocean, social and political organization is governed by the *fa'amatai* system, in which each extended family holds a chiefly title. These chiefs, the *matai*, represent their families' interests in village councils (*fono*) and matters relating to the family or village on a larger scale.

Among the Five Nations of the Iroquois Confederacy, across the area now known as upstate New York, a Clan Mother is responsible for selecting a new chief and ensuring that he performs his duties well. If he does not act in the best interests of the people, she can also remove his authority. In this matrilineal society, the hereditary title of Clan Mother is passed down among sisters and then daughters, but the title of chief is not hereditary.

How COVID-19 Reveals Class Divisions

Studies of political anthropology often focus on class differences, identifying the ways in which wealthier people consistently have access to the resources they need while others do not. These class differences become glaringly obvious in the aftermath of a natural disaster, such as a hurricane, flood, or earthquake, in which lower income communities suffer the greatest losses. In 2020, the COVID-19 pandemic evolved into a global disaster, with lower socio-economic communities, often largely comprised of people of color, at the highest risk for exposure, infection, and death.

The pandemic taught us a new term, "**essential workers**," which describes those people performing (mostly) low-wage but important jobs that keep communities functioning. Also called "critical infrastructure workers," they might be employed in any number of positions, from fields like energy, agriculture, and transportation to childcare, retail food distribution, and automotive mechanics. At the height of the pandemic, people required to physically go into work ran the greatest risk of exposure to the virus, while those people with jobs that could be transferred to a remote or online environment could remain in their homes with a far lower risk of exposure. Another term, "**frontline workers**," describes another category of people that cuts across class lines – mostly in the medical field – who ran a high risk of infection while working to keep people alive.

The vast majority of essential workers are wage laborers, who lack the job security of professional positions. They are more likely to belong to a lower socio-economic class, to have fewer medical resources in their communities, and to have underlying health conditions (see Chapter 8 for a discussion of the health impacts of structural racism). Due to these factors, the effects of COVID-19 on communities of color have been devastating. In fact, US life expectancy rates dropped significantly in 2020, with Black and Latinx Americans losing 2.1 and 3 years respectively. While no one is risk-free from the virus, some communities have paid (and continue to pay) a higher price than others. Essentially, COVID-19 has waged a "class war" in societies across the globe.

In many societies, colonial contact drastically altered the traditional political and social organization of chiefdoms. In some cases, in order for the colonial governments to better administer the Indigenous lands that they had claimed, they created new roles for "chiefs" that did not exist before. Among Indigenous peoples of Australia, for instance, British governors would choose Indigenous elders and appoint them "Kings." They would be easily identified by a metal plate hanging around their necks from a chain, stamped with their name, perhaps several images, and the word "King." Prior to contact, Indigenous society in Australia was uncentralized. As hunter-gatherers, Indigenous groups were egalitarian, and decisions

Figure 12.4
CHANGING OF THE GUARD, PARLIAMENT BUILDING, HUNGARY.
State societies often situate their governmental offices in imposing buildings with towers, walls, and gates. Security forces guard the entrances, such as this guard patrol at the Hungarian Parliament building in Budapest.
Credit: Dennis Jarvis/CC BY-SA 2.0

were made by consensus. European settlement upended the political organization of the area, as it has in colonial areas around the world.

States

State societies are industrial and heterogeneous, with a strong centralized government. State societies contain the largest populations seen among the different forms of political organization. A state society usually contains diverse groups within its borders.

A state has a formalized central government with the authority to use force to control its citizens. A written code of laws formalizes right and wrong and encourages socially sanctioned behavior. When laws are broken, there is also a codified set of punishments that correspond to the severity of the offense. The government uses an official court system to determine innocence or guilt and to impose punishments.

Because a state society is very large, a bureaucracy is necessary to administer to all the needs of its people. There are lower-level administrators, who report to higher-level governors. All of these leaders report to a central authority.

The highest authority may be an individual ruler such as an emperor, king, or queen, or it may be a collective group, such as a congress or parliament (with a ruling president or prime minister). A state society is also referred to as a **stratified society,** in which certain members have access to power, authority, and prestige, while other members (often a much larger number) are excluded. This creates sometimes radically different levels of access to power and even to basic resources for different groups of people within a state society.

A central governing body may demand taxes or tribute from its citizens, which can then be used for improvements in infrastructure or operations. Due to its power of authority, it may also demand labor from its citizens. The stability of a central government may be dependent upon how taxes are collected and used, and whether the citizens are treated justly. If a government authority demands too much of its citizens, resentment and discontent may turn into violence or revolution.

SOCIAL INEQUALITY

Social stratification, or the ranking of members of a society into a hierarchy, is not a natural feature of social organization. Many societies, such as foraging band societies, follow carefully constructed social rules so that individual members do not have more status than others. Resources abound in their environments, and cooperation and sharing are built into their daily lives; gossip and social ostracism result if they do not abide by these expectations.

Nevertheless, once a society settles in one location and begins to amass possessions, a stratified society results. Stratification is characterized by unequal access to resources. High-ranking members of a stratified society own or have access to more possessions and opportunities than low-ranking members. These wealthy individuals gain status in the form of power and prestige. Therefore, people with the most power tend to be concentrated in a small group at the top of the social pyramid. Stratification may also result when a society is colonized, resulting in a social and political hierarchy in which non-Indigenous colonizers make rules and mete out punishments for members of the now-subordinate Indigenous society.

Depending on the society and how it is structured, social hierarchies may or may not allow **social mobility** – that is, the ability to move upward or downward within the system. There are two basic types of social stratification: class and caste (see Table 12.1). **Class** stratification is based upon differences in wealth and status. Through a combination of work and opportunity, members of one class can move up into a higher class, or they can lose their status and move downward. This is referred to as **achieved status,** because it is based on personal actions.

In the United States, for instance, a person may be able to move from one class to another based on hard work, resulting in higher income and status. This is the model of the "American Dream," which many immigrants aspire to reach. In reality, it is often quite difficult to rise through the system, since one's cultural and physical environments play a large part in determining opportunities.

TABLE 12.1

Comparison of Class and Caste

Class	Caste
Determined by wealth and status (achieved status)	Determined by birth (ascribed status)
Allows social mobility	Does not allow social mobility
Influences occupation and marriage	Determines occupation (to an extent) and marriage

Caste, on the other hand, is a hierarchical system based on birth. The caste system does not allow movement from one group to another. An individual's status in society is **ascribed**, or fixed at birth, and cannot be changed. Caste may dictate a person's social standing, occupation, and who they may marry. Historically, social ranking based on ascribed status is found in societies all over the world and throughout history. North Korea's *songbun* caste system uses the status of one's ancestors to dictate a person's social and political mobility.

Although ascribed status exists in many forms all over the world, it is most commonly associated with Hindu India. The Hindu caste system is an ancient ranking system that separates people into categories based on their birth into a particular set of occupations. There are four major *varnas* or divisions: Brahmin (priests), Kshatriyas (warriors and rulers), Vaisyas (merchants, farmers, craftspeople), and Shudras (laborers, servants), as well as a category outside the system called Dalits (untouchables). The caste system in India was outlawed after India's independence from Britain in 1948. Nevertheless, social practices based on caste (such as marriage) and severe discrimination against Dalits still persist.

Gender Inequality

When looking at the question of equality between males and females in society, it may be tempting to regard male dominance as "natural." This is because in nearly all societies, men own the family's land and other resources, even when a society is matrilineal and matrilocal. This generally results in higher status and more prestigious positions for men than women. This is not to say that women cannot hold positions of power, because many women throughout history and today are leaders who hold prestige and authority. However, men tend to have more access to power, prestige, and privilege than women. The dominance of men and subordinate status of women in society is referred to as **gender stratification**.

Figure 12.5
AFGHANISTAN NATIONAL WOMEN'S VOLLEYBALL TEAM.
Female Muslim athletes on the Afghanistan National Women's Volleyball team compete internationally. The team wears lightweight and breathable head coverings ("sports hijabs"). In non-Muslim majority countries, the hijab is often controversial for political or social reasons. However, the acceptance of head scarves in all areas of social life allow Muslim women a degree of freedom to pursue their interests without shame or harassment.
Credit: US State Department/CC BY-ND 2.0

Because social expectations and status vary across cultures, it is clear that gender inequality is not "natural," but connected closely to social norms concerning power. In examining the question of male dominance, anthropologist Dr. Ernestine Friedl (1978) discovered a consistent connection between power and the distribution of food resources. She notes that in societies in which females grew and controlled access to food, women had relatively equal status to men. In societies in which men controlled food resources, especially meat, men dominated women socially and politically, and perpetrated more abuse on women. Gender stratification exists across cultures where men control the distribution of food and/or other resources.

Another example of gender stratification can be seen in the daily expectations of men's and women's behavior. If these expectations are similar, then the society regards both sexes as equal. In gender-stratified societies, however, women are faced with restrictions on their behavior from which men are largely exempt. Women's behavior may be closely monitored, with limits on where, when, and with whom a woman is seen. Her clothing may be restricted or commented upon by others, while men's clothing is not. Unmarried women, especially, are believed to represent the family's honor; therefore, the family and community members regard her actions with suspicion. Even suspected dishonorable behavior may be punished harshly.

A woman's dress code is one way that gender inequality may be expressed externally. To outsiders, the veil that Muslim women wear throughout the Middle East,

Asia, Africa, and elsewhere appears to indicate extreme gender inequality. Head and body coverings are seen as a sign of restriction and oppression imposed upon them by their families or husbands. It is true that in some countries in which women veil their heads, faces, or bodies, women experience severe **gender discrimination**.

For instance, in areas of Afghanistan where the Taliban rule, women are forbidden not only from showing their faces or bodies in public, but also from laughing or singing, seeing a male doctor, and indeed, leaving the house without being accompanied by a male relative. This type of severe gender discrimination affects women's mental health as well: in the 1990s, researchers from the American Medical Association found that 78 percent of women living in a Taliban-controlled area were clinically depressed, and 73 percent of women reported suicidal thoughts (Rasekh et al., 1998).

Nonetheless, wearing the veil is a complex and nuanced issue. Even though some women may feel restricted by their veils, it is also true that many women *choose* to wear the veil without experiencing pressure from their husbands or family members. For these women, wearing the veil is primarily a sign of devotion to their religious beliefs. It is also a way to take part in public life with a desired measure of modesty. Head and body coverings provide these women a measure of privacy, even safety, which they would not otherwise enjoy.

Environmental Inequality: Access to Water

Social stratification reaches into all aspects of human society, even our most fundamental needs, such as food and water. Water is especially crucial since a person cannot live more than a handful of days without it. Throughout history, human communities have been built around access to water, both for their own survival and for the successful growth of food crops and the keeping of domesticated animals. It is not an exaggeration to say that for the living species on our planet, water equals life.

Although clean water is a fundamental human need, vast inequalities exist regarding who has access to it. Although our blue planet has plenty of water, only 2.5 percent of it is fresh. Sixty-eight percent of that fresh water is locked into glaciers. The remaining fresh water is available underground in aquifers and groundwater, and above ground in rivers and streams. Because of the small percentage of water available for human use, and the unequal distribution of it throughout nations, the UN Food and Agriculture Organization (2013) predicts that two-thirds of the world's population will face water shortages by 2025. The countries – and people – suffering most from these shortages will be across the developing world.

It is estimated that more than a billion people do not have regular access to clean drinking water. The situation is dire in regions all over the world, but nowhere is it more threatening than in Africa, where 19 of the world's 25 countries suffering from very limited access to safe drinking water are located. Millions of the poorest inhabitants of these nations die annually from diarrheal infections caused by drinking water from open sources contaminated by human and animal waste. The vast majority of these people are children under five years old.

Although clean drinking water is a basic human need, it is no longer treated as a human right. **Privatization** of water makes it difficult for marginalized and rural people to access clean water. When local water supplies are privatized, governments grant the rights to the water supply to a private corporation, which then purifies the water, bottles it, and sells it back to the local people. This can be an extreme hardship when people have very little income. Often, the head of household must make the decision between purchasing clean water or food and medical supplies. The inequality of power affects people's daily lives, even in meeting their most essential needs.

Gender, status, and ethnic differences can impact the water rights of an individual, family, or community. For example, in sub-Saharan Africa, 60 percent of the water collected daily is done by girls aged 15 years and under. Sometimes walking to and from the water source takes up to six hours a day. Girls with this responsibility often cannot attend school due to the time it takes to bring fresh water to their households. In some places, women are harassed or assaulted as they carry water.

North America is no stranger to water issues. First Nations peoples across Canada struggle with water access for their **reserves**. A national study released in 2011 assessed 571 First Nations reserves (with nearly 500,000 inhabitants total) to evaluate their water systems, including the supply of clean drinking water and treatment of wastewater (Aboriginal Affairs and Northern Development Canada, 2011). Looking at the sources, design, operation, and monitoring of water systems in each location, the report concluded that 73 percent of the reserves had medium- to high-risk water problems, or they would in the immediate future. Moreover, nearly half of all homes on the reserves lacked sewage pipes and relied on outhouses, even in the winter. The lack of updated infrastructure has led to outbreaks of illness, including skin rashes, infections, and gastrointestinal problems, due to high levels of bacteria and chemicals.

Water Inequality in Agriculture

Feeding people is a thirsty business. Growing food uses the majority of our water supplies across the planet. It takes massive quantities of water to feed people: an average of 450 gallons to grow a pound of rice, 400 gallons for a pound of sugar, 130 gallons for a pound of wheat, and a shocking 2,650 gallons to produce a pound of coffee (Pearce, 2006).

Rivers and their tributaries provide the water needed to irrigate fields to grow food. As a result, some of the world's great rivers no longer reach the ocean: the

Figure 12.6

COLLECTING WATER IN A REFUGEE CAMP.

More than a quarter of a million refugees fled the conflict in northern Mali in 2012, some of them settling in camps across the border in Niger. These camp residents are collecting clean water provided to them by a United Nations refugee agency.

Credit: Sean Smith/ CC BY-S.A. 2.0

Nile in Egypt, the Yellow River in China, the Indus in Pakistan, and the Rio Grande in the western United States and Mexico. Also disturbing is the lack of water in farmers' wells, caused by an overuse of groundwater. Every year farmers dig deeper – in certain places more than a mile – to find the water needed to irrigate their crops. Fresh water is disappearing in places where the ability to feed people depends upon it.

It follows that food preferences play a role in a country's water requirements. Meat products, especially the raising of beef cattle, require more water to produce per calorie than any other food. It takes 20 times the amount of water per calorie to raise beef than cereal grains (such as rice or wheat). Therefore, those nations with growing economies where meat consumption has exploded, such as China, have much higher water requirements than others. Individuals with more financial resources and higher social status consume more meat products and therefore drive the market to produce more.

VIOLENCE AND WAR

Violence – either within groups or between groups – often arises out of power inequities and hierarchies. When groups clash, the size of the group and goals of the conflict shape the confrontation. Violence in smaller horticultural or pastoral societies may take the form of a **raid**, in which members of one group aim to steal or recover items, animals, or people from another group in the same society. Raids are short-term incursions with a specific goal in mind.

Ongoing violent relations between two groups in the same society are called a **feud**. Feuding often begins when a member of one group kills a member of another. This begins a long-term hostile relationship in which revenge is the goal. Feuding often occurs between extended families, who continue to avenge the murders of their kin. It also can occur between groups who share fictive kinship, such as members of an urban gang. Unlike a raid, which is over in a few hours, a feud can last for generations until the two sides agree on a truce.

Warfare is different from raiding and feuding in that it is on a much larger scale. Generally, the weapons and transport of armies are more technologically advanced. Societies can divide internally into civil wars, in which different groups within the same society go to war with one another. Civil wars may begin based on religious or ethnic issues.

Ethnic **sectarian violence** occurs when societies divided by ethnicity and religious beliefs explode with tension that has built for decades or even centuries. Some of these ethnic-based uprisings escalate into violence or full civil wars, such as occurred in the 1990s between the Hutus and Tutsis in Rwanda, Central Africa, when the extremist Hutu majority committed near **genocide** of the Tutsis, killing 70 percent of the Tutsi people living in Rwanda at the time. The early twenty-first century has witnessed conflicts between Sunni and Shia Arabs in the Middle East escalate into a long, violent, and oppressive war.

War may also be declared by one society or nation on another. The goals of war are much larger in that one side attempts to kill as many people or destroy as much property as possible until the other side calls for a truce. Industrialized nations fight over natural resources such as land, water, or raw materials. Today, some federal budgets, such as that of the United States, allocate billions of dollars to develop weapons and military technology, support the different branches of their nation's military, equip troops, and extend their nation's influence in countries in different parts of the world.

How does this type of war develop on such a massive scale? It has to do with population growth and surpluses of wealth that arise in a stratified society. Competition among state societies for access to resources is high, especially when a growing population demands them. War also arises when other methods of conflict resolution have failed, such as **diplomacy** or economic sanctions.

We may assume that war has been a part of human behavior since the beginning of our species. However, large-scale warfare cannot exist without large-scale societies. In small populations like food foragers, resources are communal. Marriage

Talking About: War and Destruction

Anthropologist Dr. Carol Cohn (1987) spent a year studying the subculture of a strategic think tank for US government defense analysts who plan nuclear strategy. She wanted to find out how people can plan the business of destruction – in other words, "think about the unthinkable." Through a process of enculturation, Cohn learned the language necessary to discuss military strategy, which she calls "technostrategic." As she became fluent in this highly specialized language, she was surprised to find that she had lost the ability to think about the human costs of war.

Abstraction and **euphemisms** focus all discussion on weapons and strategy. Cohn found that the use of several types of metaphors of domesticity allow the analysts to connect in positive ways to their work. First, the euphemisms invoke hygiene and medical healing: they talk about *clean bombs* (bombs that release power but not radiation) and *surgically clean*

strikes (bombing that takes out weapons or command centers only). Second, images of country life and recreation are used: missiles are located in *silos* as if on a farm, piles of nuclear weapons loaded in a submarine are called *Christmas tree farms*, and bombs are referred to as *re-entry vehicles*, or *RVs*. Third, the weapons are talked about as if they were responsible for their own actions: for instance, the pattern in which a bomb falls is called a *footprint*, as if the bomb was dropping itself, like a foot in the sand. This image removes human accountability for the action.

In addition, Cohn discovered male-gender attribution to the missiles. Beyond the expected phallic imagery, bomb detonations were frequently described sexually, comparing the explosion to an orgasm. Moreover, missiles are spoken about as if they were infants or little boys. The implication is that the analysts hope the bomb will be powerful and aggressive (like a boy) and not mild or

alliances between horticultural villages make it impractical to fight other villages since relatives may live there. Avoiding conflict and confrontation is important in small-scale societies where cooperation is crucial for survival.

Large-scale warfare arises with centralized states and surpluses of food and resources. This coincides with population growth and the rise of cities. Surpluses become attractive to official leaders, who then can organize their people to fight. An army or other large-scale military force is given the authority to use force against other nations.

Not all societies value aggression as a means to solve problems. Egalitarian societies, such as foragers and horticulturalists, must cooperate for reasons of survival. Egos and arguments only divide the group, making protection and pooling of

timid (like a girl). After the first successful test of the hydrogen bomb in 1952, one pleased atomic scientist wrote to another, "It's a boy" (Cohn, 1987, p. 701).

Cohn began her fieldwork interested in how nuclear defense analysts discuss massive destruction and human suffering day in and day out as part of their job. She quickly found they don't. Military strategy demands a language that focuses on weapons only in a quest for scientific rationality. But the costs of embracing this language privilege a distanced and aggressive (i.e., "masculine") view over any others. Human costs cannot be discussed; these are "feminine" concerns. To her surprise, Cohn discovered that once she was a speaker of this language, she could no longer express her own values, since they were outside of the "rational" discourse. Not only could she not articulate her ideas using this language, but also, she was written off as a "hippie" or "dumb" if she tried.

Her work carries with it an important question: What does any language allow us to think and say?

Figure 12.7 **TITAN II MISSILE.**
The Titan II missile, seen here in its resting place at the Titan Missile Museum in Arizona, was the type of missile on alert for several decades during the Cold War. An ICBM (Intercontinental Ballistic Missile) like this one had the ability to launch a nine-megaton thermonuclear warhead to a target 6,000 miles away in approximately 30 minutes.
Credit: Geoff Stearns/CC BY-SA 2.0

resources harder. Therefore, these groups develop cultural norms that lessen the inevitability of social tension. One way to remove the source of tension is for an individual or group to leave. Foragers such as the Ju/'hoansi, for instance, can join a neighboring band and live with relatives either temporarily or permanently.

Other groups have developed a set of social norms that limit possible sources of tension. For instance, among the Buid of Mindoro Island, Philippines, nonaggression is the most valued trait in a person. Social expectations reinforce this behavior to the extent that men do not face one another when speaking. Rather, they direct their comments to the larger group, which lessens the possibility of annoyance or defensiveness. Other cultural practices that minimize the risk of hostility between individuals include harvesting crops with all workers facing the same way to minimize conversation; avoiding economic debt to one another; placing little value on bravery; and rearing children without punishment. The members of the group are conditioned to avoid competition, individual leadership (ego), or authority in an effort to keep the group stable and nonviolent.

SUMMARY

This chapter explored political organization, which acts as a structure that holds society together and dictates the hierarchy of power. Mirroring the Learning Objectives stated in the chapter opening, the key points can be summed up as follows:

- Political organization regulates people's behavior through a combination of the use of power, authority, and prestige with culturally sanctioned punishments and rewards.
- In uncentralized political systems, such as bands and tribes, informal leaders use charisma and experience to lead. They rely on the members of the community to support the functioning of the social system through gossip, negotiation, and supernatural threats. In centralized political systems, such as chiefdoms and states, official leaders use power and authority to keep order.
- Other forms of social and political hierarchies, such as gender or environmental, exist, in which marginalized members of society do not have the same kinds of access to power as others. Thus those marginalized people lack social status or access to resources, such as the basic right to water.
- While the majority of societies experience violence, some small-scale societies value cooperation over competition and manage to avoid it. However, when populations grow large and more complex, the conditions necessary for war begin to emerge and settled societies fight for the resources of the other.

Review Questions

1. What are the differences between power, prestige, and authority?

2. How do sanctions and rewards work to control people's behavior within a society?

3. What are the differences between uncentralized and centralized governing systems?

4. What are the characteristics of the four types of political systems?

5. What are the differences between class and caste societies?

6. What are the effects of gender stratification?

7. What are the different forms of violent conflict, and in which types of societies are they usually seen?

Discussion Questions

1. In your view, what would be a good use of power for a leader?

2. How do people come to accept power inequality or differential access to resources as normal?

3. Although we generally classify Western nations using categories of class, not caste, do you see evidence for caste-like discrimination as well? What are the differences?

4. Due to the unequal demands on water for growing food crops and raising animals, what would you suggest as an optimum diet for human beings in the future?

Visit **www.lensofanthropology.com** for the following additional resources:

| SELF-STUDY QUESTIONS | WEBLINKS | FURTHER READING |

13

SUPERNATURALISM

LEARNING OBJECTIVES

In this chapter, students will learn:

- *reasons for the development of supernatural belief systems.*
- *the earliest evidence for supernatural beliefs.*
- *what functions religious belief serves in society.*
- *about the roles of deities, ancestor spirits, and spirits of nature.*
- *about the different roles that religious practitioners play in society.*
- *how oppressed peoples resist the imposition of a new set of beliefs.*
- *about the intersections of religious beliefs and other forms of cultural expression.*

Leaping into the fire as a sacrificial offering teaches lessons about religious origins and right and wrong behavior.
#Supernaturalism

INTRODUCTION

Humans are unique in that they are compelled to make meaning of their lives. People create and use symbols to connect to others in their communities, shape their world, and define their identities. One area of life in which symbols carry great meaning, both personally and for society, is in the realm of religious or supernatural beliefs.

Faith in spirits, gods, or unseen forces guides individual behavior in powerful ways, and it serves important functions for the social group. The teachings of religious belief systems often underpin many aspects of social life, such as the structures of power and punishment. They also extend beyond the realm of organized religion into family life, ideas about health and healing, people's relationship to the natural world, and other areas of life. Every society also has nonbelievers among its members, and their secular, scientific, or humanist value systems serve to guide their behavior in similar ways as well.

DEFINING RELIGION

Of course, beliefs vary widely throughout the world, as do the experiences and practices of religion. Because of this enormous diversity, anthropologists may not agree on a single definition of religion. It all depends on how anthropologists

Figure 13.1
GOLDEN BUDDHA, BURMA.
The golden Buddha at the Maha Myat Muni Temple in Mandalay, Burma, has a two-inch-thick layer of gold leaf that has been applied over many years by the hands of male pilgrims to the site.
Credit: Daniel Chit

approach their research. In this text, we define **religion** simply as a set of beliefs and behaviors that pertain to supernatural forces or beings, which transcend the observable world.

Religious belief systems have four components: (1) they share an interest in the supernatural (whether beings, forces, states, or places), (2) they use ritual, (3) they are guided by myths, (4) and they are symbolic. The term **supernatural** doesn't imply "unnatural" or "strange." As a word made up of the root meaning "beyond" the "natural," it simply refers to those things outside of a scientific understanding that we cannot measure or test. For practitioners, faith requires no evidence in order to believe, only acceptance.

Ritual is a symbolic practice that is ordered and regularly repeated. It provides people with a way to practice their beliefs in a consistent form, connecting them to others in the same community. Examples of ritual range from a Unitarian Universalist church service to a Buddhist meditation, from the sacrificing of a hen by a shaman in Guatemala to making the sign of the cross before a sports match to ask for divine assistance.

Myths are sacred stories that explain events, such as the beginning of the world or the creation of the first people. They serve to guide values and behaviors. It's important to clarify that these stories are referred to as "myths" not because we regard them to be untrue, but because they are outside recorded history and based on faith. An example of a myth is the Finnish creation story of how the world came to be. In one version of this story, the demigod Väinämöinen, the son of the Sky's daughter and the Sea, came down to rest in the ocean, where a beautiful bird laid eggs. When the eggs dropped, they formed the land, sky, sun, moon, and stars.

Finally, religion is symbolic because it is based on the construction of meaning between a person and their beliefs, and among people within a community who share these same beliefs. Although the belief system itself is symbolic and imbued with meaning, individual symbols that represent that belief system are also important in connecting a believer to the religion. Symbols can take many forms: objects, signs, words, **metaphors**, sound, gestures, ritual clothing – anything that acts as a mediator between the believer and the supernatural realm.

With the guiding principle of cultural relativism, an anthropological approach does not question whether one religion is more valid than another. We examine religious beliefs both from an **emic** – or insider's – perspective and from an **etic** – or outsider's – perspective. That is, anthropology attempts to learn how people think, act, and feel about their belief systems. Then, we analyze and interpret these aspects to produce a deeper and broader understanding.

Reasons for Supernatural Belief Systems

Attempting to trace the earliest evidence for religious beliefs poses a challenge. Like many aspects of culture, beliefs do not fossilize and lie buried in strata for us to uncover ("Look, there's a belief, right next to that stone tool!"). However, cultural practices may leave physical evidence that can be found by archaeologists, who seek to understand cultural practices through physical remains.

Early Evidence

The earliest evidence of religion is linked to burial sites, since the idea of burial is an early marker of culture and community. Before foragers began burying their dead, they would simply leave a corpse behind and move on to a new location. Moving after a group member has died was a practical choice because it avoided exposure to the decomposition process and to scavengers who might be attracted to the body. The idea of burial represents a radical change in this thinking process. Even the earliest burials may have had something to do with the possibility of preparing or assisting the body (or its essence/soul) for existence in an afterlife.

Although laying the dead in the ground is the most common mortuary practice, cultures throughout the world practice other forms of releasing a person into the spiritual realm. Therefore, there may be ancient funerary rites for which we have no physical evidence. For instance, many cultures cremate the remains of their loved ones. Hindu tradition requires a body to be cremated on a pyre of wood while family members are in attendance. Some cultures have similar practices but use very different symbolism to guide the practice. For instance, both Zoroastrians and Tibetan Buddhists invite scavenging vultures to remove remains of the dead by placing them in high, open places. Tibetan Buddhists believe this is the most generous and compassionate way to return the body to the circle of life. For Zoroastrians, this practice prevents the world of the living from being contaminated by the dark forces of the dead.

Religion and Early Cave Art

There may be some very basic connections between religion and art in the evolution of modern human beings. First, they may have developed around the same time in human history, around 40,000 years ago. Second, both are understood to be signs of behavioral modernity. In other words, along with cooking and language, the existence of religion and art are signs that *Homo sapiens* had reached a modern stage of social and cultural development. Third, after 40,000 years ago, early humans appear to have made art for spiritual reasons, to attempt to manipulate or communicate with forces beyond their control. Representing an image visually may be the most fundamental way to try to connect with the supernatural realm.

Some of the earliest art can be seen in sites with cave paintings (**pictographs**) and engravings (**petroglyphs**), as discussed in Chapter 5. There are many reasons early humans might have painted and carved the walls of caves, including for rituals or ceremonies, for recreating the lives of the people who lived in the region, for documenting game animals, or for aesthetic reasons. Multiple sites have abstract paintings such as geometric designs, the meanings of which are difficult to decipher. The most common interpretation of cave art is that it has religious significance.

Some of this art on the walls of caves seems to provide evidence for the use of **magic**, or the use of powers to contact and control supernatural forces or beings. Magic seeks to manipulate the outcome of events. For instance, in South Africa, a series of ancient caves show pictographs of geometric grids, zigzag lines, dots, and spirals. Based on ethnographic research among the San people who have inhabited that area for thousands of years, these shapes are similar to the patterns seen by a person deep in trance or as a result of ingesting hallucinogenic drugs (Lewis-Williams, 1998). Interestingly, the same shapes are also similar to the visions seen by sufferers of migraine headaches. These links connect this cave art to the universal physiological alterations of the human brain under those conditions.

In the northern hemisphere, cave pictographs at Lascaux in France depict hundreds of large game animals, many superimposed upon one another. One interpretation of this layered painting is that the animals were painted on a sacred spot in the cave in order to practice magic. The location of the paintings might have been a particularly powerful spot, and so many paintings were placed there to harness the same power.

Interpretations of some representational animal art see the animals as painted before a hunt to ensure success. If true, this would be evidence of **imitative magic**, or creating something to represent real life, then manipulating it in a way that imitates the desired effect. In other words, "like produces like": painting an animal may signify the wish to encounter and kill it on the hunt. Another example of imitative magic would be a shaman creating an effigy that resembles an actual person, and then sticking it with a pin to cause pain in the person's body.

Functions of Religion

One can imagine the prescientific notion that natural forces such as weather, the sun's orbit, the changing of the seasons, or solar eclipses were caused by unseen supernatural forces. In fact, humans' big brains compel us to seek understanding and knowledge. Our ancestors wanted explanations for these kinds of natural phenomena. Anthropologists believe that supernatural beliefs provide explanations for those aspects of life for which we have no logical answer, thereby fulfilling an intellectual function.

Figure 13.2

These Muslim men are washing their feet before entering a mosque for prayers in Istanbul, Turkey. In some traditions, it is important to show respect by cleansing the hands or feet, or both, before worship.

Credit: Barry D. Kass/Images of Anthropology

Religion also helps humans cope emotionally with those anxiety-producing events that we cannot control, such as accidents, illness, or death. Prayer, offering, and sacrifice are ways for a person to seek help from supernatural beings or forces. Participation in ritual practices allows a person who is suffering to feel involved in achieving a positive outcome. For these reasons, belief in supernatural beings and forces can provide emotional relief.

However, the opposite can also be true. In his ethnography *The Winds of Ixtepeji*, Michael Kearney (1972) writes about the Zapotec town of Ixtepeji, Oaxaca, Mexico, in which the air is filled with malevolent forces and spirits. These *mal aigres* ("bad airs," colloquial) can enter one's home when a door is opened. People can also manipulate them to hurt others. Imagine living in a world where the very air you breathe may be ready to kill you! Not surprisingly, the townspeople have developed cultural and psychological defenses against these perceived threats. As Kearney relates, they live in a society that is characterized by distrust and paranoia.

Differences in belief systems can also lead to distrust and paranoia. For instance, after the attack on the New York City World Trade Center on September 11, 2001, Muslim and Sikh people in the United States became the targets of fear, hate, and violence, a phenomenon known as **Islamophobia**. Suspicion grew quickly of Arab people and Sikhs who are not Arab but wear turbans as a symbol of their faith. This suspicion spread largely due to an inaccurate and unfounded link between radical Islamists involved in *jihad*, or armed struggle, and the vast majority of peaceful Muslims and people of other faiths throughout the world. Islamophobia also has had the result of socially sanctioning racial prejudice and policies that exclude people of the Muslim faith. Box 13.3 explores Islamophobia and xenophobia in more detail.

Religion, then, has many functions for society and for individual believers. Some, as in the example above, result in dividing people based on **ideology**. Nonetheless, there are many ways in which faith-based belief systems bring people together, including by (1) creating community, (2) instilling values, (3) renewing faith, (4) providing reasons for life's events, and (5) solving problems.

1. *Creating Community*: Religious ceremonies and rituals bring community members together so that individuals feel support from the group. There are many types of rituals that bring cohesiveness to a group, whether they are performed with others or alone. Services (such as those in a temple, mosque, or church) allow individual members to physically come together regularly, creating a community of worshippers. Some religious practitioners create altars in their homes, whether to gods, spirits, or their own departed ancestors. Although they may worship alone, other members of that community use the same types of altars, creating cohesion.

Some religious rituals mark life's important transitions from one social or biological role to another, such as at puberty, first menses, marriage, childbirth, or death. Anthropologists call these **rites of passage**. The three stages of a rite of passage take an individual on a journey from separation, through transition, to the final stage of reincorporation and acceptance. For example, puberty rituals, such as circumcision, may mark a boy's passage into manhood. Among the Maasai of East Africa, the Emuratta circumcision rite identifies a young man as responsible enough to protect a camp territory, to learn the skills of an adult male, to lead others, and to fight as a warrior. Months of preparation and training precede the arduous circumcision ceremony, during which no anesthesia is administered. The young man must not flinch or cry out as an elder warrior performs the surgery. Once he successfully passes the initiation, the young man is given gifts of cattle to begin his new role in Maasai society.

2. *Instilling Values*: Religious texts and oral tales teach ethics to guide behavior. Elements of religious education may come from written texts such as the Qur'an (Islam), Torah (Judaism), Bible (Christianity and Judaism), and Bhagavad Gita (Hinduism). In cultures without a written tradition, values are passed orally through poems, myths, legends, and tales. Practitioners may learn the rules of moral behavior through these texts and stories. They also learn what punishments may ensue from a failure to follow them.

A group's oral stories provide guidelines for correct action as well. Myth is a category of story, outlined earlier in this chapter, which describes the sacred origins of the world and its people. It also expresses morals and a guide to "right" behavior.

The Aztec creation myth of ancient Mexico recounts a story of the beginning of the Fifth Age of the World (*el Quinto Sol*). Before the current world was created, the gods gathered together. They were discussing the best way to create the sun anew to provide life for the world, and they concluded that sacrifice was the way to achieve this. However, none of the gods wanted to be sacrificed. Finally, two gods offered themselves: a proud and strong god (Tecciztecatl) and the humblest and poorest god, the God with Boils (Nanahuatzin). At the last minute, the strong god lost his nerve, but the lowly god calmly offered himself up, becoming the sun. Ashamed, the strong god followed, but a more powerful god kicked a rabbit at him in protest, dimming his light. The strong god, now weakened, became the moon, which is said to have the shape of a rabbit on its face. This origin myth teaches its followers that being strong and conceited are wrong, while acting humbly and for the benefit of others is right. It also provides a foundation for the religious practice of human sacrifice, without which the sun would cease, and day and night would end.

Figure 13.3
CUBAN SANTERÍA DANCER.

This Cuban dancer allows the Orisha, or deity, Yemayá to inhabit her body during a performance. The dancer wears blue to represent Yemayá's connection to the sea and motherhood. Yemayá is one of the most important deities of the Santería religion.
Credit: James Emery/CC BY 2.0

3. *Renewing Faith*: Certain regular rituals elevate the mood of participants and bring on a state of happiness or transcendence. This may include such elements as song, call-and-response, hand clapping, trance states, or dance. For instance, Islamic Sufi dancers of the Mevlevi sect in Turkey perform a form of moving meditation in which they spin in circles. Practitioners, called whirling dervishes, experience closeness to the divine by abandoning the self in a trance-like dance. The Sema, or worship ceremony, is highly regulated, from the dervishes' clothing to the movements of the feet and hands.

Some revitalization activities use the threat of danger to rejuvenate faith in their belief system. An example of this is the religious snake handlers of the Pentecostal Holiness or Church of God churches across parts of North America. The handling of venomous snakes is one way members of these sects provide evidence that the Holy Spirit has saved them. The dangerous nature of this ritual generates excitement and transcendence for the community participating and witnessing the event. Some states have outlawed the practice due to high-profile deaths from snakebites during worship.

4. *Providing Reasons*: Belief systems provide explanations for life's events. Humans want to understand why things happen – for instance, why bad things happen to good people. Religious traditions also provide reasons for behaviors, such as why certain foods can or cannot be eaten by people of certain religious communities. Practitioners may not know the origins of these restrictions; however, many anthropologists believe that some of the major food taboos are linked to the environmental pressures found in places where religions first developed. The Hindu taboo on eating beef and Muslim and Jewish taboos on eating pork are explored in Box 13.1.

Many belief systems teach that everything that happens in life is predetermined. Therefore, when a misfortune occurs, a believer might take solace in the idea that "everything happens for a reason." A divine plan that life events fit into is less frightening and chaotic than one in which accidents happen for no reason at all.

The Wape people of Papua New Guinea believe that ghosts, demons, and witches inhabit their environment. When the main food source of the Wape was produce from horticulture, any meat brought from a forest hunt was greatly anticipated. However, the forest is a dangerous and forbidding place to the Wape. Vengeful spirits of recently dead ancestors populate it. A hunter may not find any game animals or have any luck killing the ones he sees due to the will of these spirits. Therefore, if a hunting party returns to the village with no meat, the explanation is clear: angry ancestor spirits chased the animals away.

5. *Solving Problems*: Since many societies attribute the causes of events to supernatural beings and forces, they also seek help from them when problems need to

BOX 13.1 **Food Matters: Religious Food Taboos**

Religious rules and practices support a group's environment. That is, the guidelines laid down in religious stories and texts serve to maintain balance in the ecosystem. For example, spirits of the forest would never require their human worshippers to burn down trees, making the area uninhabitable. This is especially true in terms of sacred food taboos.

Why do Jews and Muslims avoid eating pork? Why do Hindus avoid beef? One cultural system considers pigs "dirty" and disease carrying and therefore not to be eaten. On the other hand, another cultural system says that cows are sacred and pure, connected to the divine Mother, and therefore can't be eaten. Elaborate symbolism supports these ideas in stories, prayers, ritual ceremonies, and texts.

In seeking the origin of these beliefs, anthropologist Dr. Marvin Harris examines the environmental conditions in which the religions developed. As mentioned earlier in the text, his approach is called **cultural materialism**, in which the external pressures of the environment dictate cultural practices. This is not the only framework with which to explain cultural choices, such as diet, but it is one that focuses on the interaction of people and their environments.

Harris argues that the pig was not well adapted to the dry, hot grasslands of the Middle East where the early Abrahamic religions developed. It was used to shadier, wetter climates in which it could keep a cool body temperature (pigs have no sweat glands, making it hard to live in a desert-like environment). Not only must humans provide shelter and water to keep them cool, but pigs also compete for resources, eating the same foods that humans live on. On the other hand, cows, sheep, and goats live happily on pasture, leaving grains for human consumption.

As farming expanded, suitable habitats for pigs decreased. It became too costly to raise pigs for meat, which is a pig's only real product. You can't milk a pig easily, and it's hard to imagine trying to wear clothing made of pig hair like you can with the wool of a sheep. Thus, they became codified as unsuitable to eat. In this way, the ban on pork among Jews and Muslims supported the expansion of farms and the raising of pasture animals, which were "good to eat" (Harris, 1985).

In Hindu cultures, cow meat was prohibited not because of potential harm to humans, but the opposite. The strong and hardy zebu cattle of India provide so many benefits alive that killing them for beef would undercut the entire system of agriculture. Their main role is to pull the plow, creating opportunities for Indian agriculture where neither other animals nor human labor suffices. Farm cattle are fed kitchen scraps or oil patties; therefore, they do not compete with people for food resources. Even in times of drought or a failed harvest, keeping cattle on the farm ensures some long-term security.

In addition, cows provide an unlimited supply of milk and other products at the center of the Indian diet, such as ghee (clarified butter), cheese, and yogurt. The giving nature of the cow is revered and protected. For all these reasons, it is easy to see why early Indian Buddhists, Hindu Brahmin priests, and, more recently, Indian leader Mohandas Gandhi condemned the killing of cows for their flesh.

be solved. Prayer is one of the most common ways that individuals request assistance, either in a communal setting or alone. Many ritual behaviors such as prayers and ceremonies are done with the purpose of solving an immediate problem, such as asking for rain during a drought, consulting the astrological charts for an auspicious day for marriage, or praying for the health of a loved one. Even mundane activities merit divine cooperation, such as lighting a candle before taking an exam, or touching a statue for luck before driving a car.

Chinese rulers consulted oracle bones for ways to appease the ancestors before a hunt, the harvest, or wars with neighboring groups. The oldest evidence of Chinese script is found on these bones, which are flat pieces of ox scapulae (shoulder blades) and the underside of turtle shells. The writing on these pieces of bone shows that rulers during the Shang period (from the sixteenth through the eleventh centuries **BCE**) regularly consulted fortune tellers specializing in **divination** to answer their questions and solve problems. It was thought that ancestors of the Shang royal family would communicate through the heating and cracking of the bone. The diviner would read the messages left by the cracks, and then inscribe the bone with the answers.

SACRED ROLES

Fundamental to supernatural beliefs is the culturally accepted existence of beings or forces that exist beyond the observable world. **Supernatural beings** are personified or embodied gods, demons, spirits, or ghosts. Like humans, they may have genders (masculine, feminine, or gender fluid). Beings may be known (ancestors) or unknowable (all-powerful gods beyond human comprehension). They may exist in the everyday world, such as in trees and rocks, or in a world beyond human comprehension.

Supernatural forces, in contrast, are disembodied powers that exist in the world. These powers may bring good or bad luck. They may exist in the air, water, or other natural features of the environment, or they may be confined to an item, such as a lucky charm or talisman to ward off evil. Because culture is fluid and changing, multiple belief systems may be used simultaneously to understand the spirit world of any society.

Figure 13.4
ORACLE BONES.
Ancient Chinese diviners used ox scapulae or the undersides of turtle shells to read a person's future. It was believed that the ancestors would guide the questioner toward the right decisions.
Credit: Cambridge University Library

Deities

Deities are distant, powerful beings. People ask them for help with life's problems, assuming they are concerned with human issues and can alter the course of events. **Gods and goddesses** are found most often in societies with a hierarchical social organization, since a society's belief systems reflect its social organization. A society's gender roles are also reflected in the gender composition of deities, in that a male-dominated, hierarchical society may worship a masculine, authoritarian god. Societies in which women do much of the labor may worship both male and female deities. Over time, societies may change their understandings of gender roles, but codified religious tenets may not change at the same pace.

Worship of one god or goddess is called **monotheism**. Monotheistic religions posit a single, all-knowing, and all-powerful deity as the absolute ruler of the universe. Judaism, Islam, and Christianity are three modern religions that stemmed from a single religion of pastoral peoples. This pastoral society's worship of a single god evolved into these three world religions. The all-powerful deity is expressed as either the masculine Yahweh in Judaism, God as Jesus Christ in Christianity, or as the neutral/genderless Allah in Islam. Although there are important female figures in each of these traditions, the subordinate role of women in pastoral society is reflected in the few leadership roles for women in these religions, even today.

Polytheistic religions, on the other hand, worship two or more gods and goddesses in a **pantheon**. In the Native Hawaiian belief system, the goddess Pele is one of the most prominent deities. She resides in the volcanoes and is associated with volcanic activity. When a pantheon exists, gods and goddesses control certain aspects of the world. The Hawaiian pantheon includes Pele's brothers and sisters, such as Kā-moho-alii, shark god and keeper of the water of life, and Hiʻiaka, spirit of the dance.

Ancestral Spirits

A belief in **ancestral spirits** comes from the idea that humans are made of two aspects, the body and the soul (essence or spirit), which separate upon death. The physical body may eventually disappear, but the soul continues to exist among the living. Spirits of one's family members may continue to live in their house or community, inhabit the physical environment, or live in another realm but visit on certain days of the year. Ancestors can be pleased or angered, which may have an impact on the health or success of the living.

The Mexican holiday *Días de los Muertos* (Days of the Dead) reflects this duality of existence in body and soul. This celebration honors family members who have passed away. It merges aspects of the ancient Aztec belief system with the Gregorian or Christian calendar, imposed upon the Aztecs during the Spanish conquest of

TABLE 13.1

Excerpt from *Cantares Mexicanos #20* by Aztec Poet Nezahualcoyotl

Nahuatl	English
Tiazque yehua xon ahuiacan.	We will pass away.
Niquittoa o ni Nezahualcoyotl. Huia!	I, Nezahualcoyotl, say, enjoy!
Cuix oc nelli nemohua oa in tlalticpac?	Do we really live on earth?
Yhui. Ohuaye.	Yhui. Ohuaye. (refrain)
Anochipa tlalticpac. Zan achica ye nican…	Not forever on earth, only a brief time here …
Tel ca chalchihuitl no xamani,no teocuit-latl in tlapani, no quetzalli poztequi:	Even jades fracture, even gold ruptures, even quetzal plumes tear:
Anochipa tlalticpac. Zan achica ye nican…	Not forever on earth: only a brief time here …
Ohuaya, ohuaya.	Ohuaya, ohuaya. (refrain)

Source: Curl, 2005

Mexico. Ancient Aztec poems recorded before 1550 CE stress that death is a natural part of the cycle of life (see Table 13.1).

Over the two-day holiday, deceased family members are believed to return to their homes. Fireworks may be shot off or petals strewn on the ground from the graveyard to town to help guide the spirits of children. Families construct altars with yellow and orange marigolds (*flores de cempoalxóchitl*, also called *cempasuchiles*), on which they place photos, food, and drinks, along with personal items (such as cigarettes or cards) for individual family members. Through the burning of copal incense, the deceased are believed to be able to enjoy these sensory pleasures. Families in rural areas may also spend the entire night in the graveyard, decorating their family graves, listening to music and singing, and sharing a feast with their neighbors.

Ancestor veneration also reinforces the social values regarding family and kinship. In traditional Chinese society, the spirits of deceased ancestors remained among the living, residing in the family shrine. Family members would regularly clean the shrine and provide offerings to please the deceased. Just as children were expected to obey and provide for their parents during their lifetimes, they were obliged to do the same after death. In fact, a woman who joined her husband's family would not be considered a full member of her husband's lineage until she died. At that time, her gravestone would be placed in the family shrine, where she would be venerated along with her husband's ancestors.

Spirits of Nature

Preindustrial peoples' lives are intimately connected to the natural world in which they live. Therefore, **spirits of nature** inhabit the world around them in the earth, sky, and water. The physical environment, whether it be forest, desert, plain, or tundra, is filled with supernatural beings and forces that can influence the lives of people there.

Because the spirits reside in the everyday environment, believers have a more equal relationship with them. In other words, in contrast to all-powerful beings, spirits of nature may be negotiated with and potentially won over. The goal of a Ju/'hoansi healer going into trance is to convince the god who has brought on the patient's sickness to relinquish its hold. This is experienced as conversation rather than prayer.

There are two main belief systems under the umbrella term *spirits of nature*. The first is **animism**, or a belief that spirit beings inhabit natural objects. Any aspect of a group's natural environment may be personified by spirit beings that are involved with human lives on a day-to-day basis. The Hawaiian goddess Pele is thought to physically embody the volcano Kīlauea on the big island of Hawaii. When Kīlauea erupts, it is because Pele is angry. Small things can annoy her as well, such as when

Figure 13.5
EVIL EYE TALISMAN.
The Turkish *nazar* is a talisman, worn or hung to protect the owner from the "evil eye." This curse is found in belief systems across the Middle East, Mediterranean, North Africa, South Asia, and Spain, as well as in areas of Spanish colonization, such as Mexico.
Credit: Brian Jeff Beggerly/CC BY-SA 2.0

visitors remove rocks of her lava from the island. A curse is said to follow those thefts until the stones are returned and she can be appeased.

We might also consider the animist thought in our own lives, in minor or mundane ways. For instance, there is animist thought in the way North Americans sometimes personify their cars by giving them names, attributing a breakdown to the will of the car, or pleading with the car to reach the next gas station before running out of gas. We may also use animist thought for solace after the death of a loved one. For instance, one may attribute the appearance of a dragonfly seen on the day of someone's passing to the person's spirit now inhabiting the dragonfly.

The second type of spirit belief is **animatism**, or the belief that supernatural forces reside in everyday things. The forces are impersonal – not spirit beings, but powers – that have control over people's lives. Supernatural forces can reside anywhere in the natural world, such as in the air, earth, or water. These forces may be helpful or harmful.

Religious specialists may be able to harness this power for human purposes. For instance, among Polynesian peoples, an object such as a fishing spear may be imbued with *mana* (power) to ensure a successful catch. Practitioners of Buddhism may meditate under the protection of the sacred fig tree (the Bodhi Tree) at Bodh Gaya in northern India while seeking to attain enlightenment. Contemporary pagans in societies all over the world emphasize the spiritual essence of the natural world and the sacred unity of all living things.

Special items that have concentrated power, such as charms to ward off evil, may be carried or worn for luck. A lucky penny, a religious charm on a necklace, and a horseshoe are all examples of animatistic charms. The Turkish *nazar* is a talisman that depicts an eye of blue-and-white glass. It is carried to protect the bearer from the "evil eye," a supernatural force caused by envious stares that can result in sickness, whether intentionally or not. In many cultures, it is thought that the evil eye is cast when a woman without children feels envy of another's child. An infant wearing the nazar will be protected, as it captures and neutralizes the force.

Another animatistic force is the power of *n/um*, the Ju/'hoansi healing force. The Ju/'hoansi believe that all people possess n/um, but some cultivate the power to harness it in trance. As mentioned above, the goal of the healer is to reach the god responsible for the patient's illness and negotiate with them to give the person back. To achieve this, the healer will enter a trance through the percussive claps and song of the group gathered to witness. Once the healer is deeply into trance, their n/um is thought to boil up and down the spine. At this point, the healer will lay hands on the sick person and attempt to communicate with the god controlling the illness. If the healer is successful, the n/um will seize the illness and suck it up into the healer's spine, sending it flying out of the healer's mouth with a cry. Since trance sends the healer into an altered physiological state with a lowered heart rate, family members and other healers must aid the trancing healer until they regain full consciousness.

RELIGIOUS PRACTITIONERS

Priests/Priestesses

Priests and priestesses are full-time religious practitioners. They are often found in societies that are based on hierarchical status, in which there is a major gap between those with power and those without it. Although they may be divinely called to this profession, priests must earn their position through a process of certification bestowed by the religious hierarchy. A priest specializes in carrying out the required rituals of the religion. This may include conducting services, interpreting sacred texts, or carrying out particular duties for members of the religious community.

Shamans

Shamans are part-time religious practitioners who specialize in communicating with spirits, ancestors, or deities. They are more likely to be found in societies in which social and political life is more egalitarian than hierarchical. People who are called to the practice of shamanism may experience visions or dreams, after which they are given the gift of healing. They may also have survived long illnesses or near-death experiences. Because shamans are people who communicate with supernatural beings and forces, any inexplicable personality traits often signal that a person has one foot in this world and another in the world of the spirits. After a person with these gifts is identified, they will be trained by more experienced practitioners to become a full shaman.

Shamans make contact with the spirit world in several different ways. One way is through artistic or symbolic means, such as the Diné (Navajo) practice of creating sandpaintings that call the gods to aid in healing. Diné people understand sickness as an imbalance between the worlds of humans and the Holy People. In order for a patient to regain health, that person must be reconnected to the gods. As they perform healing ceremonies, Diné chanters (those who perform the healing ceremony) seek to restore harmony between the human and spirit worlds.

Another way shamans make contact with the spirit world is through **trance**, or an altered state of consciousness. The mechanism a person takes to enter into trance is culturally specific. Some societies may use drumbeats and hand claps.

Eveni shamans of northeastern Siberia, dressed in furs, skins, feathers, and antlers, enter into trance with the help of a drum. They send their soul flying to the realm of the spirits on the back of a reindeer. In this other realm, they are prepared to negotiate on behalf of their people, and even fight if necessary. The shaman asks for blessings from the spirits before returning to the taiga (snow forest) after the journey, and then spends the rest of the evening dancing and feasting with the community (Vitebski, 2005). The word *shaman* originally stems from the Tungusic language family of Siberia.

Some shamans may use hallucinogenic substances; the Yanömami, for instance, snort *ebene* (crushed *Virola* tree bark) to provoke visions. Yanömami shamans have the ability to communicate with spirits called *xapiripë*, who manifest themselves as tiny lights. The xapiripë can heal sickness, help hunters find game animals, and protect community members. Their protection, as manipulated by shamans, is especially important now, given the encroachment of non-Yanömami settlers and the destruction of the forest, as discussed in Box 13.2.

Because shamans have an influence on the outcome of events, we say that they are practitioners of magic. As such, shamans have a strong voice in determining the outcome of community issues. Even a member of the community who might be marginalized for odd behavior in a different society ends up with the power to speak for the spirits. With this power, a shaman can pass judgments on community members who have transgressed norms.

RELIGIOUS RESISTANCE

Belief systems, like all other aspects of culture, are subject to change and modification over time, whether by internal or external pressures. Conquest and colonialism impose a dominant society's religious belief system on the groups whose lands are being occupied. Communities often exert their agency by resisting these imposed and enforced changes to the core values and symbols of their society. They may attempt to merge the two systems or resist by inventing a new tradition.

When the Spanish explorers led by Hernán Cortés conquered the Aztec forces of Cuauhtémoc and Motecuhzoma in Mexico City in 1521, they imposed the Catholic religion with one god on the Aztec people, who worshipped a pantheon of gods. One way for the Aztec people to hold on to some of their beliefs while outwardly assimilating to the new religious system was to merge them, in a synthesis anthropologists refer to as **syncretism**. Syncretic beliefs bring the old and new belief systems together in ways that make sense to people who are forced to undergo a complete revision of their worldview. Tonantzin, the Aztec mother goddess, was reimagined as the Catholic Virgin Mother. Huitzilopochtli, the god of war and sacrifice and the most revered Aztec god in the pantheon, merged into the idea of

BOX 13.2 **Disappearing Forest of the Yanömami**

Public awareness of the deforestation of the Amazonian rainforest began in North America in the 1980s, when the extent to which logging, mining, and development had changed the forest environment came to light. Decades later, rainforests all over the world are still losing land to both legal and illegal activity.

What effects come from deforestation? Loss of the forest removes animal and plant habitats, leading to extinction. The loss of biodiversity creates challenges for the organisms that remain, because it upsets the food cycle. Since medicine comes from plants, the loss of yet-unknown plant life hinders the development of medical advances. In addition, erosion kills the microorganisms that keep soil alive, leading to vast areas of infertile land. The loss of trees and resulting decomposition of leftover tree trunks account for approximately 11 percent of human-caused carbon dioxide emissions globally (Food and Agriculture Organization of the United Nations, 2018). All the emissions, plus the loss of the forest as a carbon sink, contribute directly to the buildup of greenhouse gases, the warming of the oceans, and the alarming effects of climate change.

As one might imagine, the loss of forest has been tragic for the Yanömami and other Indigenous peoples of Brazil and Venezuela. In 1973, the Brazilian government built a Trans-Amazonian Highway, opening up interior land to commercial exploitation and settlers. The influx of workers and settlers had a devastating impact on the health and lives of the Yanömami. Nearly 20 percent of the population died from new diseases such as smallpox and malaria, for which they had no immunity. Hundreds of Indigenous people, including women and children, have been beaten and killed by *garimpeiros* (non-Yanömami prospectors). By 1990, 70 percent of Yanömami land had been taken from Indigenous control for use in commercial activities, leaving the Yanömami people with only 30 percent of their original land (Bier, 2005).

According to Davi Kopenawa (2013), a Yanömami shaman and spokesperson, shamans of the Amazonian rainforest recognize the terrible destruction of their ancestral lands. Nonetheless, they work harder than ever to extend their influence and protection to the entire rainforest and, generously, to non-Yanömami people. Kopenawa affirms that "shamans do not only repel the dangerous things to protect the inhabitants of the forest. They also work to protect the white people who live under the same sky. This is why if [the shamans] die, the white people will remain alone and helpless on their ravaged land.... If they persist in devastating the forest, all the unknown and dangerous beings that inhabit and defend it will take revenge" (p. 404).

the Catholic God. When oppressed Aztec people went to worship, they could still connect to meaningful symbols, only in a new form.

Whole societies forced to undergo major religious conversions as part of the colonization process might seek ways to resist and change their fate. One of the ways in which these actors had agency was to create a **religious revitalization**

movement, through which people could appeal to their old gods for help and deliverance. Revitalization movements generally begin with a charismatic leader who reports having visions or other communication with deities or spirits.

One well-known revitalization movement is the **Ghost Dance**, which began with the Northern Paiute (Numa) and spread to Native American nations across the West and into the Great Plains. Many Native communities used circle dances for ritual and prayer. After American settlers encroaching on their lands interrupted their traditional lifeways, Native peoples sought answers and an end to their suffering. When a Paiute prophet named Wovoka preached that a type of five-day circle dance could lead them back to happiness and to reuniting with their ancestors in Heaven, the idea caught on and spread. Wovoka claimed that God had told him that if all Natives performed the dance correctly, all evil would be gone from the world, leaving them with peace.

Cargo cults are another form of revitalization movement in which acts are performed to hasten the return of happiness and material wealth. Beginning after European contact with islands in the Pacific in the 1800s, groups of Indigenous people began to believe that the wealth enjoyed by the European invaders actually was destined for them. If they practiced the right supernatural rituals, then ships would come in, bringing all the "cargo" they desired.

On Vanuatu, a Melanesian island, a specific cargo cult developed centering on a mythical American serviceman named John Frum (John "from" America). This practice began in the 1940s, after the United States military had occupied the island. Practitioners believed that if they returned to their traditional customs and rejected Western ones, John Frum would bring their desired "cargo" on an airplane, and all non-Indigenous people would leave the island. Rituals celebrating John Frum include flag raising, marching, and maintenance of a painted landing strip on which the cargo will land. John Frum Day is celebrated annually on February 15th and has become the ideological focal point of a modern-day political party.

SUPERNATURAL BELIEFS AND CULTURAL EXPRESSION

Although the practice of religion is a very personal experience, it is also embedded in wider cultural practices. Religious beliefs are expressed in symbols such as images and iconography, and in music, dance, rituals, and patterns of behavior. For this reason, religious beliefs and the arts are closely connected as expressive systems. That is, the inner experience of an individual may be expressed in personal ways and shared in communal ones. Religious expression appears in many areas of life, including the modification of one's body.

Figure 13.6
BODY-PIERCING RITUAL AMONG TAMIL HINDU.

This young Tamil Hindu man is participating in the Panguni Uthiram festival, in Tamil Nadu, India, during which he pierces his cheeks with a spear to show his devotion. Like participants in the Thaipusam festival, devotees pierce body parts and also carry offerings and images of the gods in chariots.
Credit: Avena Matondang/ Images of Anthropology

Body Modification and Religion

People have manipulated their bodies for religious reasons for thousands of years. A person's physical devotion may take many forms: painting the face or body, fasting, or shaving one's head. Sikhs are required to allow their hair to grow; men wrap it in a turban, but the turban is optional for women. Male infant circumcision is mandated in Judaism through a ceremony called a *bris milah*, or Covenant of Circumcision.

In times of intensive worship, such as ceremonies, festivals, or other meaningful events, believers may undergo voluntary painful and arduous physical trials, such as walking long distances on the knees or self-flagellation (whipping). Tamil Hindu worshippers insert hooks in the skin of their backs and spears through their cheeks at several important festivals throughout the year. During the annual Thaipusam festival, Tamil Hindus show their devotion to the Hindu god Murugan. Devotees pierced by hooks and spears make a six-mile trek, called *kavadi*, through the streets to the sacred Batu Caves.

Humans have also been permanently marking the skin with tattoos to harness healing forces or protective powers for thousands of years. These sacred tattoos both refer to the symbols of a religious belief system and produce a magical outcome. In this way, they are similar to religious language and writing, which speak of a belief system while invoking the power of the belief system.

In 1991, an ancient mummy was found in thawing ice in the Ötztal Alps, on the Austrian–Italian border. Named after the site of his death, Ötzi the Ice Man may have been attacked and murdered there 5,300 years ago. Subsequent analysis of his body has provided a wealth of information about his life, including the fact that he suffered from a host of ailments. Ötzi has more than 50 tattoo marks at 12 different sites on his body where he would have experienced physical pain. The placement of the tattoos, mostly along his back, shows that they would have been applied by another person, likely a religious healing specialist, attempting to ease the pain. This type of tattooing marks him as a person who may have had wealth or status, since he had access to the art of a healer.

Sacred tattoo designs are also placed on the body for magical protection and power. The Thai, Shan of Burma, and Khmer of Cambodia share the tradition of *sak*

BOX 13.3   **Talking About: Islamophobia and the "Chinese Virus"**

Cultural anthropology uses the term *ethnocentrism* to refer to the idea that the things we do are normal, good, and right, and that the different beliefs and behaviors of others are abnormal or wrong. Ethnocentrism can become dangerous when it turns into hateful speech and actions against a group of people solely because they practice a particular religion or come from another place. Islamophobia, or the fear of and hostility toward Muslims, especially to justify actions or policies against them, has spiked in the US since the events of **September 11th**, 2001. This type of **xenophobia**, or fear of foreigners, is not new, however. It has been used throughout history to turn public sentiment against groups of people through the use of stereotypes and misinformation.

Rhetoric in the US after September 11th equated Muslims with terrorists, creating the narrative that Muslim men are aggressive and violent and hate Americans, and that Muslim women are oppressed and therefore need to be "freed" (Clay, 2017). The US saw a huge spike in hate crimes against Muslims that year before dropping again in 2002. Over a decade later, the inflammatory anti-foreigner rhetoric of Donald Trump as he entered the US presidential race in 2015–16 would cause the number of assaults against Muslims to surpass even those of 2001. European countries have also experienced a similar rise in anti-immigrant and anti-foreigner rhetoric, violence, and policies. Language matters, and it has a direct impact on people's health, safety, and security.

"Othering" of foreigners by characterizing them as violent, dirty, or disease carrying is a linguistic strategy employed to gain support for discriminatory policies. Nations have used this strategy ever since nations first existed. In the US, several historical moments stand out in which the administration has deployed xenophobia as a strategy against Asians and Asian Americans. In particular, the US used xenophobic rhetoric to enact an anti-immigration law (the Chinese Exclusion Act) barring Chinese people from entering the country for 60 years, from 1882 to 1943. And the repercussions of the racist slogans and characterizations from the Japanese internment camps of World War II and the Vietnam War are still felt in the US today.

The tactic to refer to the COVID-19 virus in the early months of 2020 as the "Chinese virus" is a more recent example of this. By choosing to characterize it as a virus that is somehow intrinsic to Chinese people, the Trump administration built this same type of narrative: that foreigners are dirty and should be banned from infecting our nation. Racist and violent acts against Asian American people of all ethnicities spiked as this narrative took hold, with California alone reporting over 800 acts of racist violence against Asian American people from April to June 2020.

Dr. Manjusha Kulkarni, executive director of the Asian Pacific Planning and Policy Council, succinctly summarizes the effects of this kind of speech: "So what does that say about the spread of the contagion of racism?" she asks. "It actually moves much more quickly than the disease" (as cited in Strochlic, 2020). Whether against a religious group or citizens of a nation, language can be a powerful political weapon.

Figure 13.7
SAK YANT TATTOO.
Sak yant tattoos are talismans worn for protection. Traditionally, the tattoo was applied with inked bamboo needles tapped into the skin, but today it is more often applied with long steel needles. A code of conduct accompanies the yantra tattoo by which the wearer should abide, including Do Not Steal, Do Not Lie, and Do Not Speak Poorly of Anyone's Mother, Including Your Own.
Credit: Andres Guerra

yant, or *yantra* tattooing. Buddhist monks or yantra specialists apply the designs on young Thai men, who wear them for protection. These tattoos have a long history, beginning in the first century BCE with Khmer warriors, who tattooed their entire bodies so they could be invisible to harm. Today, members of street gangs and soldiers in the military also wear yantra tattoos as talismans to ward off misfortune.

A number of cultures practiced facial tattooing for spiritual reasons that were also linked to social practices. Ainu women of northern Japan and Russia wore lip tattoos that were applied before marriage. They were thought to repel evil spirits that could enter the woman's body through the mouth. Bearing the lip tattoo also signified that a woman would have a place among her ancestors in the afterlife.

SUMMARY

This chapter examined how belief systems guide people's behavior in society by providing a symbolic framework for aspects of cultural life. Mirroring the Learning Objectives stated in the chapter opening, the key points can be summed up as follows:

- Evidence for early religious practices focuses on burials, especially those with grave goods.

- Supernatural beliefs help individuals and entire religious communities explain events and cope emotionally with things they cannot control. In addition, beliefs function in various ways to guide people's behavior, create cohesion within groups, and maintain ecological practices that support their own success in a given environment.
- Different types of sacred beings and forces inhabit the worlds of different types of societies. For instance, hierarchical societies will often worship deities, societies with a strong moral code for the respect of elders will revere ancestors, and small-scale groups that rely on the natural world for resources will populate the natural environment with beings and forces.
- Priests, priestesses, and shamans intervene on behalf of the spirit world and may relay messages to and from those inhabiting the world beyond.
- Throughout history, many traditional societies have been forced to adopt the religion of a society that has come to dominate their region. Rather than surrender their deeply held beliefs, people develop revitalization movements, bringing hope that supernatural beings will help things return to the way they were.
- Because supernatural beliefs are central to social structure, they are found in many other forms of cultural expression, including body modification.

Review Questions

1. What does it mean to say that human culture is founded on symbolic systems?

2. How and why do anthropologists think the earliest religions developed?

3. What kinds of political systems tend to correlate with the veneration of deities, ancestors, and spirits in nature?

4. What are the functions of religious belief in society, both on an individual and a social level?

5. What are some differences between the roles of priests/priestesses and shamans?

6. What are the goals of a religious revitalization movement?

7. How do religion and cultural expression overlap?

Discussion Questions

1. Use the five functions of religion stated above to describe the functions of your own belief system. If you do not subscribe to a formal religion, then describe your system of morals and values.

2. What are some differences between scientific and spiritual systems of healing? What are some similarities?

3. Have you personally undergone a medical treatment that relied on traditional knowledge of medicine? How was the experience different from Western scientific medical treatments?

Visit **www.lensofanthropology.com** for the following additional resources:

SELF-STUDY QUESTIONS   **WEBLINKS**   **FURTHER READING**

14

ANTHROPOLOGY AND SUSTAINABILITY

LEARNING OBJECTIVES

In this chapter, students will learn:

- *the connections between anthropology and sustainability.*

- *useful definitions of sustainability that resonate with the anthropological perspective.*

- *how anthropologists have approached the study of peoples and ecosystems throughout the history of the discipline of anthropology.*

- *some of the current frameworks in environmental anthropology, including the study of Traditional Ecological Knowledge (TEK) and ethnoecology.*

- *the importance for anthropologists of some of the major issues in sustainability studies.*

- *ways that anthropologists can help inform the discussion about a sustainable future.*

INTRODUCTION

One of the key concerns of anthropology across the subfields is the way humans adapt to their environments. Biological anthropologists may examine foraging strategies among primates, or the ways people have adapted to climate change. Archaeologists study the remains of people's lives situated within a particular ecological and regional context. Cultural anthropologists may focus on how people create and modify their beliefs and behaviors to adapt to particular environmental pressures. In sum, anthropology has been concerned with issues of **human ecology** since the discipline's inception and can provide a long-term view of human adaptations. This chapter examines the intersections between the fields of **sustainability** and anthropology, and it looks at how anthropologists can be instrumental in finding sustainability solutions.

Both anthropologists and sustainability researchers put people at the center of their research. That is, both fields focus on how people live: what works for them in terms of adaptive strategies and what doesn't work (both today and in the past). Both fields also recognize that for any issue there are multiple ways to understand and engage with it. That is, both emic and etic perspectives are valid. With so many examples of human societies throughout time faced with similar challenges, anthropology provides a wealth of detailed knowledge from the inside of how these societies have solved their problems – or not.

In addition to human–environment interactions, anthropologists are interested in the connections between local and global processes. Cultural anthropology provides a close-up view of local processes through fieldwork, and then a larger analysis allows them to be situated in the global context. Archaeology can discover how global changes in climate led to the dispersal or demise of peoples. Biological anthropology may highlight physiologic or demographic changes in human populations resulting from long-term environmental change. Therefore, the anthropological lens is uniquely suited to inform global projects that seek sustainability solutions.

Anthropologist Dr. Sally Ethelston (2006) tells the story of Miriam, who lives with her family in Manshiet Nasir outside Cairo, Egypt. Originally a squatter

Figure 14.1
A GROUP OF ZABBALEEN BOYS IN MOQATTAM VILLAGE, CAIRO.
These boys live in Moqattam, a Zabbaleen village, where they help recycle Cairo's trash.
Credit: Ayoung0131/CC BY-SA 3.0

settlement, Manshiet Nasir is home to thousands of *zabbaleen* (garbage collector) families who make up a poor Christian Coptic minority in a predominantly Muslim area. Zabbaleen men collect Cairo's garbage, and women sort it by hand, including hospital waste, food waste, and rusted and broken metal and glass. The community collects approximately 3,000 tons of Cairo's waste each day and recycles about 80 percent in order to make a living. A thick, slimy layer of garbage covers the streets due to inadequate sewage systems.

Miriam and other Zabbaleen people, forced to the edges of Egyptian society, suffer under these risky and unhealthy conditions daily in order to make ends meet. However, by looking at the larger trends, one can understand the forces creating these unsustainable circumstances: population growth, scarce resources such as water and land, and social and economic restrictions on women. In order to impose further controls on women, the official Egyptian narrative blames female fertility for the unsustainable population growth.

However, when anthropologists examine the larger social and political pressures, they see something different. Rather than placing blame on women for reproduction, anthropologists identify connections among areas of gender equality, reproductive rights, use of resources, land development, and population. When local problems are placed into a more global context, we can ask the right questions about how to reach solutions.

Anthropologists believe that people and culture cannot be separated from their environments. The idea of interconnectedness between nature and culture means that these areas of life are intimately tied together. However, anthropologists do not believe that traditional peoples lived in some sort of primitive state of organic balance with their environments. Sometimes this myth is invoked to provide an ideal model of a sustainable society: if modern people could return to the "simpler" and more "natural" ways of our ancestors, then the problem of sustainability would be solved. While we can certainly identify practices in traditional societies that conserve the environment, each cultural practice is bound up in a complex web of beliefs and behaviors that may be fundamentally different from our own. There is no "going back" to some simple ideal. We can only move forward.

HISTORY OF HUMAN-ENVIRONMENTAL RELATIONS

Anthropological research underscores the interconnectedness of life. Today, a web of complex relationships around the globe links people and products. Corporate decisions made in an office in Paris or Beijing can set off a chain of events that involve people, environmental resources, and politics in Sri Lanka or Bolivia. The chain is largely invisible to the consumer, who knows little to nothing about where a product was made, by whom, and under what conditions. Modern marketing stresses every decision as a personal one, based only on identity and personal choice. For most consumers, owning an item that expresses something about who we are – or who we want to be – becomes the most important consideration, over cost or the process of production.

The way early humans perceived their place in nature was markedly different than it is today. As discussed in Chapters 6 and 10, before the development of agriculture, bands of people hunted and gathered, planted small horticultural plots, or practiced a pastoral lifestyle. These adaptations required intimate knowledge of the ecosystem within which they lived. Economies were, for the most part, local. To survive, it was crucial to know where to gather or hunt, when to plant and harvest, or where and when to take animals to pasture. People's foodways placed them within the natural world, as part of it, albeit a part that learned to manipulate it for human needs. People may not have had "natural" instincts about sustainability, but they certainly had a more equitable and reciprocal relationship with nature for the vast majority of human history.

With the rise of intensive cultivation approximately 9,000 years ago, human societies began to change the nature of their relationship to the land. Intensive agricultural techniques require more labor, technology, and inputs into the soil than small-scale horticulture. Although productivity increased, allowing societies to feed growing populations, large-scale cultivation changed the ecological balance. Large plots of land needed to be cleared for planting. Farmers domesticated animals for food and labor, requiring close contact with animal waste. The use of draft animals allowed deeper plowing, but also released into the air new pathogens that impacted human health.

Industry began to grow exponentially several hundred years ago. Nations' resources and wealth developed at a faster rate than ever before. For the first time, goods flowed around the globe from industrialized countries, especially from those in the northern hemisphere. These goods were often produced outside these countries, in non-industrialized nations. These developing nations provided natural resources and raw materials but saw little of the profit. The problems of social, economic, and political inequities stemming from this period lie at the root of many of the sustainability issues of our time. The exploitation of underdeveloped areas for the profit of corporations in developed nations creates great inequities.

After 1950, in a period economists call the Great Acceleration, demands for fuel, food, timber, water, and other natural resources exploded. This was primarily due to the growth of populations and human consumption. With this era began the highest level of deforestation and destruction of the world's ecosystems ever seen on the planet. Global resources seemed limitless, and little attention was paid to conservation. Some scholars see this period as the beginning of the **Anthropocene**, the geological and environmental era in which humans have drastically and undeniably altered the planet as a whole. (Others place the start of the era during the Industrial Revolution or the development of agriculture; the debate is ongoing.)

Anthropologist Dr. David Maybury-Lewis (2006) argues that the real change around this time occurred within human societies and personal values. When we left behind small-scale, cooperative communities, society's priorities shifted from collective needs to individual needs. Maybury-Lewis argues that Western society especially glorifies the individual's rights and desires. It releases the individual from the complex bonds of family and kinship that rooted people in their communities for nearly all of human history. He sees evidence for this refocusing of social values in the changes that took place in child-rearing (a move to independence training), social status (now conferred by power and money, not compassion or generosity), and the structure of the modern nuclear family (free from the obligations of extended families).

All these changes in society and culture have led to incredible advancements. Limitless creativity has released ambition, competition, and achievement as never before. Unfortunately, it has also isolated individuals from not only other people

TABLE 14.1

Human Impacts on the Biosphere

1.	Evidence for global warming due to human production of CO_2 and other greenhouse gases is now unequivocal.
2.	Between 5 and 20 percent of the approximately 14 million plant and animal species on earth are threatened with extinction.
3.	The Living Planet Report compiled by the World Wide Fund for Nature (WWF) reports that from 1970 to 2012, the average number of vertebrate species on earth has been reduced by 58 percent. In 40 years, the planet has lost more than half of its species of mammals, amphibians, reptiles, birds, and fish. Specifically, there has been a reduction in land species by an average of 38 percent, marine species by an average of 36 percent, and freshwater species by an average of 81 percent (WWF, 2016).
4.	Overharvesting has devastated both ocean and inshore fisheries. The population of large predatory fish has been reduced by more than 90 percent from preindustrial levels.
5.	More than two million people globally die prematurely every year due to outdoor and indoor air pollution and respiratory disease.
6.	Per capita availability of fresh water is declining globally, and contaminated water remains the single greatest environmental cause of human sickness and death.
7.	We currently use up the resources of 1.6 earths. (This is possible because we use resources faster than they can regenerate. For example, trees are cut down faster than they can grow, and carbon is released into the atmosphere and oceans faster than it can be absorbed.)

Source: Adams & Jenreneaud, 2008; WWF, 2016.

but also the natural world. Modern industrial societies have developed a sense of ownership and entitlement over the land, air, and water. Maybury-Lewis sees this shift in social and cultural values leading directly to our modern environmental crisis. Table 14.1 lists some of the most severe impacts humans have made on the planet.

DEFINING SUSTAINABILITY

A general definition of *sustainability* is the ability to keep something in existence, to support or continue a practice indefinitely. Clearly, this is problematic when applied to the earth, as it does not have limitless resources. Our planet is an example of a **closed-loop system**, or a system that has finite resources and cannot sustain indefinite growth. Therefore, the most general definition of the term sustainability cannot be applied accurately to life on earth.

To address this dilemma, sustainability scholarship focuses on the well-being of people, now and in the future. The most commonly used definition is the one originally developed by the 1987 Brundtland Commission of the United Nations. The commission described sustainability as "meeting the needs of the present without compromising the ability of future generations to meet their own needs" (UN, 1999).

There are echoes of this definition in a much-quoted passage on considering the impacts of our actions today on the next seven generations. Perhaps falsely attributed to Chief Seattle of the Duwamish and Suquamish Nations, the original context of this quote may reside in the Great Binding Law of the Iroquois Nation (Haudenosaunee):

> In all of your deliberations in the Confederate Council, in your efforts at law making, in all your official acts, self-interest shall be cast into oblivion. Cast not over your shoulder behind you the warnings of the nephews and nieces should they chide you for any error or wrong you may do, but return to the way of the Great Law which is just and right. Look and listen for the welfare of the whole people and have always in view not only the present but also the coming generations, even those whose faces are yet beneath the surface of the ground – the unborn of the future Nation. (as cited in Murphy, 2001, para. 28)

Modern Iroquois leaders continue to invoke this idea. Oren Lyons is a Faithkeeper of the Turtle Clan of the Seneca Nations, one of the five original nations of the Iroquois Confederacy. He is an activist, professor, author, and leader who has won awards for his work on Indigenous rights and development. Lyons says, "We are looking ahead, as it is one of the first mandates given us as chiefs, to make sure every decision that we make relates to the welfare and well-being of the seventh generation to come" (as cited in Vecsey & Venables 1980, pp. 173–4). The idea of the "seventh generation" is a powerful reminder that people today are stewards of the future resources of the planet.

The United States Environmental Protection Agency (EPA) uses an environmentally focused description of sustainability, in that it considers all the features of human society that depend on the environment. The EPA (n.d.) writes, "Sustainability is based on a simple principle: Everything that we need for our survival and well-being depends, either directly or indirectly, on our natural environment. Sustainability creates and maintains the conditions under which humans and nature can exist in productive harmony, that permit fulfilling the social, economic, and other requirements of present and future generations. Sustainability is important to making sure that we have, and will continue to have, the water, materials, and resources to protect human health and our environment" (para. 1–2).

The approaches of both the Brundtland Commission and the EPA resonate with the anthropological perspective. Anthropology is a holistic science that seeks to interpret the connections among all aspects of human life. Anthropologists focus on whether and how people get their needs met – just as these definitions do.

The availability of and access to environmental resources is linked to food-getting and work patterns, which affect social life, economic stability, family structures, and child-rearing. People make meaning of these aspects of life through culture: symbolism, belief systems, myth, and artistic expression.

A stable and productive environment is fundamental to meeting people's needs. Furthermore, a healthy environment supports human "welfare and well-being." From this perspective, the goals of anthropology are very similar to the goals of sustainability: both are interested in understanding what works for human beings on this planet.

Components of Sustainable Development

Sustainable development is often described as having three components. These different components were named the **three pillars of sustainability** at the Rio Earth Summit in Brazil in 1992. As shown in Figure 14.2, the pillars are social, environmental, and economic. Ideally, each of these aspects of human life must be supported and in balance if we are to reach the goal of a sustainable world.

Environmental sustainability is the ability of the environment to renew resources and accommodate waste at the same rate at which resources are used and waste is generated. It implies that human practices should protect and preserve those aspects of the physical environment that sustain life. This includes not only our major life-giving ecosystems – such as the land and soil, atmosphere, freshwater resources, and oceans – but also all natural resources, from the smallest biomes to the most complex living systems.

Social sustainability is the ability of social systems (such as families, communities, regions, or nations) to provide for the needs of their people so that they can attain a stable and healthy standard of living. Aspects of social sustainability include equity, justice, fair governance systems, human rights, quality of life, and diversity. A socially sustainable society would be one in which people can rely on a dependable infrastructure for health, order, education, and employment, while also feeling interconnected to others in social and cultural life.

Figure 14.2
THE THREE PILLARS OF SUSTAINABILITY.
The three pillars of sustainability (social, environmental, and economic) are shown in this diagram as equal components of a healthy and stable system.

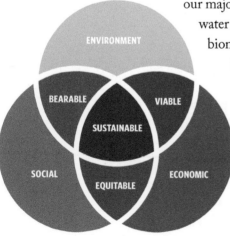

Economic sustainability is the ability of the economy to support indefinite growth while ensuring a minimum quality of life for all members of society. However, there is an inverse relationship between economic growth and environmental conservation. That is, economic development generally causes environmental degradation (but not always in the same region). Therefore, sustainable economic development would address overconsumption in the developed world and find ways to manage resource use and the environmental impacts of growing economies.

The concentric model of sustainability uses the three-pillar approach in a way that responds more accurately to the realities of life on earth. This model emphasizes the importance of the environment, for without a productive and healthy environment, the social and economic realms of life would not be able to function.

Where the three pillars overlap, certain goals need to be met in order to increase the likelihood of a sustainable outcome. For instance, as seen in Figure 14.2, where the environment interfaces with society, life should be "bearable." Where society interfaces with the economy, life should be "equitable." Finally, where economic and environmental issues overlap, life should be "viable."

While the three-pillar model provides a good place to begin sustainability discussions, alternate models have also been proposed. One of these models shows the fundamental importance of the environment to provide for sustained life on earth, as discussed above. If the environment is depleted, social structures will collapse and there will be no economic output. Therefore, our first priority should be to protect the environment, as it is the foundation of social and economic life. Figure 14.3 represents an environment-centered approach.

How might the models of sustainability translate into practice on the ground for those interested in pursuing sustainability projects? Rather than sustainability science remaining in the universities, corporate accounting offices, and governmental policy offices, steps toward sustainability in the local context can engage the full participation of the people the work aims to help. This engaged approach to sustainability research is similar to that already used by applied and public anthropologists involved in **participatory action research** (PAR). In PAR projects, the community's needs and goals are identified through a process of participant observation and consultation. Then, external factors that impact the needs of the local community can be analyzed. Finally, the collaborative team – including community members – identifies solutions. Since projects trying to support change (e.g., cleaner water, higher harvest yields, the implementation of sustainable farming techniques) often require funding from outside agencies, the reports

Figure 14.3
CONCENTRIC MODEL
OF SUSTAINABILITY

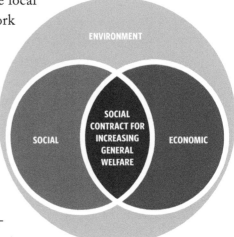

Tragedy of the Commons

In the mid-twentieth century, ecologist Dr. Garrett Hardin (1968) illustrated the nature of the sustainability problem with a scenario. He called it the **tragedy of the commons**. The *commons* may refer to any publicly shared resource, such as water, land, or air. In Hardin's original analogy, it refers to an open pasture shared by herdsmen and their cattle. The inevitability of the tragedy comes from overpopulation.

At first, there is enough space in the commons for all the cattle, and enough pasture for all to graze freely. However, each herdsman (let's assume that in this society the herding is done by men) wishes to grow his herd by adding another head of cattle. The benefits to him personally are obvious, since he will have a larger herd. The cost, on the other hand, is that there will be less pasture for all animals to share. He chooses to grow his herd because the benefits of adding cattle are his alone. On the other hand, the costs of losing a little pasture due to the addition of one more animal are distributed among all the herders. To this individual herdsman, the benefits outweigh the costs. Unfortunately, each herdsman has the same private goal: to maximize his herd and his profit. Eventually, with this mindset, the tragedy occurs: the pasture becomes overgrazed, and there are no resources left for anyone.

The analogy can be applied to human use of natural resources. If each person acts in their own best interest, then the depletion of resources follows. Hardin argues that people will naturally act selfishly when they weigh the pros and cons of the situation.

Applying this model to humans' use of natural resources, how do policy-makers try to avoid what they see as the inevitable depletion of resources? Hardin argues that a degree of financial coercion is necessary, especially in charging more money to individuals for the resources used and pleasures enjoyed. Governmental agencies take this approach, seeing "tragedy" as the inevitable result of collective resource use. Therefore, **privatization** of resources by corporations and government regulation is seen as the only way to prevent total ecological destruction. Unfortunately, conservation does not always result from these policies.

Under certain circumstances, the problem of the commons may actually be self-regulating. Anthropologists cite many examples of groups of people that regulate their own joint use. Individual users of an area may voluntarily cooperate to constrain or conserve the use of resources, making it clear that the tragedy of the commons model is not universally applicable.

For instance, among rural South Indian farming villages, anthropologist Dr. Robert Wade (1988) found that farmers cooperated when they had the power to affect outcomes. With an understanding of how their actions affected overall success, farmers cooperated to share irrigation for the benefit of all. Part of their success was a strict set of social guidelines governing behavior that prevented any individual farmer from "free riding" to the detriment of another. When a farmer didn't adhere to these guidelines, there were negative social repercussions imposed by the other members of the group.

Individual users of an area may voluntarily cooperate to constrain or conserve the use of resources. People do not always operate in the way Hardin's model predicts, because it does not consider the potential for an internalized code of conduct that stems from different cultural norms and values.

ANTHROPOLOGICAL APPROACHES TO SUSTAINABILITY STUDIES

Since the early twentieth century, anthropologists have sought to understand the relationship between people and their environments. This relationship is referred to as *ecological* because it stresses this fundamental connection. Several different frameworks or models have been used, each building on – and reacting to – the ideas that came before. The following section introduces some of the major theories and how they seek to understand the human–environment relationship.

Cultural Ecology

Dr. Julian Steward (1955/2006) is largely considered the first anthropologist to develop a **paradigm** based on the interactions of people in their particular environments. His ideas came out of a major debate in anthropology on how much of a people's culture developed in direct response to environmental pressures. This idea is known as *determinism*, which argues that the limitations of the environment determine people's behavior.

Steward was the primary proponent of the theory of **cultural ecology**, which began in the 1940s. During periods of fieldwork among the Shoshone (*Newe*) of the Nevada Great Basin area, he saw how the specific environmental pressures of the Great Basin ecosystem limited the possibilities of food procurement. In particular, the Shoshones' traditional plant-based foraging diet had been drastically reduced by settlers' introduction of sheep and cattle.

Steward believes that food-getting practices directly affect social and economic life, creating a central set of behaviors that he calls the **culture core**. To identify features of the culture core, an ethnographer would need to examine the technology used in food procurement, the patterns of social life directly linked to those practices, and then, finally, how other aspects of life were influenced by social patterns.

BOX 14.1

Standing Rock Sioux Reservation and the Dakota Access Pipeline

In late 2016, social media made the world aware of a controversial project called the Dakota Access Pipeline (DAPL). The 1,172-mile (1,886-km) pipe system was designed to move crude oil from the Bakken oil fields in North Dakota through three states to an oil tank farm in Illinois. Along its route, the pipeline comes within 492 feet (150 meters) of the Standing Rock Sioux Reservation.

The line passes under several central water resources, including the Missouri River and Lake Oahe. The pipeline not only poses a risk of leaks into the water supply, but also passes directly through treaty lands owned by the reservation. Oil leaks are standard occurrences for pipelines, with approximately 700 oil leaks in North Dakota from May 2016 to May 2017 reported by the North Dakota Department of Health (2018).

In an effort to protect sacred sites, including burial grounds, and the reservation's water supply, the Standing Rock Sioux of North Dakota, Meskwaki of Iowa, Oglala Sioux and Cheyenne River Sioux of South Dakota, and other tribal nations showed early opposition to the route of the pipeline. In April 2016, a Standing Rock tribal elder established a sacred camp at the site of construction.

By the end of the year, several more camps had sprung up to house some 1,000 people, both Indigenous people and environmental activists, opposed to the pipeline route. Video footage recorded the violent tactics of the security forces protecting the construction sites: protesters beaten by soldiers in riot gear, bitten by attack dogs, struck with water cannons resulting in broken bones and hypothermia, and locked in small cages.

The American Anthropological Association (2016) published a statement declaring its solidarity with Oceti Sakowin Oyate (the Great Sioux Nation) as the pipeline "violates the cultural and collective environmental human rights of the Tribe to life, land, cultural preservation, health, clean water, and a clean environment" (para. 2). Archaeologists and historical preservation specialists testified along with tribal elders in September 2016 in an effort to preserve 82 sacred sites and graves in the Cannonball River section. The area was bulldozed shortly after. Political figure and social justice advocate Reverend Jesse Jackson called the disregard for tribal sovereignty an act of "environmental racism" (as cited in Thorbecke, 2016).

For instance, the Shoshone hunted and gathered as their foraging strategy. Both hunting and foraging take place in groups, divided into men's and women's work. The sexual division of labor influences political life, in that control of the protein resources gives men – the hunters and fishermen – more power than women – the gatherers. Simply put, Steward argued that the production of food was central to the organization of Shoshone society.

In January 2017, the US government approved construction and removed the last of the protesters from the camps. The Dakota Access Pipeline began transporting about 500,000 gallons of oil daily in the summer of 2017. But tribal governments sued, demanding that the US Army Corps of Engineers undertake a thorough environmental impact assessment. Tribal leaders are now hoping that the newly inaugurated administration in 2021 will shut down the flow of oil in the pipeline, decreasing risks to their land, while the assessment is conducted.

Figure 14.4 **STANDING ROCK CAMP BANNER.**
This banner at the Oceti Sakowin Camp includes the Sioux words "Mni Wiconi," or "Water Is Life."
Credit: Becker1999/CC BY 2.0

At the time, the cultural ecology model established itself as a model of multilinear cultural evolution in opposition to the predominant model of universal (linear) evolution. As noted in Chapter 1, the theory of universal evolution argued that all cultures went through the same steps as they modernized. This implied a hierarchy of development from simple to complex. Steward rejected that notion. He focused instead on the particulars of each culture: their environmental limitations and resources, their patterns of behavior pertaining to subsistence, and any area of social life influenced by these patterns.

Ecological Anthropology

Steward's cultural ecology model eventually transformed into a field called **ecological anthropology**. This framework is similar to a cultural ecology model, in that culture and social organization are the outcomes of a group's adaptation to the particular challenges of its environment. While cultural ecology focuses on culture as the unit of analysis, ecological anthropologists define the population as the unit of study. Within ecological anthropology, a framework called systems theory is used to measure the inputs and outputs of the system.

Systems Theory

The **systems theory** model examines a particular geographic area inhabited by people as a closed-loop system. It understands the population to be in a state of equilibrium within its environment of finite resources. Research focuses on the flow of energy and matter, as well as information, in an attempt to quantify how the system functions. This framework borrows ideas from biological studies of natural ecosystems.

Anthropologist Dr. Roy Rappaport (1968) used a systems theory approach to study the Tsembaga Maring of Papua New Guinea in the mid-twentieth century. His work focused on how the Maring functioned within their ecosystem, in which everything was connected: plants, animals, and humans. Rappaport saw how the products of human culture created measurable effects in the flow of material goods through the ecosystem. For example, a ceremonial sacrifice of pigs created the sudden availability of tons of pork to be consumed. He noted that the timing of pig-slaughtering rituals was linked to a regular cycle of war, when injured warriors would require extra animal protein for strength. Systems theory assumes that cultural practices, such as animal slaughtering rituals, exist to fulfill the needs of human life, such as nutrition.

Political Ecology

Researchers looking to understand the relationship between ecology and power developed a related framework called **political ecology**. Using this model, anthropologists and other scholars focus on the complex relationships between the environment, economics, and politics. These studies focus especially on the developing world, where people who are marginalized lack access to or control of resources.

Issues studied by political ecologists include ecological justice, such as deforestation of an inhabited area of the rainforest; human rights, such as the right to clean drinking water; and cultural identity, such as the right to pursue traditional modes of hunting. There are also many forms of resistance that Indigenous and other minority groups mount in the face of externally made decisions that may be detrimental to their health and well-being. Therefore, political ecologists also focus on environmental agency and activism in its many forms around the world.

Environmental Anthropology

Today, hundreds of practicing and teaching anthropologists interested in these issues belong to the Anthropology and Environment Society, a section of the American Anthropological Association. Within this section, one might find members who identify themselves as ecological anthropologists, political ecologists, or human geographers. Many, however, use the term **environmental anthropologist** to represent their broad interests. There are several frameworks that environmental anthropologists may use to inform their research, including Traditional Ecological Knowledge and ethnoecology.

Traditional Ecological Knowledge (TEK)

While it is important to avoid stereotyping the "simple" lives of Indigenous peoples, there is much to be learned from the knowledge and practices that a group has developed and handed down over generations. The study of **Traditional Ecological Knowledge (TEK)** seeks to understand the collective and cumulative knowledge that a group of people has gained through living in their particular ecosystem. (TEK may also be called Indigenous knowledge or local knowledge, especially outside Canada.)

Studies that use local knowledge to inform their research seek to understand the broad and deep knowledge local groups have of the interrelationships among people, plants, animals, and nature. In particular, projects might focus on traditional methods of food procurement, such as hunting, gathering, trapping, fishing, and farming. Projects might also examine ways in which people manage their local ecosystem, such as through forestry, water use, and soil management. These practices are bound up in spiritual and cultural understandings of the world as well. When applied to policy-making, these understandings may be used as tools in not only short-term resource management but also in plans for the long-term sustainability of a given area.

However, TEK has the potential to be misused by external agencies. For instance, there have been many examples since the 1990s of government agencies using TEK data as simply another piece of externally calculated scientific data. This removes the local knowledge from its deeply complex cultural context. Attempting to apply a cultural "fact" as a discrete piece of data is similar to removing an artifact from an archaeological site and trying to make some sense of it in the absence of contextual clues. Furthermore, if Indigenous peoples do not have control over how this information is embedded into new contexts, such as in external reports used to develop policy, then it ceases to be of use to them. It may even be detrimental to them and end up excluding their further input in future land management debates.

Figure 14.5
GLACIER BAY.
To local peoples, glaciers are sentient and social beings that play a role in community life. Researchers who use discrete bits of data and represent them as traditional knowledge lack the full cultural context. Some of the cultural context may be crucial to conservation.
Credit: Luis A. González

The fundamental conflict between Indigenous and scientific understandings of the natural world is exemplified by Tlingit and Athapaskan understandings of glaciers in the Mount Saint Elias range in the Yukon and Alaska. In *Do Glaciers Listen?* (2005), Julie Cruikshank explores the tension between scientific concepts of nature and Indigenous concepts of nature. In science, nature is something separate from the social world of people and culture. In other words, glaciers are areas of pristine wilderness that can be studied and measured. For the local Indigenous residents, on the other hand, the glacier is a social being, enmeshed in their community histories.

Kitty Smith, a local woman, relates an oral narrative to Cruikshank describing the story of a man and his son, camped out at a "bad place" called *Dadzik*. An excerpt from this story illustrates local understandings of what happens when a person cooks with heat too close to a glacier.

> One old man said, "Take care of that grease.
> The wind is blowing that way, toward the ice.
> I smelled grease when I was walking around.
> Ah! Right there! Look at the grease! Be careful of the grease …"
> Then it started to get light – just like there was a light on top of the ice.
> It starts getting warm. People start to sweat.
> "Well," that man said, "we've got to get away."
> They're close to that glacier.
> They started.
> That little kid, his dad held him.
> Meat, blankets, everything they left behind – just to get away.
> When they began to move down, they could see eyes, just like the sun, on top of the ice.
> Two eyes came out when they were going down.
> It's getting hotter and hotter – too much heat, just like a stove. (p. 113)

In the tale, when the man and his son return in the morning to collect their things, the heat from the glacier has cooked all of their meat.

In the local view, glaciers are responsive members of the world. They can be pleased or angered depending upon the actions of humans, with results for the local community. In this case, local people know that glaciers are angered by the use of hot grease and will retaliate by producing severe heat in return through the ice, sometimes causing major calving of the glacier. Indigenous and scientific understandings are interwoven in their worldview.

Ethnoecology

Researchers who identify themselves as **ethnoecologists** tend to emphasize traditional peoples' knowledge of flora and fauna. Like other environmental anthropologists, they look at the interactions a group of people has with its natural environment. In particular, ethnoecology tends to focus on Indigenous concepts of plants and their uses for food, medicine, or ritual.

Understanding the unique ways in which a group classifies its environment can provide anthropologists with a key to understanding the group's **worldview**. For instance, in early ethnoecological studies, anthropologist Dr. Harold Conklin (1986) discovered that the Hanunóo people of Mindoro Island, Philippines, use a fundamentally different color classification scheme than ours, based on their forest environment. In the forest, light/dark and fresh/dry characteristics play an important part in perception. Therefore, the Hanunóo use four major categories of color: "darkness" (dark colors including black, deep blues, greens, and purples); "lightness" (light colors including white and other pale hues); "redness" (colors of dry plant life such as reds, oranges, and yellows); and "greenness" (colors of fresh plant life such as light greens and browns).

Of course, Hanunóo people can see all the variations in the color spectrum that any other person sees, and express them as well, using modifications to their basic color scheme. The four-category division simply reflects how they interpret the colors in a way that is relevant to their lives. Ethnoecologists use this kind of data to understand the way that people understand their role in the world around them.

ISSUES IN SUSTAINABILITY STUDIES

Population Growth

Underscoring the urgency of the sustainability discussion is the rate at which the world's population is growing. For most of human history, the world's population remained more or less steady, with slow growth resulting from the beginnings of agriculture. With the effects of the Industrial Revolution in the eighteenth and nineteenth centuries, the world's population began to leap forward in exponential growth. Many factors contribute to lower mortality rates and higher life expectancies. Beginning in the twentieth century, the world's population began to explode, jumping from 1.6 billion in 1900 to nearly 7.5 billion in 2018. The UN Department of Economic and Social Affairs (2017) projects that we will reach 9.5 billion by about 2050. The world's population is growing faster than resources can support it.

As you look at Table 14.2, note the years between milestones of one billion more people on earth. Each increment of one billion more people on earth has come at a faster rate. The population growth rate is currently holding steady at about one billion more people on earth every 12 to 16 years.

The direct relevance of population growth for sustainability is that it results in fewer resources for the majority of the world's people (consider Miriam's story at the beginning of the chapter) and greater environmental depletion. This is especially true when increased population leads to higher consumption and economic growth. In other words, the more populations expand and economic growth increases, the more natural and nonrenewable resources need to be exploited. This effect, on a global scale, has the worst impacts on people in societies producing the goods to support growth.

In particular, the growth of high-income countries and their economies has the most severe effect on environmental resources in low-income countries. This is because most natural resources for energy and products are sought outside high-income countries. For example, to produce smartphones and other digital products, minerals must be extracted from mines across the world. In the Democratic Republic of the Congo (DRC), armed terror groups largely control the extraction of minerals. The groups, run by local warlords, are known for stripping land and controlling natural resources. Because much of their mining enterprise is illegal, the environmental destruction is accompanied by other social, political, and economic problems, such as arms trafficking and sexual violence against women. Demand for goods in highly industrialized, high-income countries profoundly affects the lives of people producing the raw materials.

TABLE 14.2

World Population Growth

Population (billions)	1	2	3	4	5	6	7	8	9
Year	1804	1927	1959	1974	1987	1999	2012	2026	2042
Years between Milestones	–	123	32	15	13	12	13	14	16

Source: http://www.census.gov/popclock/.

Nevertheless, the population crisis has also initiated scientific advancements meant to increase food production in low-income countries. For instance, in the 1990s, Chinese plant geneticists began producing over 80 varieties of "super rice," each of which is suited to less-than-ideal growing conditions. Small-holder rice farms in the Philippines suffer from saltwater flooding and intermittent drought, with no means of washing out the salt deposits from the rice farms. The green super rice variety developed for soils with high salinity has allowed Filipino farmers to increase their yields. In countries across Asia, different **genetically modified** strains of super rice ease some of the risks of unpredictable weather events due to climate change.

Globalization of Food

Industrial food production, distribution, and consumption are enmeshed in global processes. These interconnected systems and processes are referred to as **globalization**. In the modern global economy, people in places all over the world are linked in a complex, yet largely invisible, chain of producers and consumers.

Globalization is the integration of economic, social, political, and geographic boundaries and processes. Global studies scholar Dr. Manfred Steger (2003) defines it as "a multidimensional set of social processes that create, multiply, stretch and intensify worldwide social interdependencies and exchanges while at the same time fostering in people a growing awareness of deepening connections between the local and the distant" (p. 34). Because inequalities exist structurally, the poorest people generally suffer the most inequity as a result of globalization. This is seen clearly in our food system. As shown by Pelto and Pelto (2013), there are three major transformations that have affected food access since the industrialization and globalization of the food system:

1) Food production and distribution are embedded in an increasingly intertwined and rapidly growing network of global interdependency.
2) In high-income nations, this leads to better nutrition due to a wider availability of diverse foods.
3) In low-income nations, the elite benefit while most people who were dependent on local production and methods suffer economically and nutritionally.

Due to interdependent food networks, urban dwellers in modern industrial society have access to a smorgasbord of world cuisines and international foodstuffs. Nutritious and ripe produce is available in the middle of winter, something our ancestors couldn't have imagined. Eating a wide range of healthy products supports the health of those who can access those foods.

BOX 14.2 **Food Matters: Chocolate Production**

Serious costs are borne by laborers in industries with little regulation or oversight, such as the chocolate industry. Approximately 75 percent of all chocolate the world eats begins as raw cacao in West Africa. As of 2019, two million children in Côte D'Ivoire (Ivory Coast) and Ghana work to produce and harvest the cacao bean (Whoriskey & Siegel, 2019).

According to UNICEF (2018), an organization that fights for the rights of children all over the world, a complex system of pressures forces children into chocolate production, including poverty, limited access to clean water, environmental degradation, and few child protection laws. Cacao farming communities rely primarily on the labor of the children of farmers, as well as child migrants from neighboring countries, in order to meet the demand for cacao. The children work in hazardous conditions, and many do not get paid for their labor.

In order to combat these practices, a group of organizations are focusing on the interwoven factors that drive **child labor** in cacao farming. These include economic, environmental, social, and political causes. For instance, improving women's economic status and investing in farming communities decreases the number of children who will be sent to work rather than to school.

Both the governments of Ghana and Côte D'Ivoire have been engaged this past decade in fighting child labor in the chocolate industry by providing direct resources to communities and enacting laws. For instance, in 2016, Côte D'Ivoire passed laws that raise the legal working age to 16 and make education mandatory until that age. These and other ongoing efforts also involve the world's largest chocolate companies, who are struggling to meet the goals of child labor–free chocolate. Unfortunately, the Hershey Company, Mars, Nestlé, and Godiva cannot yet say that their chocolate products are free from child labor (Whoriskey & Siegel, 2019).

On the other hand, the globalization of the market creates a set of issues for food production and distribution that should be examined critically. Importing produce from all over the world depletes the local farming economy and creates hardships for people dependent on agricultural jobs. Produce, meat, fish, and other imports in grocery stores are often more expensive than the less healthy, processed foods in the center aisles, making healthy eating difficult for those on limited income. For individuals who are **food insecure**, this means that a healthful diet is often out of reach.

In addition, developing nations engaged in the production of raw materials for food often lack the same protections as the countries in which demand originates. With fewer environmental protections, the soil and waterways of producing nations may be polluted by pesticide runoff or other waste products. With fewer labor protections, laborers may work under conditions that are maladaptive: unhealthy, illegal, or potentially dangerous. The farming of cacao beans for the chocolate industry, as described in Box 14.2, provides an example. (Also see Chapter 7 for how historical archaeology has uncovered the origins of these practices.)

One way in which consumers in high-income nations exert some control over food production and health is to adopt a deliberate form of eating that emphasizes fresh, plant-based, and/or local foods. **Vegetarianism** is a diet that emphasizes plant-based foods and restricts meat and fish, and **veganism** is a diet comprised wholly of plant-based foods and restricts products made from animals or their products, such as dairy and eggs. Either of these diets may be adopted for reasons that include supporting personal or environmental health, or rejecting the idea of animals as commodities for human consumption. Often this stems from a sense of kinship with and sensitivity to the emotional life of animals, and the rejection of industrial farming practices that cause animals to suffer.

A **locavore** diet emphasizes foods produced in one's local community, which supports the local economy, allows consumers to get to know their farmers, and provides fresh produce. Communities all over the world, especially those

Figure 14.6
PIKE MARKET PRODUCE VENDOR, SEATTLE.

A locavore diet – that is, eating what is grown in one's local area – emphasizes fresh produce and supports the local economy.
Credit: Laura T. González

BOX 14.3 **Natural Disasters and Inequitable Responses – The Case of Puerto Rico**

The ongoing effects of climate change suddenly ramped up in 2017 with an unusually active hurricane season in the Atlantic. Category 4 and 5 hurricanes made landfall multiple times in places where storms of such severity are normally seen no more than once in a generation. One of the places that was hit hardest by these megastorms was the Caribbean island of Puerto Rico. Hurricane Irma passed just north of the island on September 6, leaving one million people without power. Two weeks later, on September 20, Hurricane Maria, a strong Category 4 hurricane, passed directly over the entire island.

When Hurricane Maria slammed across Puerto Rico with up to 175 mph winds, it destroyed the island's entire electrical grid and left utter devastation in its path. It was the fifth strongest storm to ever hit United States territory. More than 1,000 people lost their lives to the storm, both directly and indirectly.

Puerto Rico is an island of 3.4 million people, who have been American citizens since 1917. However, as a US commonwealth, Puerto Rico lacks a voting representative in Congress and citizens cannot vote in a presidential election. This creates unique challenges for the island, which has little leverage in the federal

Figure 14.7 **PUERTO RICAN BOY AFTER HURRICANE MARIA.**
One month after Hurricane Maria devastated Puerto Rico, 88 percent of residents were still without power to run refrigerators to store food and medicine, medical supply machines such as oxygen and dialysis, air conditioning and fans (important in this tropical region), or to flush toilets. This young boy's home has lost its roof and walls.
Credit: Courtesy of US Department of Agriculture

outside urban centers, eat the foods grown in their local areas. Nevertheless, when living in an urban environment, the sheer variety of imported foods available makes eating from the local area a deliberate choice. Locavores may adopt a vegetarian, vegan, or **flexitarian** diet, in which they seek out fresh and ethically sourced foods without following a strict set of rules. By adopting eating habits in which meat, especially industrially produced meat, is purchased less or not at all, these consumers help reduce the environmental degradation that meat farming can produce.

government. Just prior to Hurricane Maria, Puerto Rico declared bankruptcy with a $123 billion debt. Infrastructure, especially power and water, is in dire need of upgrades. However, without a voting member of Congress, the US government is not pressured to prioritize the needs of Puerto Ricans.

The White House was strongly criticized for not mobilizing quickly enough to provide aid and not directing enough resources toward the restoration of infrastructure. One month after the disaster, 88 percent of residents still were without power, and 29 percent of people were without running water in their homes. Four months later, in January 2018, it was reported that one-third of the island was still without power. The humanitarian aid organization Oxfam, which rarely provides aid to wealthy countries such as the United States, announced in October 2017 that it would work with leaders in San Juan directly. This is due to what they saw as a "slow and inadequate response" from the US government (Oxfam, 2017).

Climate change has devastating effects on people of island nations, especially those suffering the consequences of inaction of the world's governments, including the United States. In many cases, major disasters resulting from climate events create refugees – people who can no longer get their basic needs met in their home communities. Not everyone has the option to start over in a new country, however. The poorest citizens of the world may be left waiting for aid while homeless, hungry, and without clean drinking water or medicine. "To deny climate change ... is to deny a truth we have just lived," said Roosevelt Skerrit, prime minister of Dominica, after Maria devastated that island (UN News, 2017). Islands like these are paying a heavy price for the effects of climate change, primarily driven by large industrial nations. In the case of Puerto Rico, having no voting representatives means the US citizens on the island are at the bottom of a two-tiered system with little accountability from the US government. Climate change separates and divides – with the poorest nations suffering the heaviest burden.

HOW CAN ANTHROPOLOGISTS HELP?

More than any other academic discipline, anthropology uses a long-range, holistic perspective that connects the dots between local and global systems. Therefore, anthropologists are uniquely suited to engage with sustainability issues on multiple levels. In particular, these include using the methods and theories of anthropology, disseminating information, and, for those who teach, engaging students and campuses.

First, the basis of anthropological practice and theory is to learn about people. Anthropologists spend years among people in developing nations – people who are largely invisible to consumers in the developed world. Experience in the field alongside people who suffer the inequities of global systems provides an intimate understanding of their struggles. With both local and global knowledge, anthropologists can help provide the kinds of solutions that are most needed.

Anthropologist Dr. John Bodley (2008) makes the case that anthropologists should be an essential part of the discussion on sustainable development. The advantages of our evolutionary and cross-cultural approaches provide a deep and broad perspective on societal change. He argues that it is the scale and scope of our societies that have created the major problems of today: environmental problems, hunger, poverty, and conflict. A reduction in **culture scale** would help to mitigate some of these problems in our globally linked, commercial societies. Bodley concludes that the long-term success of small-scale tribal societies should serve as a guiding model. As the experts on 100,000 years of tribal societies, anthropologists have much to say about successfully living in regional ecosystems supported by local economic markets and in small-scale social communities.

Second, armed with the knowledge of human history and societies, anthropologists can bring this insight to the larger public, helping people understand the effects of their actions. The discipline of anthropology has a great responsibility to publish and present information in ways that will reach the general public. Statistics and charts give facts, but the actual stories of a people's struggles tie us emotionally to their plight. Consumer behavior can change rapidly when knowledge leads to compassion.

Pioneering anthropologist Dr. **Margaret Mead** did just that: she brought knowledge of human societies to the public. Mead worked among islanders of the South Pacific and Indonesia, exploring major questions of the early twentieth century such as race, gender, child-rearing, and the **nature versus nurture** controversy. She was able to bridge the gap and translate her conclusions into formats to which the general public had access.

Although later researchers have questioned the validity of some of her data, Mead's lasting legacy is the way in which she was able to reach a wider public. As a curator in the American Museum of Natural History in New York for 50 years, she designed exhibits shown to visitors. She appeared on radio and television programs and taught at several colleges. Importantly, she published regularly in *Redbook*, a women's magazine. In this way, Mead reached the general public with critical information about human beings and culture, breaking down stereotypes and paving the way for the women's rights movement in the 1960s.

Finally, the majority of anthropologists teach at colleges and universities. In this setting, anthropologists have the potential to effect major change, especially among students and the campus community. Sustainability Studies is now a major at many institutions, training students in an interdisciplinary way to work with development solutions. Many campuses have created offices of sustainability and lead the way in green building for new campus structures, water conservation on campus, and recycling or composting programs. Some urban college campuses are even building organic farms to offer degrees in urban agriculture and supply their cafeterias with fresh, organic produce. Anthropologists appear in these initiatives as leaders, alongside biologists, environmental scientists, and ecologists.

At Emory University in Atlanta, Georgia, anthropologist Dr. Peggy Barlett (2005) helped develop a series of sustainability workshops to engage faculty on her campus. Based on the successful Ponderosa Project at Northern Arizona University, Emory's Piedmont Project seeks to connect the campus community with the local geography and ecosystem of the Piedmont region. Workshop participants hear speakers on local sustainability issues, spend time outdoors in the Piedmont forests, and draw on their own expertise as inspiration for including sustainability topics in their classes. Barlett concludes that the success of this model results from not only learning new information but also reconnecting physically with nature – with the resulting emotional shift that participants experience. This type of intellectual and experiential workshop now serves as a model for campuses everywhere.

SUMMARY

This chapter explored the relationship between the field of anthropology and the challenges and scope of sustainability. Mirroring the Learning Objectives stated in the chapter opening, the key points can be summed up as follows:

- A sustainable future on our planet requires an understanding of how humans have coexisted and manipulated their environments and continue to do so.
- Anthropology is interested in the human–environment relationship: what has worked and what has failed over a long span of time and in every corner of the globe. Therefore, anthropology can inform discussions about sustainability.
- Sustainable development has at least three aspects: social, economic, and environmental. The environmental aspect is the most fundamental, for without healthy air, soil, or water, human society would collapse.
- Earlier in history, humans had a very different interaction with their local ecosystems as small bands of foragers. As populations grew, we drew more resources from the environment. Demand for resources increased

exponentially in the middle of the twentieth century, and the environment has suffered proportionally.

- There have been multiple theoretical approaches to these questions within anthropology, all looking at the relationships between people and their ecosystems.
- Some of the most pressing questions include the issues of population growth and food security.
- Anthropologists can provide much of this data from their ethnographic and archaeological fieldwork, their biological research, and applied anthropology projects in which they seek to solve problems for people.

Review Questions

1. What are some of the unique contributions that the study of anthropology can make to the discussion of sustainability solutions?

2. How have the ways that anthropologists approach ecological studies changed over the years?

3. How is the loss of biodiversity encouraging current studies of ethnoecology and Traditional Ecological Knowledge?

4. Why is population growth a fundamental problem for sustainability?

5. What are the pros and cons of the globalized food economy?

Discussion Questions

1. The health of our planet's oceans is suffering due to human practices. How has global use of the ocean resulted in a "tragedy of the commons"?

2. Can you think of ways that alternative food movements are aligned with Bodley's idea of "culture scale"?

3. Have you had a personal experience that caused you to think differently about nature or our place in it?

Visit **www.lensofanthropology.com** for the following additional resources:

| SELF-STUDY QUESTIONS | WEBLINKS | FURTHER READING |

GLOSSARY

The numbers that follow the definitions indicate the chapters where the term is discussed.

Acheulean a cultural tradition based largely on specific kinds of stone tools and associated primarily with *Homo erectus* and *Homo ergaster* 5

achieved status a social role a person achieves due to work and opportunity 12

adaptive radiation a process by which one species occupies a new ecological niche, quickly increasing its population and diversifying into new species 2, 3

affiliative friendly 2

affinal related by marriage 11

agency the capacity of a person to think for themselves and control their life choices 8

agriculture a farming technique that can support a large population using advanced tools and irrigation, and requiring more preparation and maintenance of the soil; also known as intensive cultivation 1, 6, 10

allele an alternate form of a gene 3, 11

allele frequency the relative proportion of a particular allele occurring, compared to other alleles that could be selected 3

ancestor veneration worship of one's ancestors 13

ancestral spirits the essence of one's family ancestors who have remained in contact with the mortal world 13

androgynous gender neutral; in the middle of the gender spectrum 11

androphilia romantic or sexual attraction to males 11

animal husbandry the use and breeding of animals for purposes that benefit humans 10 animatism the belief that spiritual forces inhabit natural objects 13

animism the belief that spiritual beings inhabit natural objects 13

Anthropocene a proposed geological period to describe the years in which humans have had a significant impact on the earth's environment, observable in the geological record; there is no consensus on the validity of the term to describe the geological period or when such a period began 2, 14

anthropological perspective evolutionary, holistic, and comparative methods applied to the study of humans 1

anthropology the study of human biology and culture, past and present, through evolutionary, holistic, and comparative perspectives 1

applied anthropology a branch of anthropology in which the researcher uses knowledge of anthropological methods, theory, and perspectives to solve contemporary human problems 1, 8

archaeological record the material remains of the human past and, in some cases, the description of the human past based on the material remains 5

archaeological site any location where there is physical evidence of past human activity 1, 5

archaeology the branch of anthropology focusing on the study of humans through the remains of their physical activities 1

arranged marriage the practice in which parents find a suitable husband or wife for their child 11

artifact any portable object showing evidence of being made or used by people 1, 5

ascribed status a social role of a person that is fixed at birth 12

asexual an individual without sexual desires 11

assigned sex at birth the sex label (usually male or female) that a medical practitioner gives to an infant at birth by assessing their outer appearance 11

Atlantis a fictional island civilization described by Plato; both island and civilization are destroyed in Plato's account 7

atlatl spear-thrower 5

authority having legitimate power by law 12

balanced reciprocity a form of exchange in which the value of goods is specified, as is the time frame of repayment 10

band a small egalitarian society of food foragers who live and travel together 6, 10, 12

base camps discrete areas with physical evidence that people were temporarily occupying a place for resource processing or habitation 5

BCE "Before the Common Era," a secular calendar notation equal to BC 3, 12, 13

Beringia a large, unglaciated landmass connecting North America and Asia during the last ice age 6

Big Man an informal leader who possesses authority based on prestige and persuasive power, found in Melanesian societies 6, 12

bilateral descent the act of tracing one's genealogy through both the mother's and the father's line 11

binary having two parts; in gender studies, it refers to a two-gender system of masculine males and feminine females 11

biodiversity the variety of life on earth including plants, animals, and microorganisms; the diversity of living organisms in a given ecosystem, area, or the world 10

biological adaptation a physical adaptation that allows an organism to survive better in its environment 8

biological anthropology the branch of anthropology focused on human biology, including evolution and contemporary variability 1

bipedalism moving mostly by using two legs 1

bisexual an individual attracted romantically or sexually to both males and females or to two genders 11

Black Lives Matter a movement that began in 2013 as a grassroots activist campaign to make visible the experiences of Black Americans subjected to police brutality; has since grown into a global movement that seeks to confront and end systematic racism and violence against Black people 1, 8

body modification the practice of altering the body for reasons of identity, attractiveness, or social status 11

brachiation a form of locomotion primarily by arm-over-arm swinging 2

bride-price a form of marriage compensation in which the family of the groom is required to present valuable gifts to the bride's family 11

bride service a form of marriage compensation in which the groom is required to work for the bride's family 11

bushmeat meat from wild animals, usually referring to animals from forested regions of Africa 2

Cahokia a World Heritage Site near St. Louis, Missouri 1

capitalist system a system of economics in which a country's industry is controlled by private and corporate ownership in order to make a profit 10

cargo cult a religious revitalization movement in Melanesia that uses ritual to seek help and material wealth 13

cargo system a political and religious system among the Maya in which adult males must serve the community in a volunteer position for at least one year; a leveling mechanism 10

carrying capacity the number of people who can be sustained with the existing resources of a given area 6, 10 caste a hierarchical system based on birth; most commonly associated with Hindu India **12**

caste system a system of social stratification in India in which a person is born into a hereditary group traditionally linked to certain occupations **11**

Catarrhini a primate infraorder, including Old World monkeys, apes, and humans **2**

cave art art painted or incised on cave walls, including petroglyphs and pictographs **5**

CE "Common Era," a secular calendar notation equal to AD **3, 7, 9, 10, 11, 13**

centralized system a political system with a centralized governing body that has the power and authority to govern **12**

ceramics baked clay **6**

Cercopithecoidea a primate superfamily, including Old World monkeys **2**

chemical inputs synthetic additives, such as pesticides and fertilizers, that raise the yield of crops in industrial agriculture **10**

chiefdom a type of political organization found in settlements ranging from a few thousand to tens of thousands of people; characterized by social inequality and hereditary leadership, and based on horticulture **6, 12**

child labor the exploitation of a child's labor for business or industry, especially when interfering with the ability of the child to attend school and when the work is physically, emotionally, or morally inhumane **14**

child marriage the practice in which parents marry young daughters to older men who offer to provide for them **11**

cisgender individuals who internally experience their gender identity as aligned with their assigned sex **11**

city a settlement supporting a dense population with a centralized government, specialization, and socio-economic hierarchy **6, 7, 10**

civilization a type of society characterized by state-level political organization, a system of writing, at least one city, and monumental architecture **6, 7**

clan a social division that separates members of a society into two or more groups; also called a moiety **11**

class a form of social stratification based on differences in wealth and status **12**

closed-loop system a system that has finite resources and cannot sustain indefinite growth **14**

coastal migration route probably the route of the first migrants to the Americas, along the coast of Alaska and British Columbia **6**

code switching the practice of moving easily between speech styles or languages in a conversation or single utterance **9**

colonialism the domination and subjugation of Indigenous peoples by Europeans and their descendants 1

commercial archaeology *see* cultural resource management

community a group of people who share a physical location; people who live, work, and play together 8

confined animal feeding operations (CAFOs) industrial farming enterprises in which large numbers of animals are prepared for human consumption; the basis of conventional meat production 10

consanguineal related by blood 11

conventional describes food-growing processes in industrial societies in which pesticides and other chemicals are used 10

cooperative societies bands and other small-scale human groups that rely on sharing resources for survival 10

coprolite preserved feces, usually referring to human feces 5

COVID-19 the acronym given to the disease caused by a new coronavirus first identified in 2019; COVID-19 created a global pandemic 1, 3, 7, 8, 9, 12

creole a stable language that forms over time when two languages have sustained contact; often the result of a pidgin language's natural development 9

cultural adaptation a belief or behavior that allows an organism with culture (especially humans) to better thrive in its environment 8

cultural anthropology the branch of anthropology focusing on contemporary cultures 1

cultural appropriation the use of an element of a minority or oppressed culture by a dominant culture in an inappropriate context, such as the way Euro-Americans use elements of Native American culture for fashion, logos, and mascots 1

cultural ecology a framework of understanding culture by examining the limitations of the environment and food-getting practices 14

cultural landscape a distinctive geographic area with cultural significance 5

cultural materialism a framework for understanding society that is directly related to whatever adaptations are necessary to survive in its environment 12, 13

cultural model a widely shared understanding about the world that helps us organize our experiences in it; determines the metaphors used in communication 9

cultural relativism the idea that all cultures are equally valid, and that every culture can be understood only in its own context 1, 8

cultural resource management (CRM) doing archaeology in advance of development projects; also known as commercial archaeology 1 culture the learned and shared things that people think, do, and have as members of a society 1

culture core a set of features of culture that are similar in societies practicing the same food-getting strategies; an aspect of the cultural ecology model 14

culture scale the scope or reach of culture; implied is the idea that smaller-scale societies are more sustainable than larger-scale societies 14

dating by association dating artifacts and sites by their association with other artifacts, ecofacts, or geological features of known age 4

deities gods and goddesses 13

dental arcade the shape of tooth rows, such as parabolic (wider at back than front) or u-shaped 4

dental formula the kind and number of teeth, usually described for one quarter of the mouth 2

dependence training a set of child-rearing practices that supports compliance to the family unit over individual needs 8

descent group a social group of people who trace their descent from a common ancestor 11

diasporic spread to different parts of the world, especially used in reference to ethnic or cultural groups 8

diastema a gap between teeth 4

diplomacy the relations and negotiations between nations 12

discourse written and spoken communication 8

discrimination actions taken as a result of prejudice 8

divination the art of reading the future 13

DNA deoxyribonucleic acid; a molecule that contains the genetic instructions for living organisms 3

domestication shaping the evolution of a species for human use 6, 10

dowry a form of marriage compensation in which the family of the bride is required to present valuable gifts to the groom's family or to the couple 11

dowry death deaths of women in the homes of their in-laws due to unmet dowry demands 11

ecofacts botanical remains, animal remains, and sediments in archaeological sites that have cultural relevance 5

ecological anthropology a framework of understanding culture that uses systems theory to understand a population as a closed-loop system 14

economic sustainability the ability of the economy to support indefinite growth while ensuring a minimum quality of life for all members of society 14

economics how goods and services are produced, distributed, and consumed in a society 10 egalitarian describes a society in which every member has the same access to resources and status; non-hierarchical 2, 10

emic an insider's view; the perspective of the subject 8, 13

emigrant a person who leaves their own country permanently in order to live in another one 8

empire a kind of political system with one state being territorially expansive and exerting control over others, such as the Roman and Inka empires 7

enculturation the process by which culture is passed from generation to generation 8

endogamy the practice of marrying within one's social or ancestral group 11

entomophagy the practice of eating insects for food 8

environmental anthropologist an anthropologist interested in the relationships between people and the environment 14

environmental sustainability the ability of the environment to renew resources and accommodate waste at the same rate at which resources are used and waste is generated 14

epigenetics the study of how parts of the genome may become activated or deactivated as an organism develops 3

ergonomics the science of designing things so they create little or no physical stress on the human body 1

essential workers people who governments around the globe have deemed crucial for the functioning of society amid the COVID-19 pandemic 1, 12

ethnic cleansing violent and aggressive intergroup conflicts in which one group attempts to commit genocide of the other due to ethnic differences 12

ethnicity a term used to describe the heritage, geographic origin, language, and other features of a person 8

ethnocentrism the belief that our own customs are normal while the customs of others are strange, wrong, or even disgusting 8

ethnoecologist a person who studies the interactions a group of people has with their natural environment, focusing especially on the use of flora and fauna 14

ethnographer a cultural anthropologist who studies a group of people in a field setting 8

ethnographic research the process of studying culture, undertaken in a field setting 1, 8

ethnography the written or visual product of ethnographic (field) research 1, 8

ethnolinguistics the study of the relationship between language and culture; a subset of linguistic anthropology 9

etic an outsider's view; an objective explanation 8, 13

eugenics a pseudoscience of "race improvement" 8

euphemisms polite or socially acceptable words and phrases that are used in place of ones that are considered unpleasant or offensive **12**

exogamy the practice of marrying outside one's social or ancestral group **11**

extended family a family unit consisting of blood-related members and their spouses; a mix of consanguineal and affinal kin **11**

externalized controls rules that regulate behavior by encouraging conformity to social norms; may be negative (punishments) or positive (rewards) **12**

extinction that which occurs when a taxonomic group, usually a species, ceases to exist, either because it could not adapt to changing circumstances or because it evolved into a new species **3**

family of choice a group of people who consider themselves to be family members, even though they may not be affinal or consanguineal kin; originated in the LGBTQ+ community **11**

family of orientation blood-related family members, including parents, siblings, grandparents, and other relatives **11**

family of procreation the family unit created by marriage or partnership, including spouses/partners and their children **11**

fandom a group of people who collectively are fans of something or someone; considered to be a community of people **9**

faunal remains animal remains in paleoanthropological and archaeological sites, usually restricted to bones and hair but also possibly including fur, nails, claws, horn, antler, skin, and soft tissue **5**

feature a nonportable object or patterning created by people and recognized archaeologically, such as a fire hearth **5**

feud ongoing violent relations between two groups in the same society **12**

fictive kinship including non-blood relations in the family with all the expectations of blood-related family members **11**

flexitarian describing a diet that is primarily vegetarian but does not exclude infrequent consumption of meat **14**

folivory a diet focusing on leaves and other rough plant foliage **2**

food foragers people who utilize the food resources available in the environment; also called hunter-gatherers **10**

food insecure not always having access to food; not knowing where one's next meal might come from **14**

food producers people who transform the environment with the goal of food production, using farming and/or animal husbandry **6, 10**

food production transforming the environment with the goal of producing food using farming and/or animal husbandry **6, 10**

food security the availability of and access to safe and nutritious food **1**

foodways the methods, knowledge, and practices regarding food in a particular society 10

foraging utilization of food resources available in the environment; also known as food foraging or hunting and gathering 1, 6, 10

foramen magnum the hole at the base of the skull through which the spinal column enters the brain 4

forced marriage the practice in which parents demand their child marry someone they have chosen 11

fossil in anthropology, any preserved early human bones or teeth 4

fossil record in anthropology, the assemblage of early human remains, or the interpretation of human evolution based on human remains 4

frontline workers a subset of employees, usually in the medical field, who are at greater risk for contracting COVID-19 because of the nature of their work; this type of work usually entails direct contact with people, including COVID-19 patients 12

frugivory a diet focusing on fruit 2

garbology the study of contemporary garbage using the methods of archaeology 7

gender the set of social meanings assigned by culture to a person's identity as male, female, both, or neither, as well as the expression of that identity in social behavior 11

gender discrimination the apparent or real dominance of men and subordinate status of women in a society 12

gender dysphoria the stress that can result from feeling that one's gender identity does not match one's assigned sex at birth 11

gendered speech different speech patterns based on the cultural expectations of the different sexes 9

gender expression the outward display of a person's gender in clothing, mannerisms, and language use 11

gender fluid gender identity that lies somewhere along the gender spectrum; nonbinary or changing 11

gender identity a person's internal experience of identity as male, female, both, or neither 11

gender queer gender identity that lies somewhere along the gender spectrum; sometimes used as an umbrella term to signify "not cisgender" 11

gender roles the culturally appropriate or expected roles of individuals in a society 11

gender spectrum the variety of gender identities that exist on a continuum 11

gender stratification the hierarchical division of males and females in society 12

gene a unit of heredity 3

gene flow the movement of genes between populations 3

generalized reciprocity a form of specialized sharing in which the value of a gift is not specified at the time of exchange, nor is the time of repayment 10

genetically modified (GM) altered at the level of the gene; refers particularly to food crops that have been modified by introducing genes from another organism to enhance or create desired traits in the species 10, 14

genetic drift the random factor in evolution, including changes in allele frequencies by chance rather than selection 3

genetic engineering the manipulation and modification of DNA to alter an organism 3

genetics the study of a particular gene or groups of genes, especially as it relates to inheritance 3

genocide the deliberate killing of an entire ethnic group 12

genome the complete genetic makeup of an organism 3

genomics the study of genomes 3

genotype the genetic makeup of an organism 3

genus a taxonomic category, above the level of species 2, 3

Ghost Dance a religious revitalization movement started among the Northern Paiute that used a five-day circle dance to seek help from the supernatural realm 13

globalization the integration of economic, social, political, and geographic boundaries in complex chains of interconnected systems and processes 14

glycemic index a measure of the rise in blood glucose (sugar) after eating 10

gods and goddesses distant and powerful supernatural beings; deities 13

gradualism the idea that evolutionary change is a long, slow process 3

Great Rift Valley an area in East Africa where many important paleoanthropological sites are located 4

group a loose term for people who share culture; they often live in the same region 8

guilt culture a culture that focuses on one's own sense of right and wrong and the punishment that can result from breaking the rules 12

gynophilia romantic or sexual attraction to females 11

habitation sites areas with physical evidence indicating that people were living there, at least temporarily 5

Haplorhini a suborder of Primates, including monkeys, apes, and humans 2

hearth a discrete area where people controlled a fire 5

heterogeneous sharing few identity markers 8

heterosexuality the romantic or sexual attraction or sexual behavior between partners of the opposite sex 11 hijra a third gender found in India and Pakistan in which male-bodied or intersex individuals adopt female mannerisms and dress 11

holistic describes the viewpoint that all aspects of biology and/or culture are interrelated 1

Homininae the biological family to which humans belong; some also consider Homininae to include gorillas, chimpanzees, and bonobos 1

Hominoidea a superfamily of the infraorder Catarrhini; Hominoidea includes apes and humans 2

homogeneous sharing similar identity markers 8

Homo sapiens the genus and species to which modern humans belong 1

homosexuality the romantic or sexual attraction or sexual behavior between partners of the same sex 11

honorifics linguistic ways to show honor or respect 9

horticulturalists food producers who cultivate the land in small-scale farms or gardens 6, 10

horticulture land cultivation in small-scale farms or gardens 1, 6, 10

household a domestic unit of residence in which members contribute to child-rearing, inheritance, and the production and consumption of goods 11

human at a minimum, Homo sapiens, although most anthropologists define it as any member of the genus Homo or biological family Homininae 1

human ecology the study of the complex relationships between humans and their environments 14

Human Terrain System US Army program involving anthropologists deployed with military units in active conflict zones 8

hunter-gatherers people who utilize the food resources available in the environment; also called food foragers 10

Hylobatidae one of the three families of the superfamily Hominoidea; Hylobatidae includes gibbons and siamangs, otherwise known as "lesser apes" 2

hypothesis a possible explanation that may be tested using scientific methods 3

ice-free corridor the space between two ice sheets covering most of Canada during the last ice age, providing a possible route from Beringia to areas south 6

ideal behavior how people believe they behave or would like to behave; the norms of a society 8

identity markers cultural characteristics of a person such as ethnicity, socio-economic class, religious beliefs, age, gender, and interests 8

ideology refers to beliefs and values, including religion 1, 13

imitative magic a form of magic in which a practitioner creates something to represent real life, then manipulates it in a way that imitates the desired effect; the magical idea that like produces like 13

imperialism a state or political entity's attempt at exerting control over another territory or cultural group, often by force and/or violence 9

incest taboo prohibition against sexual relations between immediate family members 11

independence training the set of child-rearing practices that foster a child's self-reliance 8, 11

index sign an emotional expression that carries meaning directly related to the response 9

Indigenous describes people who can trace their ancestry in an area into the distant past; Indigenous peoples in North America include Native Americans, First Nations, American Indians, Métis, and Inuit 1

industrialism methods of producing food and goods using highly mechanized machinery and digital information 1, 7, 10

informants study subjects of an anthropologist; also referred to variously as collaborators, field subjects, or associates 8

inheritance of acquired characteristics the (incorrect) idea that characteristics acquired during one's lifetime could be passed on to offspring 3

Inka ancient culture of the Peruvian Andes; the spelling preferred over "Inca" by modern scholars 7

insectivory a diet focusing on insects 2

intensive agriculture a farming technique that can support a large population using advanced tools and irrigation, and requiring more preparation and maintenance of the soil 10

internalized controls impulses that guide a person toward right behavior based on a moral system 12

intersex having a combination of physiologic or morphological elements of both sexes 11

Inuit Indigenous peoples inhabiting Arctic Canada, Alaska, and Greenland; in Alaska, the term *Eskimo* remains commonly used, while in Canada and Greenland, it is a racial slur 7, 10

Islamophobia fear of and prejudice toward Islam and people of the Islamic faith (Muslims) 13

judgment sample a method of choosing informants based on their knowledge or skills 8

key associate *see* key informant

key informant a person with whom the ethnographer spends a great amount of time because of the person's knowledge, skills, or insight 8

kinesics the cultural use of body movements, including gestures 9

kinship family relations; involves a complex set of expectations and responsibilities 11

Kula Ring a system of balanced reciprocity in which gifts circulate among trading partners in the Trobriand Islands 10

language a symbolic system expressing meaning through sounds and/or gestures 9

language registers different styles of speaking within a single language 9

lens of anthropology a particular way to view the world through the perspectives, ideas, methods, theories, ethics, and research results of anthropology 1

leveling mechanism a social and economic obligation to distribute wealth so that no one member of a group accumulates more than anyone else 10

lineage a line of descent from a common ancestor 11

linguistic anthropology the branch of anthropology focusing on human languages 1

linguistic relativity principle the idea that the language one speaks shapes the way one sees the world 9

lithic scatter an accumulation of lithic (stone) flakes left behind from making stone tools 5

locavore describing a diet based on locally produced foods 14

Lomekwian a proposed new tool tradition or industry based on what appears to be 3.3-million-year-old tools at the Lomekwi locality in Kenya, Africa 5

Lower Paleolithic the cultural time period from about 2.6 million to about 500,000 years ago 5

magic the use of powers to contact and control supernatural forces or beings 13

maladaptive a cultural practice leading to harm or death; not productive for a culture's survival in the long run 8

Margaret Mead pioneering figure in early cultural anthropology; one of the first female anthropologists to undertake long-term fieldwork 14

market economy an economic system in which prices for goods and services are set by supply and demand 10

marriage the practice of creating socially and legally recognized partnerships in society 11

marriage compensation gifts or services exchanged between the families of a married couple 11

mass extinctions widespread and rapid extinctions, usually restricted to situations in which at least half of all living species on earth become extinct 3

matrilineal descent tracing one's genealogy through the mother's line 11

matrilocal a residence pattern in which a husband moves to his wife's household of orientation 11

Mesoamerica the term used in anthropology to describe the area now encompassing Mexico and Central America 7

Mesolithic the time period from about 15,000 to 10,000 years ago; also known as the Middle Stone Age 6

Mesopotamia in ancient times, the area around modern-day Iraq 6, 7

metaphor a comparison to something as a way to suggest a similarity 13

#MeToo a social movement that began in 2006 with activist Tarana Burke, who wished to draw attention to and assist women of color who had been sexually assaulted; the movement has since become international, inspiring women of all walks of life to come forward with their stories of sexual harassment and assault 1

midden a discrete accumulation of refuse 5

Middle Paleolithic the cultural time period from about 500,000 to 40,000 years ago 5

money anything that is used to measure and pay for the value of goods and services 10

monoculture a technique used in industrial farming in which a single crop is planted on many acres 10

monogamy the marriage practice of having a single spouse 11

monotheism a religious belief system worshipping a single god or goddess 13

morpheme the smallest part of a word that conveys meaning 9

multipurpose money commodities that can be used for other practical purposes besides simply as money 10

mutation an error in the replication of DNA 3

myth a sacred story that explains the origins of the world or people in it 13

nature versus nurture the debate over which aspects of human life are fixed in one's genetic makeup, and which are learned through culture 14

negative reciprocity a deceptive practice in which an exchange is unequal; an exchange in which the seller asks for more than the value of the item 10

Neolithic the time period from about 10,000 to 5,000 years ago; also known as the New Stone Age 6

neolocal a residence pattern in which a couple moves to their own household after marriage 11

nobles high-status members of a society wherein rank is often inherited 10

nomadic moving within a large area frequently to access food resources 10

nuclear family a family unit consisting of two generations, most often parents and their children 11

nurture kinship non-blood relationships based on mutual caring and attachment 11

nutrition transition a shift in diet and activity level that accompanies modernization and results in obesity and related health problems 10

Occam's razor the notion that the simplest explanation is usually the best; also known as Occam's rule 7

Oldowan a cultural tradition based largely on specific kinds of stone tools and associated primarily with *Homo habilis* and *Homo rudolfensis* 5

olfaction sense of smell 2

omnivorous a diet including a wide variety of plants and animals 2

ontogeny the development of an individual from conception to maturity 2

paleoanthropology the study of early humans, using both archaeology and biological anthropology 1

paleo diet a fad diet based on the notion that people should be eating "like our ancestors" before the domestication of plants and animals 6

paleoenvironment the ancient environment 5

Paleolithic the time period from about 2.6 million to 15,000 years ago; also known as the Old Stone Age 5

Pan the genus to which chimpanzees and bonobos belong 2

pandemic an outbreak of disease that spreads over large portions of the globe and affects a significant proportion of the population 7

pansexual an individual not limited in romantic or sexual attraction by sex or gender 11

pantheon a set of gods and goddesses in a religious belief system 13

paradigm a set of concepts; a model 14

paralanguage the ways we express meaning through sounds beyond words alone; a subset of semantics 9

participant observation a research method used in anthropology in which an ethnographer lives with a group of people and observes their regular activities and behaviors 1, 8

participatory action research an applied anthropological method of field research and implementation of solutions; relies on close collaboration with the target community 8, 14

pastoralism a way of life that revolves around domesticating animals and herding them to pasture 1, 6, 10

patrilineal descent tracing one's genealogy through the father's line 11

patrilocal a residence pattern in which a wife moves to her husband's household of orientation 11

peasants low-status members of a society who farm for a living 10

personality the unique way an individual thinks, feels, and acts 8

petroglyph an inscription on stone 5, 13

phenotype the physical expression of a genotype; what an organism looks like 3

phonemics the study of how sounds convey meaning 9

phonetics the study of the sounds in human speech **9**

physical anthropology *see* biological anthropology

phytoliths plants turned to stone **5**

pictograph a painting on stone **5, 13**

pidgin a language formed when two groups lack a common language; uses limited vocabulary and grammar for mutual comprehension **9**

Platyrrhini a Primate infraorder, commonly known as New World monkeys **2**

political ecology a framework for understanding culture that focuses on the complex relationships between the environment, economics, and politics **14**

political organization the way a society maintains order internally and manages affairs externally **12**

polyandry the marriage practice of having two or more husbands at the same time **11**

polygamy the marriage practice of having two or more spouses at the same time **11**

polygyny the marriage practice of having two or more wives at the same time **11**

polysexual an individual attracted to people of multiple genders or sexes **11**

polytheistic describes a religious belief system in which multiple gods and goddesses are worshipped **13**

Pongidae one of the three families of the superfamily Hominoidea; some use Pongidae to refer to the three genera of Pongo, Gorilla, and Pan, while others use it only for Pongo **2**

Pongo the genus to which orangutans belong **2**

popular culture mainstream culture in a society, including mass media, television, music, art, movies, and books **1**

potassium argon dating dating technique based on measuring the ratio of potassium and argon in volcanic sediments **4**

potlatch a Pacific Northwest Coast gift-giving ceremony with great cultural significance **10**

pottery articles made of baked clay, especially containers **6**

power the ability to compel another person to do something that they would not do otherwise **12**

pragmatics the context within which language occurs **9**

prehensile the ability to grasp **2**

prehistory the time before written records were kept in a given area **1**

prejudice an unfavorable bias toward something or someone; a preformed opinion not based on fact **8**

prestige the positive reputation or high regard of a person or other entity merited by actions, wealth, authority, or status **12**

priest (priestess) a full-time religious practitioner **13**

Primate one of the taxonomic orders of the class Mammalia 1, 2

primatology the study of primates, usually in the wild, using the framework of anthropology 1

privatization selling ownership of public resources to private companies 12, 14

prognathism having a protruding face 4

proxemics the cultural use of space, including how close people stand to one another 9

pseudoarchaeology the study of the human past, but not within the framework of science or scholarly archaeology 7

punctuated equilibrium the idea that evolutionary change may alternate between periods of slow, gradual change and short periods of significant change 3

qualitative a type of research strategy or perspective focusing on observations, actions, symbols, interviews, and words to gather data 1

quantitative a type of research strategy or perspective focusing on methods designed to produce data in the form of numbers 1

race a term used to describe varieties or subspecies of a species; inaccurately used to refer to human differences 1, 8

radiocarbon dating a dating technique based on measuring how much carbon-14 is in preserved organic remains 4

raid violence in which members of one group aim to steal or recover items, animals, or people from another group in the same society 12

random sample a method of choosing informants randomly 8

ranked society a society in which prestige and authority are inherited through families 12

real behavior how people actually behave, as observed by an ethnographer in the field 8

reciprocity a set of social rules that govern the specialized sharing of food and other items 10

redistribution an economic system in which goods and money flow into a central entity, such as a governmental authority or a religious institution, which then sorts and allocates those goods back into the community 10

religion a set of beliefs and behaviors pertaining to supernatural forces or beings that transcend the observable world 13

religious revitalization movement a process by which an oppressed group seeks supernatural aid through the creation of new ritual behaviors 13

reserves land set aside for Canada's First Nations peoples; in the United States, these areas are called reservations 12

resistance acting in opposition to a dominant power using cultural meanings and symbols 8 resource processing sites areas where physical remains indicate that people were harvesting and/or processing resources 5

rhinarium the fleshy part at the end of the nose of Strepsirhini and many other animals 2

rites of passage rituals marking life's important transitions from one social or biological role to another 11, 13

ritual a symbolic practice that is ordered and regularly repeated 13

rock art paintings on rock (pictographs) and inscriptions on rock (petroglyphs) 5

Rosetta Stone a slab of basalt on which are inscribed three different kinds of writing; provided the key for deciphering Egyptian hieroglyphs 7

salvage ethnography ethnography done with a sense of urgency to record cultures based on the assumption that the cultures are rapidly disappearing 1

sanction punishment that results from breaking rules 12

savannah an environment common in tropical and subtropical Africa, consisting of wild grasses and sparse tree growth 4

scarification inscribing scars on the body as a marker of identity 11

science a framework for investigating and understanding things that includes a specific set of principles and methods 1, 3

sectarian violence a form of violence inspired by different sects (ideological or religious groups) in the same society 12

semantics the study of how words and phrases are put together in meaningful ways 9

September 11th in 2001, the day on which a group of extremists affiliated with a group known as al-Qaeda hijacked four planes in the United States; under the direction of terrorist leader Osama bin Laden, these coordinated attacks killed thousands of Americans in New York, Pennsylvania, and Washington, DC 13

serial monogamy the marriage practice of taking a series of partners, one after the other 11

sex the biological and physiological differences of human beings based on sex chromosomes, hormones, reproductive structures, and external genitalia 11

sexual division of labor the division of tasks in a community based on sex 10

sexuality romantic or physical attraction to another person 11

sexual orientation the nature of one's romantic or sexual attraction to another person 11

sexual selection when mates are chosen based on characteristics unrelated to survival 3

shaman a part-time religious practitioner 13

shame culture a culture in which conformity to social expectations stems from wanting to live up to others' expectations 12

sharia law scriptural guidelines for Muslim religious adherents to follow 11

sign in communication, something that stands for something else 9

silent language the very specific set of nonverbal cues such as gestures, body movements, and facial expressions that is acquired by speakers of a language 9

snowball sample a method of finding informants through association with previous informants 8

social capital the idea, developed by Pierre Bourdieu, that skills and resources give an individual status by virtue of their membership in a social group 10

social density the frequency and intensity of interactions among group members in a society 10

social distance the degree of separation or exclusion between members of different social groups 10

social mobility the ability of members of a society to rise in social class 12

social stratification the ranking of members of society into a hierarchy 12

social sustainability the ability of social systems (such as families, communities, regions, or nations) to provide for the needs of their people so that they can attain a stable and healthy standard of living 14

society people who share a large number of social or cultural connections; in the animal world, a group of animals born with instincts that cause them to occupy a particular place in the group hierarchy 8

sodality a group that brings people together through common concerns, age, or interests 12

special-purpose money items used only to measure the value of things and otherwise lacking a practical purpose 10

specialization the development of certain skills that others in the group do not share; characteristic of complex societies 10

speciation the evolution of new species or the process of creating species 3

species a population that can mate and produce fertile offspring in natural conditions 2, 3

speech verbal communication using sounds 9

speech community a group that shares language patterns 9

spirits of nature unobservable beings and forces that inhabit the natural world 13

state a type of political organization in a highly populated, industrial society with strong centralized government 6, 12

stratified society a type of society based on a hierarchy in which certain groups have access to resources and power while others do not 12

Strepsirhini a suborder of Primates, including lemurs and lorises 2

subculture a group of people within a culture who are connected by similar identity markers; may include ethnic heritage or common interests 8

subsistence food procurement; basic food needs for survival 1, 10

superfamily a taxonomic category; a subdivision of infraorder 2

supernatural describes those aspects of life that are outside a scientific understanding and that we cannot measure or test; religious 13

supernatural beings personified or embodied beings, such as deities or spirits, that exist beyond the observable world 13

supernatural forces disembodied powers, such as luck, that exist beyond the observable world 13

sustainability the ability to keep something in existence; to support a practice indefinitely 14

swidden (slash-and-burn) cultivation a farming technique in which plant material is burned and crops are planted in the ashes 10

symbol something that stands for something else with little or no natural relationship to its referent; a type of sign 8, 9

syncretism a synthesis of two or more religious belief systems 13

syntax the study of how units of speech are put together to create sentences 9

systems theory a model of understanding an ecosystem that assumes the ecosystem is a closed-loop system with finite resources 14

taphonomy the study of what happens to organic remains after death 4

taxa a category in the system of biological classification of organisms 2

taxonomic order the subdivision of a taxonomic class in the biological classification system; Primates are an order of the class Mammalia 2

technology the tools, skills, and knowledge used by people to survive 1, 10

Terracotta Army approximately 8,000 life-size warriors made of terracotta in China, presumably guarding the tomb of an emperor 7

terraced describes a farming technique that uses graduated steps on hilly terrain 10

theory in science, an extremely well-supported idea; an idea that has not been disproven 3

third gender a gender role accepted in some societies as combining elements of both male and female genders 11

three pillars of sustainability a model of sustainable development with three components: sustainability of the environment, society, and the economy 14

Traditional Ecological Knowledge (TEK) the collective and cumulative knowledge that a group of people has gained over many generations living in a particular ecosystem 1, 10, 14

Traditional Use Studies (TUS) anthropological studies focusing on the ways in which Indigenous peoples have used their lands and resources and continue to do so 1

tragedy of the commons the idea that individual actors sharing a natural resource will inevitably act in their own best interest, eventually depleting the resource 14

trance an altered state of consciousness with diminished sensory and motor activity and lack of memory recall 13

transgender a term used to describe people who internally experience and/or express their gender identity as different from their assigned sex at birth 11

transhumance a pattern of seasonal migration in which pastoralists move back and forth over long distances to productive pastures 10

tribe a type of political organization with an uncentralized power structure, often seen among horticulturalists and pastoralists 6, 12

two-spirit an Indigenous person in North American who identifies as a third gender occupying a role between males and females, with characteristics of each 11

uncentralized system a political system with no centralized governing body in which decisions are made by the community 12

uniformitarianism the idea that the processes that created landscapes of the past are the same processes in operation today 3

unilineal descent tracing one's genealogy through either the mother's or the father's line 11

unilinear theory an evolutionary model that proposed societies progressed from savagery through barbarism and then to civilization; now entirely discredited 1

Upper Paleolithic the cultural time period from about 40,000 to 12,000 years ago 5

utterance an uninterrupted sequence of spoken or written language 9

veganism a diet containing no animal products, including no dairy or eggs 14

vegetarianism a meat-free diet 14

vocalizations intentional sounds humans make to express themselves, but not actually words 9

voice qualities the background characteristics of a person's voice including pitch, rhythm, and articulation 9

warfare an extended violent conflict in which one side attempts to kill as many people or destroy as much property as possible until the other side surrenders 12

World Heritage Site a site of outstanding heritage value, designated by the United Nations Educational, Scientific, and Cultural Organization (UNESCO) 1, 7

worldview the way a group understands and interprets the world; includes all aspects of its culture 14

xenophobia the fear of foreigners, or people outside of one's nation 13

REFERENCES

Abel, A. (1997). Paleohooters: Men as brute hunters, women as shapely sucklers: Who turned prehistory into "Baywatch"? *Saturday Night, 112*(6), 15–16.

Aboriginal Affairs and Northern Development Canada. (2011, July 14). *Fact sheet – The results of the national assessment of First Nations water and wastewater systems.* https://www.sac-isc.gc.ca/eng/1313762701121/1533829864884

Adams, W.M., & Jeanreneaud, S.J. (2008). *Transition to sustainability: Towards a diverse and humane world.* International Union for Conservation of Nature and Natural Resources. https://www.iucn.org/sites/dev/files/import/downloads/transition_to_sustainability_sep_08__en__2.pdf

Agbe-Davies, A.S. (2002). Black scholars, Black pasts. *SAA Archaeological Record, 2*(4), 24–8. https://documents.saa.org/container/docs/default-source/doc-publications/tsar-articles-on-race/agbedavis2002.pdf?sfvrsn=ce177399_2

American Academy of Dermatology. (n.d.). *Indoor tanning.* https://www.aad.org/media/stats-indoor-tanning

American Anthropological Association. (2012). *American Anthropological Association (AAA) code of ethics.* https://www.americananthro.org/ethics-and-methods

American Anthropological Association. (2016, September 6). *AAA stands with tribal nations opposing Dakota Access pipeline.* http://www.americananthro.org/ParticipateAndAdvocate/AdvocacyDetail.aspx?ItemNumber=20656

American Anthropological Association Executive Board. (2004, February 26). *Statement on marriage and the family.* http://www.aaanet.org/issues/policy-advocacy/Statement-on-Marriage-and-the-Family.cfm

Antoine, D. (2008). The archaeology of "plague." *Medical History, 52*(S27), 101–14. https://doi.org/10.1017/S0025727300072112

Asch, T., & Chagnon, N. (Directors). (1974). *A man called Bee: A study of the Yanomamo* [Film]. DER.

Atkins, B.T., & Rundell, M. (2008). *The Oxford guide to practical lexicography.* Oxford University Press.

Barlett, P.F. (2005). Reconnecting with place: Faculty and the Piedmont Project at Emory University. In P.F. Barlett (Ed.), *Urban place: Reconnecting with the natural world* (pp. 39–60). MIT Press.

Battle-Baptiste, W. (2011). *Black feminist archaeology*. Left Coast Press.

Bem, S. (1974). The measurement of psychological androgyny. *Journal of Counseling and Clinical Psychology, 42*(2), 155–62. https://doi.org/10.1037/h0036215

Berger, L., & Hawks, J. (2017). *Almost human: The astonishing tale of Homo naledi and the discovery that changed our human story*. National Geographic.

Berman, J.C. (1999). Bad hair days in the paleolithic: Modern (re)constructions of the cave man. *American Anthropologist, 101*(2), 288–304. https://doi.org/10.1525/aa.1999.101.2.288

Bier, S. (2005, August). *Conflict and human rights in the Amazon: The Yanömami*. ICE Case Studies. http://mandalaprojects.com/ice/ice-cases/yanomami.htm

Bodley, J. (2008). *Anthropology and contemporary human problems*. AltaMira Press.

Bos, K.I., Herbig, A., Sahl, J., Waglechner, N., Fourment, M., Forrest, S.A., Klunk, J., Schuenemann, V.J., Poinar, D., Kuch, M., Golding, G.B., Dutour, O., Keim, P., Wagner, D.M., Holmes, E.C., Krause, J., & Poinar, H.N. (2016). Eighteenth century *Yersinia pestis* genomes reveal the long-term persistence of an historical plague focus. *eLife 5*, 1–11. https://doi.org/10.7554/eLife.12994

Bourdieu, P. (1986). The forms of capital. In J. Richardson (Ed.), *Handbook of theory of research for the sociology of education* (pp. 241–58). Greenword Press.

Bryson, B. (2003). *A short history of nearly everything*. Doubleday Canada.

Campbell Galman, S. (2017). Research in pain [Graphic panel]. *American Anthropological Association Anthropology News, 58*(3), e56–e62. https://doi.org/10.1111/AN.431

Card, D., Dustmann, C., & Preston, I. (2012). Immigration, wages, and compositional amenities. *Journal of the European Economic Association, 10*(1), 78–119. https://doi.org/10.1111/j.1542-4774.2011.01051.x

Carleton, T. (2017). Crop-damaging temperatures increase suicide rates in India. *PNAS, 114*(33), 8746–51. https://doi.org/10.1073/pnas.1701354114

Carroll, J.B. (Ed). (2003). *On the origin of species by means of natural selection by Charles Darwin*. Broadview Press.

Cate, S. (2008). Breaking bread with a spread. *Gastronomica: The Journal of Food and Culture, 8*(3), 17–24. https://doi.org/10.1525/gfc.2008.8.3.17

Chapman, C.A., & Chapman, L.J. (1990). Dietary variability in primate populations. *Primates, 31*(1), 121–8. https://doi.org/10.1007/BF02381035

Clancy, K.B.H., Nelson, R.G., Rutherford, J.N., & Hind, K. (2014). Survey of Academic Field Experiences (SAFE): Trainees report harassment and assault. *PLoS ONE, 9*(7), e102172. https://doi.org/10.1371/journal.pone.0102172

Clay, R.A. (2017). Islamophobia. *Monitor on Psychology, 48*(4), 34. https://www.apa.org/monitor/2017/04/islamophobia

Coe, M.D. (2013). *Mexico: From the Olmecs to the Aztecs* (7th ed.). Thames & Hudson.

Cohn, C. (1987). Sex and death in the rational world of defense intellectuals. *Signs, 12*(4), 687–718. https://doi.org/10.1086/494362

Conkey, M.W., & Gero, J. (1991). *Engendering archaeology: Women and prehistory*. Blackwell Publishers.

Conklin, H.C. (1986). Hanunóo color categories. *Journal of Anthropological Research, 42*(3), 441–6. https://doi.org/10.1086/jar.42.3.3630047

Cordain, L. (2011). *The paleo diet: Lose weight and get healthy by eating the foods you were designed to eat* (Rev. ed.). John Wiley.

Crandall, B.D., & Stahl, P.W. (1995). Human digestive effects on a micromammalian skeleton. *Journal of Archaeological Science, 22*(6), 789–97. https://doi.org/10.1016/0305-4403(95)90008-X

Cruikshank, J. (2005). *Do glaciers listen? Local knowledge, colonial encounters, and social imagination.* UBC Press.

Curl, J. (2005). *Ancient American poets.* Bilingual Review Press.

Darwin, C. (1859). *On the origin of species by means of natural selection, or the preservation of favoured races in the struggle for life.* John Murray.

Darwin, C. (1871). *The descent of man and selection in relation to sex.* John Murray.

Deloria, V.J. (1988). *Custer died for your sins: An Indian manifesto.* University of Oklahoma Press.

de Queiroz, A. (2014). *The monkey's voyage: How improbable journeys shaped the history of life.* Basic Books.

Egbert, A., & Liao, K. (2020, December 21). *The color of coronavirus: 2020 year in review.* APM Research Lab. https://www.apmresearchlab.org/covid/deaths-2020-review

Enabled Archaeology Foundation. (n.d.). *Enabled Archaeology Foundation: Home.* Retrieved November 17, 2021, from https://enabledarchaeologyfoundation.wordpress.com/

Engelke, M. (2017). *Think like an anthropologist.* Pelican.

EPA (United States Environmental Protection Agency). (n.d.). *Sustainability.* Retrieved November 17, 2021, from https://www.epa.gov/sustainability

Ethelston, S. (2006). Gender, population, environment. In N. Haenn & R. Wilk (Eds.), *The environment in anthropology: A reader in ecology, culture and sustainable living* (pp. 113–17). New York University Press.

Evans-Pritchard, E.E. (1940). *The Nuer: A description of the modes of livelihood and political institutions of a Nilotic people.* Oxford University Press.

Fedorak, S.A. (2009). *Pop culture: The culture of everyday life.* University of Toronto Press.

Fiske, S. (2007). Improving the effectiveness of corporate culture. *Anthropology News, 48*(5), 44–45. https://doi.org/10.1525/an.2007.48.5.44

Food and Agriculture Organization of the United Nations (FAO). (n.d.). *Water scarcity.* Retrieved November 17, 2021, from https://www.fao.org/land-water/water/water-scarcity/en/

Food and Agriculture Organization of the United Nations (FAO). (2018). *REDD+ reducing emissions from deforestation and forest degradation.* http://www.fao.org/redd/en/

Franklin, M. (1997a). "Power to the people": Sociopolitics and the archaeology of Black Americans. *Historical Archaeology, 31*(3), 36–50. https://doi.org/10.1007/BF03374229

Franklin, M. (1997b). Why are there so few Black American archaeologists? *Antiquity, 71*(274), 799–801. https://doi.org/10.1017/S0003598X00085732

Franklin, M., Dunnavant, J., Flewellen, A.O., & Odewale, A. (2020). The future is now: Archaeology and the eradication of anti-Blackness. *International Journal of Historical Archaeology, 24*, 753–66. https://doi.org/10.1007/s10761-020-00577-1

Fraser, M.A. (2007). *Dis/abling exclusion, en/abling access: Identifying and removing barriers in archaeological practice for persons with (dis)/abilities* [Doctoral dissertation, American University, Washington, DC]. AUDRA. http://hdl.handle.net/1961/thesesdissertations:3295

Friedl, E. (1978). Society and sex roles. *Human Nature, 1,* 68–75.

Fuentes, A. (2012). *Race, monogamy, and other lies they told you: Busting myths about human nature.* University of California Press.

Fuentes, A. (2017). *The creative spark: How imagination made humans exceptional.* Dutton.

Galdikas, B. (1995). *Reflections of Eden: My years with the orangutans of Borneo.* Little, Brown.

Garber, M. (2012, December 12). 5555, or, how to laugh online in other languages. *The Atlantic.* http://www.theatlantic.com/technology/archive/2012/12/5555-or-how-to-laugh-online-in-other-languages/266175/

Gassaway, W.T. (n.d.). *Maize diety (Chicomecoatl).* The MET. Retrieved November 17, 2021, from https://www.metmuseum.org/art/collection/search/307633

Geertz, C. (1973). *The interpretation of cultures.* Basic Books.

Gero, J.M. (1985). Sociopolitics and the woman-at-home-ideology. *American Antiquity, 50*(2), 342–50. https://doi.org/10.2307/280492

González, L.T. (2013). Modern arranged marriage in Mumbai. *Teaching Anthropology: SACC Notes, 19*(1/2), 34–43.

Goodall, J. (1990). *Through a window: My thirty years with chimpanzees of Gombe.* Houghton Mifflin.

Gould, S.J. (1997). Nonoverlapping magisteria. *Natural History, 106,* 16–22.

Gusterson, H. (2013). Anthropology in the news? *Anthropology Today, 29*(6), 11–13. https://doi.org/10.1111/1467-8322.12071

Hall, E.T. (1990). *The hidden dimension.* Anchor Books.

Hardin, G. (1968). The tragedy of the commons. *Science, 162*(3859), 1243–8. https://doi.org/10.1126/science.162.3859.1243

Hare, G.P., Thomas, C.D., Topper, T.N., & Gotthart, R.M. (2012). The archaeology of Yukon ice patches: New artifacts, observations, and insights. *Arctic, 65*(1), 118–35. https://doi.org/10.14430/arctic4188

Harris, M. (1985). *Good to eat: Riddles of food and culture.* Waveland Press.

Hayden, B. (2003). Were luxury foods the first domesticates? Ethnoarchaeological perspectives from Southeast Asia. *World Archaeology, 34*(3), 458–69. https://doi.org/10.1080/0043824021000026459a

Heslin, R. (1974, May). *Steps toward a taxonomy of touching.* Annual Meeting of the Midwestern Psychological Association, Chicago, IL.

Hodgetts, L., Supernant, K., Lyons, N., & Welch, J.R. (2020). Broadening #MeToo: Tracking dynamics in Canadian archaeology through a survey on experiences within the discipline. *Canadian Journal of Archaeology, 44*(1), 20–47. https://canadianarchaeology.com/caa/publications/canadian-journal-archaeologyjournal-canadien-darcheologie/44/1/020-047

Humes, E. (2012). *Our dirty love affair with trash.* Avery.

Intergovernmental Panel on Climate Change (IPCC). (2014). *Climate change 2014: Synthesis report.* Contribution of Working Groups I, II and III to the Fifth Assessment Report of the Intergovernmental Panel on Climate Change. IPCC.

International Organization for Migration. (2019). *World Migration Report 2020.* https://publications.iom.int/books/world-migration-report-2020

Kearney, M. (1972). *The winds of Ixtepeji: World view and society in a Zapotec town.* Waveland Press.

Kirksey, E. (2021). *The mutant project: Inside the global race to genetically modify humans.* St. Martin's Press.

Kopenawa, D. (2013). *The falling sky: Words of a Yanomami shaman.* The Belknap Press of Harvard University Press.

Lakoff, R. (1973). Language and woman's place. *Language in Society, 2*(1), 45–80. https://doi .org/10.1017/S0047404500000051

Leakey, M., & Leakey, S. (2020). *The sediments of time: My lifelong search for the past.* Houghton Mifflin Harcourt.

Lee, R.B. (2013). *The Dobe Ju/'hoansi* (4th ed.). Cengage Learning.

Leonetti, G., Signoli, M., Pelissier, A.L., Hershkovitz, I., Brunet, C., & Dutour, O. (1997). Evidence of pin implantation as a means of verifying death during the Great Plague of Marseilles (1722). *Journal of Forensic Sciences, 42*(4), 744–8.

Lesnick, J.J. (2018). *Edible insects and human evolution.* University Press of Florida.

Levine, N. (1988). *The dynamics of polyandry.* University of Chicago Press.

Lewis, M.P. (2013). *Ethnologue: Languages of the world* (17th ed.). SIL International. Online version: http://www.ethnologue.com

Lewis-Williams, J. (1998). The signs of all times: Entoptic phenomena in Upper Paleolithic art. *Current Anthropology, 29*(2), 209–45. https://doi.org/10.1086/203629

Lightfoot, K.G. (2013). The study of Indigenous management practices in California: An introduction. *California Archaeology, 5*(2), 209–19. https://doi.org/10.1179/1947461X13Z .00000000011

Lightfoot, K.G., Cuthrell, R.Q., Boone, C.M., Byrne, R., Chavez, A.S., Collings, L., Cowart, A., Evett, R.R., Fine, P.A.V., Gifford-Gonzalez, D., Hylkema, M.G., Lopez, V., Misiewicz, T.M., & Reid, R.E.B. (2013). Anthropogenic burning on the central California coast in Late Holocene and early historic times: Findings, implications, and future directions. *California Archaeology, 5*(2), 371–90. https://doi.org/10.1179/1947461X13Z .00000000020

Linnaeus, C. (1758). *Systema naturae: Regnum animale.* Cura Societatis Zoologicae Germanicae.

Logan, A.L. (2016a). An archaeology of food security in Banda, Ghana. *Archaeological Papers of the American Anthropological Association, 27*(1), 106–19. https://doi.org/10.1111 /apaa.12077

Logan, A.L. (2016b). "Why can't people feed themselves?" Archaeology as alternative archive of food security in Banda, Ghana. *American Anthropologist, 118*(3), 508–24. https:// doi.org/10.1111/aman.12603

Logan, A.L. (2020). *The scarcity slot: Excavating histories of food security in Ghana.* University of California Press.

Maaravi, Y., Levy, A., Gur, T., Confino, D., & Segal, S. (2021). "The tragedy of the commons": How individualism and collectivism affected the spread of the COVID-19 pandemic. *Frontiers in Public Health, 9*, 627559. https://doi.org/10.3389/fpubh.2021.627559

Malinowski, B. (1929). *The sexual life of savages in north-western Melanesia.* Eugenics Publishing Company.

Marks, J. (2017). *Is science racist?* Polity Press.

Marlowe, F. (2010). *The Hadza: Hunter-gatherers of Tanzania.* University of California Press.

Mauss, M. (1966). *The gift: Forms and functions of exchange in archaic societies* (I. Cunnison, Trans.). Cohen & West. (Original work published 1925)

Maybury-Lewis, D. (2006). On the importance of being tribal: Tribal wisdom. In N. Haenn & R.R. Wilk (Eds.), *The environment in anthropology: A reader in ecology, culture, and sustainable living* (pp. 390–9). New York University Press.

McCann, A., & Brandom, R. (2012, August 23). *How to say LOL in 14 different languages*. BuzzFeed. http://www.buzzfeed.com/atmccann/how-to-say-lol-in-14-different-languages

McGovern, P.E. (2009). *Uncorking the past: The quest for wine, beer, and other alcoholic beverages*. University of California Press.

McIlwraith, T. (2012). *"We are still Didene": Stories of hunting and history from northern British Columbia*. University of Toronto Press.

McKeon, M. (n.d.). *Penobscot Nation and their river*. Indigenous Religious Traditions. Retrieved December 14, 2021, from https://sites.coloradocollege.edu/indigenoustraditions/sacred-lands/penobscot-nation-and-their-river/

McWhorter, J. (2013, February). *Txtng is killing language. JK!!!* TED. http://www.ted.com/talks/john_mcwhorter_txtng_is_ killing_language_jk.html

Mowat, F. (1987). *Virunga: The passion of Dian Fossey*. McClelland and Stewart.

Murphy, G. (2001, October 1). *Great Law of Peace of the Haudenosaunee*. Iroquois Confederacy and the US Constitution. http://web.pdx.edu/~caskeym/iroquois_web/html/greatlaw.html

Nabhan, G.P. (2002). *Coming home to eat: The sensual pleasures and global politics of local foods*. W.W. Norton and Company.

Nagle, R. (2013). *Picking up: On the streets and behind the trucks with the sanitation workers in New York City*. Farrar, Straus and Giroux.

Nanda, S. (1990). *Neither man nor woman: The hijras of India*. Wadsworth.

Nanda, S. (2000). Arranging a marriage in India. In P.R. Devita (Ed.), *Stumbling toward truth: Anthropologists at work* (pp. 196–204). Waveland Press.

National Collaborating Centre for Aboriginal Health. (2012). *The state of knowledge of Aboriginal health: A review of Aboriginal public health in Canada*. https://www.ccnsa-nccah.ca/docs/context/RPT-StateKnowledgeReview-EN.pdf

Nicholas, G. (2018, February 14). It's taken thousands of years, but Western science is finally catching up to Traditional Knowledge. *The Conversation*. https://theconversation.com/its-taken-thousands-of-years-but-western-science-is-finally-catching-up-to-traditional-knowledge-90291

North Dakota Department of Health. (2018). *Spill investigation program*. https://deq.nd.gov/WQ/4_spill_Investigations/default.aspx/#Environmental_Incident_Report

Nowell, A., & Chang, M.L. (2014). Science, the media, and interpretations of Upper Paleolithic figurines. *American Anthropologist, 116*(3), 562–77. https://doi.org/10.1111/aman.12121

O'Donnell, N. (2020, February 13). Poll: America praises female politicians, but gender bias remains a barrier to election. *NBC Bay Area*. https://www.nbcbayarea.com/news/politics/decision-2020/women-presidential-candidates-still-held-back-by-electability-doubts-lx-morning-consult-poll-finds/2233460/

Oxfam. (2017, October 2). *Statement by Oxfam America President Abby Maxman regarding Puerto Rico hurricane response* [Press release]. https://www.oxfamamerica.org/press/statement-by-oxfam-america-president-abby-maxman-regarding-puerto-rico-hurricane-response/

Paxson, H. (2012). *The life of cheese: Crafting food and value in America*. University of California Press.

Pearce, F. (2006). *When the rivers run dry: Water – The defining crisis of the twenty-first century*. Beacon Press.

Pelto, G.H., & Pelto, P.J. (2013). Diet and delocalization: Dietary changes since 1750. In D.L. Dufour, A.H. Goodman, & G.H. Pelto (Eds.), *Nutritional anthropology: Biocultural perspectives on food and nutrition* (pp. 353–61). Oxford University Press.

Phillips, T., & Gilchrist, R. (2012). Inclusive, accessible, archaeology: Enabling persons with disabilities. In J. Carmen & R. Skeates (Eds.), *The Oxford handbook of public archaeology* (pp. 673–93). Oxford University Press.

Pirog, R., Van Pelt, T., Enshayan, K., & Cook, E. (2001). Food, fuel, and freeways: An Iowa perspective on how far food travels, fuel usage, and greenhouse gas emissions. *Leopold Center Publications and Papers, 3*. https://lib.dr.iastate.edu/leopold_pubspapers/3

Polanyi, K. (1944). *The great transformation*. Farrar & Rinehart.

Popkin, B. (2001). The nutrition transition and obesity in the developing world. *Journal of Nutrition, 131*(3), 871S–35. https://doi.org/10.1093/jn/131.3.871S

Pruetz, J., & Herzog, N. (2017). Savanna chimpanzees at Fongoli, Senegal, navigate a fire landscape. *Current Anthropology, 58*(S16), s337–s350. https://doi.org/10.1086/692112

Rappaport, R.A. (1968). *Pigs for the ancestors*. Yale University Press.

Rasekh, Z., Bauer, H., Manos, M., & Iacopino, V. (1998). Women's health and human rights in Afghanistan. *Journal of the American Medical Association, 280*(5), 449–55. https://doi.org/10.1001/jama.280.5.449

Rathje, W.L. (2002). Garbology: The archaeology of fresh garbage. In B.J. Little (Ed.), *Public benefits of archaeology* (pp. 85–100). University Press of Florida.

Reckin, R. (2013). Ice patch archaeology in global perspective: Archaeological discoveries from alpine ice patches worldwide and their relationship with paleoclimates. *Journal of World Prehistory, 26*(4), 323–85. https://doi.org/10.1007/s10963-013-9068-3

Ruck, R. (1999). *The tropic of baseball: Baseball in the Dominican Republic*. Bison Books.

Sagan, C. (1979). *Transcript for "The case of the ancient astronauts."* WGBH Educational Foundation.

Sahlins, M. (1972). *Stone age economics*. Transaction Publishers.

Sampeck, K.E. (2019a). Cacao and violence: Consequences of money in colonial Guatemala. *Historical Archaeology, 53*(1), 535–58. https://doi.org/10.1007/s41636-019-00206-7

Sampeck, K.E. (2019b). Early modern landscapes of chocolate: The case of Tacuscalco. In C.R. DeCorse (Ed.), *Power, political economy, and historical landscapes of the modern world: Interdisciplinary perspectives* (pp. 105–12). SUNY Press.

Sampeck, K.E., & Thayn, J. (2017). Translating tastes: A cartography of chocolate colonialism. In S. Schwartzkopf and K.E. Sampeck (Eds.), *Substance and seduction: Ingested commodities in early modern Mesoamerica* (pp. 72–99). University of Texas Press.

Sapolsky, R. (2017). *Behave: The biology of humans at our best and worst*. Penguin.

SCAPE. (2021). *Scotland's coastal heritage at risk project (SCHARP)*. https://scapetrust.org/coastal-heritage-at-risk/

Schlumbaum, A., Campos, P.F., Volken, S., Volken, M., Hafner, A., & Schibler, J. (2010). Ancient DNA, a Neolithic legging from the Swiss Alps and the early history of goat. *Journal of Archaeological Science, 37*(6). 1247–51. https://doi.org/10.1016/j.jas.2009.12.025

Schmidt, R.A., & Voss, B.L. (Eds.). (2000). *Archaeologies of sexuality*. Routledge.

Schultz, E., & Lavenda, R. (2009). *Cultural anthropology: A perspective on the human condition* (7th ed.). Oxford University Press.

Service, E. (1962). *Primitive social organization*. Random House.

Seymour, S. (1999). *Women, family, and child care in India: A world in transition*. Cambridge University Press.

Shostak, M. (1981). *Nisa: The life and words of a !Kung woman*. Harvard University Press.

Soukup, K. (2006). Travelling through layers: Inuit artists appropriate new technologies. *Canadian Journal of Communication, 31*(1), 239–46. https://doi.org/10.22230/cjc .2006v31n1a1769

Steger, M. (2003). *Globalization: A very short introduction*. Oxford University Press.

Stegman, E., & Phillips, V.F. (2014, July 22). *Missing the point: The real impact of Native mascots and team names on American Indian and Alaska Native Youth*. Center for American Progress. https://www.americanprogress.org/article/missing-the-point/

Steward, J. (2006). The concept and method of cultural ecology. In N. Haenn & R.R. Wilk (Eds.), *The environment in anthropology: A reader in ecology, culture, and sustainable living* (pp. 5–9). New York University Press. (Original work published 1955)

Stiles, D., Redmond, I., Cress, D., Nellemann, C., & Formo, R.K. (Eds.). (2013). *Stolen apes: The illicit trade in chimpanzees, gorillas, bonobos, and orangutans*. United Nations Environment Program and UNESCO.

Strochlic, N. (2020, September 2). America's long history of scapegoating its Asian citizens. *National Geographic*. https://www.nationalgeographic.com/history/2020/09/asian -american-racism-covid/#close

Tallbear, K. (2013). *Native American DNA: Tribal belonging and the false promise of genetic science*. University of Minnesota Press.

Tannen, D. (2007). *You just don't understand: Women and men in conversation*. William Morrow Paperbacks.

Thorbecke, K. (2016, November 3). Why a previously proposed route for the Dakota Access Pipeline was rejected. *ABC News*. http://abcnews.go.com/US/previously-proposed-route -dakota-access-pipeline-rejected/story?id=43274356

Tracy, M. (2013, October 9). The most offensive team names in sports: A definitive ranking. *New Republic*. https://newrepublic.com/article/115106/ranking-racist-sports-team -mascots-names-and-logos

Tsunoda, T. (2006). *Language endangerment and language revitalization: An introduction*. Walter de Gruyter.

Turner, V.W. (1967). *The forest of symbols: Aspects of Ndembu ritual*. Cornell University Press.

Tuttle, R.H. (2014). *Apes and human evolution*. Harvard University Press.

UN (United Nations). (1999, December 16). *A/RES/42/187 Report of the World Commission on Environment and Development*. https://undocs.org/pdf?symbol=en/A /RES/42/187

UN Department of Economic and Social Affairs. (2017, June 21). *World population projected to reach 9.8 billion in 2050, and 11.2 billion in 2100*. https://www.un.org/development /desa/en/news/population/world-population-prospects-2017.html

UNICEF. (2013). *Female genital mutilation/cutting: A statistical overview and exploration of the dynamics of change*. https://www.unicef.org/reports/female-genital-mutilation-cutting

UNICEF. (2018, November) *Children's rights in the cocoa-growing communities of Cote D'Ivoire: Synthesis Report.* https://sites.unicef.org/csr/css/synthesis-report-children-rights-cocoa-communities-en.pdf

UNICEF. (2021, July). *Percentage of girls and women aged 15–49 years who have undergone FGM (by place of residence and household wealth quintile).* https://data.unicef.org/resources/dataset/fgm/

UN News. (2017, September). "To deny climate change is to deny a truth we have just lived" says Prime Minister of storm-hit Dominica. *UN News.* https://news.un.org/en/story/2017/09/566742-deny-climate-change-deny-truth-we-have-just-lived-says-prime-minister-storm-hit

UN Women. (2020). *Women in politics: 2020.* https://www.unwomen.org/en/digital-library/publications/2020/03/women-in-politics-map-2020

US Centers for Disease Control. (2011). *Fact sheet – CDC health disparities and inequality report – US, 2011.* https://www.cdc.gov/minorityhealth/chdir/2011/factsheet.pdf

Vakoch, D. (Ed.). (2014). *Archaeology, anthropology, and interstellar communication.* NASA.

Van Huis, A., Van Itterbeek, J., Klunder, H., Mertens, E., Halloran, A., Muir, G., & Vantomme, P. (2013). *Edible insects: Future prospects for food and feed security.* Food and Agriculture Organization of the United Nations. http://www.fao.org/docrep/018/i3253e/i3253e.pdf

Vecsey, C., & Venables, R.W. (1980). *American Indian environments: Ecological issues in Native American history.* Syracuse University Press.

Verhaar, J.W.M. (Ed.). (1990). *Melanesian pidgin and Tok Pisin: Proceedings of the First International Conference on Pidgins and Creoles in Melanesia.* John Benjamins Publishing.

Vitebski, P. (2005). *The reindeer people: Living with animals and spirits in Siberia.* Houghton Mifflin.

Wade, R. (1988). *Village republics: Economic conditions for collective action in South India.* Cambridge University Press.

WHO (World Health Organization). (2020, February). *Female genital mutilation.* http://www.who.int/mediacentre/factsheets/fs241/en/

Whoriskey, P., & Siegel, R. (2019, June 5). Cocoa's child laborers. *Washington Post.* https://www.washingtonpost.com/graphics/2019/business/hershey-nestle-mars-chocolate-child-labor-west-africa/

Wiessner, P. (2002). Hunting, healing, and *Hxaro* exchange: A long-term perspective on !Kung (Ju/'oansi) large-game hunting. *Evolution and Human Behavior, 23*(6), 407–36. https://doi.org/10.1016/S1090-5138(02)00096-X

Wilk, R.R. (1985). The ancient Maya and the political present. *Journal of Anthropological Research, 41*(3), 307–26. https://www.jstor.org/stable/3630596

Wood, B., & Boyle, E. (2016). Hominin taxic diversity: Fact or fantasy? *Yearbook of Physical Anthropology, 159*(s61), 37–78. https://doi.org/10.1002/ajpa.22902

Wrangham, R. (2009). *Catching fire: How cooking made us human.* Basic Books.

WWF (World Wide Fund for Nature). (2016). *Living planet report 2016: Summary.* http://awsassets.panda.org/downloads/lpr_living_planet_report_2016_summary.pdf

Wynn, T., & Gowlett, J. (2018). The handaxe reconsidered. *Evolutionary Anthropology: Issues, News, and Reviews, 27*(1), 21–9. https://doi.org/10.1002/evan.21552

Zuk, M. (2013). *Paleofantasy: What evolution really tells us about sex, diet, and how we live.* W.W. Norton and Company.

INDEX

Page numbers in italics refer to figures, tables, and maps.

family
    arrangement in North America, 270, *273*
    residence patterns, 277–80
    terms for different sides, *285*
family of choice, 280, 284
family of procreation, 272
family relations, 284–5, *285*
family roles, and marriage, 272
features, 110, 189
feces analysis, 182
Fedorak, Shirley, 25
female genital mutilation (FGM),
    201, 202
"feminine" (girl-like) gender characteristics,
    287, *287*
feuds, 319
fictive kinship, 284
fieldwork
    ethics, 208–9
    as focus in anthropology, 14
    methods, 206–8
figurines of Upper Paleolithic, 126, *126*, 127
fire, 122, 180
fire control, 119, 121–2
First Nations, 288, 293, *293*, 317
    See also Indigenous peoples in Canada
Five Nations of the Iroquois
    Confederacy, 310
flexitarian diet, 370
floating gardens of Aztecs (*chinampas*), 255
flute, 127
fluted point, in sites over 5,000 years,
    143–4
food
    and anthropology, 3, 4
    bugs as, 186, 188, 200
    and ethnocentrism, *199*, 200
    globalization, 367–70
    and language, 216
    and power, 315
    in prisons, 256–7, *257*
    taboos in religion, 333, 334
    transition to production, 144–6, 148
food carrying, and bipedalism, 91
food distribution, and industrialism, 263

food-getting
    adaptive strategies, 240–2
    economics (*See* economics of
        food-getting)
    exchange processes, 245, 247
    food producers (*See* food producers)
    foragers, 242–5, 248
    and human diet, 264–5
    major shifts in, 260
    overview, 240
food insecurity, in archaeology, 179
food processing, 263, 264
food producers
    definition and types, 241
    horticulturalists, 247–51
    in industrialism, 260–4
    intensive agriculturalists, 253–60
    pastoralists, 251–3
food production
    20,000 to 5,000 years ago, 135, *137*, 144–6,
        148–52
    and industrialism, 170, 260, 262, 265
    of insects, 188
    and social and political systems, 154–6
    and sustainability, 359–60
    and technology, 260–1
    transition to, 144–6, 148
    and water, 317–18
food security, 23
food surplus, 146, 153, 169–70
foodways
    of Aztecs, 255
    definition, 9
    of foragers, 242–3
    of horticulturalists, 248
    in industrialism, 260–2
    of intensive agriculturalists, 254
    of pastoralists, 252–3
    study, 4
foot (human), and bipedalism, 93
foragers
    conflict in, 319–21
    dead of, 328
    definition and description, 241, 242
    evidence and inferences, 149–50

Lyell, Charles, 66
Lyons, Oren, 355

Maaravi, Y., and colleagues, 204–5
Maasai, 6, *6*, *155*, 287, 331
macaque, *39*
Machu Picchu, *166*
magic, 329, 341
maize, as crop, 255
maladaptive practices, 201, 202–3
male bias, in archaeology, 112
male dominance, 314–15
male violence, 88–9
Mammalia class, 33
mammals, 33, 41
mammoths, 77, 143, 144
Manshiet Nasir settlement (Egypt), 350–1
manure, as problem, 263–4
Māori language (*reo Māori*), 236
Māori people, and *moko*, 298
maritime adaptation, in settlement of North
    America, 142–3
market economy, 259–60
market exchange, 245
Marks, Jonathan, 103
marriage
    as economic exchange, 272, 280–3
    practices and characteristics, 270–2, *271*
    spouses in, 272–7
Mars, occupation of, 227
Marseille, and Great Plague (1720), 172
"masculine" (boy-like) gender
    characteristics, 287, *287*
mass extinctions, 75, 76
material culture, 8
matrilineal descent, 285
matrilocal residence (matrilocality), 278–9
Mauss, Marcel, 245
Maya, collapse of, 167
Maybury-Lewis, David, 353–4
McDermott, Monika, 297
McGovern, Patrick E., 148, 149
McIlwraith, Thomas (Tad), 246
Mead, Margaret, 372
Meadowcroft site, *141–2*

meat eating, 50, 118
media, and anthropology in twenty-first
    century, 20–1
medicine, and cultural models, 226–8
men. *See* gender; "male" topics
Mendel, Gregor, 68–70, *70*
"men's speech," 228–9
Mesa Verde, *142*, *171*
Mesolithic, 136
Mesopotamia, 156, 163
Mesozoic, 41
metalworking, in cultural periods, 137
metaphors, 226, 320
Métis. *See* Indigenous peoples in Canada
Mexico, death and life ritual, 336–7
middens, 110
Middle Paleolithic, description and culture,
    115, *115–16*
migration routes in North America, 138–40,
    139, 182–3
milk, and lactose tolerance, 58–9, 135
Minoans, 163
Miocene, and primate evolution, 42, 44
missiles, euphemisms used, 320–1
modern evolutionary synthesis, 75
moko, 298
money, description and forms, 259–60
monkeys, and rafting theory, 43
monocultured crops, 261, *261*
monogamy, 272–3
monotheism and monotheistic religions, 336
monumental architecture, 162
moon, *26*
Morgan, Lewis Henry, 16
morphemes, 219
mortuary practices, 328
mounds in US (Cahokia), 15, *142*, *171*
multipurpose money, 260
Mumbai (India), and arranged marriages,
    283, *284*
Mundurucú communities, 277
museums, collecting for, 17
musical instruments, in Paleolithic, 127
Muslims, 330–1, 334, 345
Muslim women, *315*, 315–16